Understanding Allergic Airway Diseases

Parteek Prasher • Mousmee Sharma •
Sachin Kumar Singh • Ronan MacLoughlin •
Kavita Pabreja • Kamal Dua
Editors

Understanding Allergic Airway Diseases

Contemporary Treatment Paradigm

Springer

Editors
Parteek Prasher
Department of Chemistry
University of Petroleum & Energy Studies
Dehradun, India

Mousmee Sharma
Department of Chemistry
Uttaranchal University
Dehradun, India

Sachin Kumar Singh
School of Pharmaceutical Sciences
Lovely Professional University
Phagwara, Punjab, India

Ronan MacLoughlin
Research and Development
Aerogen
Galway, Ireland

Kavita Pabreja
Regulatory Affairs
Bioproperties Pty Ltd
Glenorie, NSW, Australia

Kamal Dua
Discipline of Pharmacy
University of Technology
Sydney, NSW, Australia

ISBN 978-981-97-1952-5 ISBN 978-981-97-1953-2 (eBook)
https://doi.org/10.1007/978-981-97-1953-2

This Springer imprint is published by the registered company Springer Nature Singapore Pte Ltd.
The registered company address is: 152 Beach Road, #21-01/04 Gateway East, Singapore 189721, Singapore

Preface

In the realm of medical science, few domains have witnessed as profound a transformation as our understanding of allergic airway diseases. From the earliest observations of wheezing and breathlessness to the intricate molecular pathways that govern these conditions, the journey has been both enlightening and challenging. The landscape of allergic airway diseases, encompassing conditions such as asthma, allergic rhinitis, and sinusitis, has not only expanded in its scope but has also revealed layers of complexity that demand a nuanced approach.

Understanding Allergic Airway Diseases: Contemporary Treatment Paradigm emerges from a convergence of clinical experience, cutting-edge research, and patient narratives. This book is not merely a compilation of facts and figures but a comprehensive exploration into the multifaceted nature of allergic airway diseases. We delve deep into the aetiology, pathophysiology, clinical manifestations, and diagnostic modalities that characterize these conditions. Furthermore, in recognition of the dynamic nature of medical science, this text provides a thorough review of the contemporary treatment paradigms that are shaping patient care today.

One cannot overstate the significance of understanding allergic airway diseases in the context of modern healthcare. With an increasing prevalence globally, the socio-economic burden of these conditions is palpable. Yet, amidst the challenges lie opportunities to innovate, collaborate, and redefine the boundaries of therapeutic interventions. This book endeavours to bridge the gap between knowledge and practice, equipping clinicians, researchers, and students with the tools they need to navigate the complexities of allergic airway diseases with confidence and competence.

As you embark on this literary journey, you will encounter a blend of foundational principles, emerging trends, and expert insights. Each chapter is meticulously crafted to offer a balanced perspective, drawing upon evidence-based practices while fostering a spirit of inquiry. Whether you are a seasoned pulmonologist, an aspiring medical student, or a patient seeking clarity, *Understanding Allergic Airway Diseases: Contemporary Treatment Paradigm* promises to be a valuable resource.

In closing, I extend my heartfelt gratitude to all the contributors who have dedicated their expertise and passion to this endeavour. May this book serve as a beacon of knowledge, fostering innovation, compassion, and excellence in the ever-evolving landscape of allergic airway diseases.

Dehradun, India Parteek Prasher
Dehradun, India Mousmee Sharma
Phagwara, Punjab, India Sachin Kumar Singh
Galway, Ireland Ronan MacLoughlin
Glenorie, NSW, Australia Kavita Pabreja
Sydney, NSW, Australia Kamal Dua

Contents

About the Editors

Parteek Prasher, Assistant Professor, Department of Chemistry, University of Petroleum and Energy Studies, holds a research experience of 8 years in medicinal chemistry and biomaterials while working in the field of anti-inflammatory and anticancer chemotherapy. The research team led by Dr. Prasher focuses on the development of rationally designed molecules for targeting various disorders, including inflammation, cancer, and antimicrobial MDR, which also includes the identification of novel pharmacophores, chemical moieties, and APIs for developing potential chemotherapeutics. With several national/international collaborations, he has published more than 100 peer-reviewed publications in journals of repute. He has been a recipient of the prestigious Junior/Senior Research Fellowship by the Council of Scientific and Industrial Research, Government of India (2012–2015). He received the "Lectureship" Award from the Government of India (2012). He is also a recipient of the Teachers Associateship for Research Fellowship (TARE) and a research grant by the Science and Engineering Research Board of the Department of Science and Technology, Government of India, for 3 years, from December 2021 to December 2024.

Mousmee Sharma, Assistant Professor, Department of Chemistry, Uttaranchal University, has research experience of more than 5 years in physical chemistry and biomaterials while working in the field of biomembranes. The recent research work of Dr. Mousmee is focused on the development of mucoadhesive materials for drug delivery across the physiological mucus barrier. She has published 50 research/review articles and 20 book chapters in journals of high repute. She is a "gold medalist" in her postgraduation.

Sachin Kumar Singh, Professor at School of Pharmaceutical Sciences, Lovely Professional University, India, holds his Master of Pharmacy in Pharmaceutical Analysis from JSS College of Pharmacy, Ooty, Tamil Nadu, India, and also holds a PhD degree in Pharmaceutical Sciences from Karpagam University, Coimbatore, India. He has completed one DST-SERB-funded project on the development of SNEDDS of polypeptide-k for the treatment of diabetes mellitus. He has published more than 200 research papers with a cumulative impact of 400 and co-authored 10 book chapters. He has two Indian patents and two Australian patents. He has reviewed more than 75 research papers upon invitation from publishers of high repute. He has graduated 8 PhD students and 30 M pharmacy students and currently graduating 3 masters and 6 PhD students. Currently, his h index is 27 with total citations > 2300. In January 2020, he received Dr. B.C. Deb Memorial Award for the popularization of science by the Indian Science Congress Association at 107th Indian Science Congress held at UAS Bengaluru. He is the first recipient from the field of pharmaceutical sciences to receive this award. He is also a recipient of the 2014 Bharat Shiksha Ratan Award given by the Global Society for Health and Education Growth, Delhi, India. His core area is improving the oral bioavailability of poorly water-soluble drugs using various formulation approaches, including nanotechnology, complexation, solid dispersion, liquisolid technology, etc.

Ronan MacLoughlin is currently an Associate Director R&D, Science and Emerging Technologies, Aerogen Limited, an Irish medical device company specializing in respiratory drug delivery. He has more than 16 years of experience in the design and development of aerosol drug delivery systems and drug–device combinations. He currently has more than 70 articles, conference presentations, and conference papers, combined with several patent applications currently under review. He holds the position of Adjunct Associate Professor in Trinity College Dublin's (Ireland) Department of Pharmaceutics and Pharmaceutical Technology, Honorary Senior Lecturer in the Royal College of Surgeons Ireland (RCSI), Chairman of the Enterprise Ireland Funded Medical Engineering Technology (MET) Gateway, and Chairman of the Industry Advisory Group for CURAM, the Science Foundation Ireland-funded medical device research center. Additionally, he is currently co-chair of the pediatric and CF working group within the International Society for Aerosols in Medicine (ISAM).

Kavita Pabreja holds a doctoral degree in drug discovery biology from Monash Institute of Pharmaceutical Sciences, Monash University, Australia. Her lecturing experience covers more than 5 years across national and international universities. Her research experience includes more than 10 years of laboratory-based research focused on understanding the downstream signaling pathways associated with GLP-1R in diabetes using new-generation sequencing (NGS) technology, RNA-Seq, and investigating the pathogenesis underlying diabetic neuropathic pain and inflammation. She published more than 21 original research and review articles in high-quality peer-reviewed journals. Currently, she is utilizing her knowledge and skills in the role of Regulatory Affairs Coordinator with a leading Australian novel vaccine research company.

Kamal Dua, Senior Lecturer, Discipline of Pharmacy, Graduate School of Health, University of Technology Sydney (UTS), has a research experience of over 12 years working in the field of drug delivery targeting inflammatory diseases. He is also a leader of Drug Delivery Research in the Centre for Inflammation at Centenary Institute/UTS, where the targets identified from the research projects are pursued to develop novel formulations as the first step toward translation into clinics. His key research areas are drug delivery and immunology, specifically addressing how these disciplines can advance one another helping the community to live longer and healthier. This is evidenced by his extensive publication record in reputed journals. His research interests focus on harnessing the pharmaceutical potential of modulating critical regulators such as interleukins and microRNAs and developing new and effective drug delivery formulations for the management of inflammation in chronic airway diseases and cancer.

Introduction to Allergic Airway Disease

1

Anurag Mishra, Mohit Agrawal, and Yogendra Singh

Abstract

The incidence of allergic airway illnesses has seen a notable escalation in recent decades, affecting around 30 percent of the population in affluent nations. The global prevalence of this condition is estimated to impact over 315 million individuals, with around 10% of asthma sufferers experiencing severe or uncontrolled symptoms. In the setting of airway allergies, immunological responses develop upon initial encounters with allergens in a manner associated with type 2 helper T-cell (Th2) functioning. The aforementioned procedure leads to the expression of interleukin (IL)-4, IL-5, and IL-13 inflammatory proteins, as well as the production of allergen-specific antibodies known as immunoglobulin E (IgE) at heightened levels. Efficient treatment techniques are required for the control and management of allergic disorders. Medicinal interventions for respiratory conditions include the use of bronchodilators and inhaled corticosteroids, which have demonstrated efficacy in mitigating inflammation and alleviating associated symptoms. However, it is important to note that these medications are unable to alter the inherent nature and course of the disease. Therefore, the need is crucial to clarify the shared major physiological variables and biological mechanisms driving AADs, because any therapy that targets these "master switches" is better than the standard approach.

A. Mishra
NIMS Institute of Pharmacy, NIMS University Rajasthan, Jaipur, India

M. Agrawal
School of Medical & Allied Sciences, K.R. Mangalam University, Gurgaon, Haryana, India

Y. Singh (✉)
Department of Pharmacology, Maharishi Arvind College of Pharmacy, Jaipur, Rajasthan, India

P. Prasher et al. (eds.), *Understanding Allergic Airway Diseases*,
https://doi.org/10.1007/978-981-97-1953-2_1

Keywords

Allergic airway disease · Allergic asthma · Inflammation

1.1 Introduction

The term "allergic airway disease (AADs)" refers to a range of chronic disorders that affect the airways and lungs and are typically induced by allergic responses to particular chemicals. Common forms of AAD include allergic asthma (AA), chronic rhinosinusitis (CRS), and allergic rhinitis (AR). The exponential rise in AA, AR and CRS incidence rates since the global advent of industrialization has been a major burden on both the economy and human society. While each AAD has its own unique set of symptoms and anatomical location where damage occurs, recent research reveals that these conditions may actually coexist in the same person and share some of the common pathophysiological pathways. In addition, inflammation and constriction of the airways are hallmarks of these illnesses, which can cause a wide variety of symptoms and greatly reduce an individual's quality of life.

The incidence of allergic airway illnesses has seen a notable escalation in recent decades, affecting around 30 percent of the population in affluent nations. AA is widely recognized as a very significant condition globally, mostly due to the substantial impact it has on individuals' functional limitations and the prolonged length of impairment it entails. The global prevalence of this condition is estimated to impact over 315 million individuals, with around 10% of asthma sufferers experiencing severe or uncontrolled symptoms. Several variables contribute to the heightened prevalence of asthma, encompassing genetic susceptibility and exposure to environmental pollutants and aeroallergens. The development of chronic airway inflammation starts with repetitive occurrences of wheezing, dyspnea, cough, and chest tightness. This process subsequently results in inflammatory structural alterations, such as increased sensitivity of the bronchial tubes to various stimuli, blockage of the airways, and remodeling of the airways caused by inflammation. AR refers to an allergic inflammation of the nasal mucosa, which manifests as symptoms such as sneezing, rhinorrhea, pruritus, and nasal congestion. Furthermore, CRS is defined by chronic inflammation of the nasal and sinus mucous membranes, which can cause a wide range of symptoms. Chronic rhinosinusitis is distinct from acute rhinosinusitis, which is an infection of the nasal and sinus cavities that clears itself quickly. According to reports, there is a global prevalence of allergic rhinitis among adults ranging from 10% to 40%. Moreover, there has been a significant surge in the number of cases observed in recent decades. Despite the recent publication of more articles on the role of AAD, such as AR and AA, in the severity and mortality of COVID-19, a meta-analysis found that having a history of AAD is not likely to worsen the course of COVID-19. In addition, individuals with COVID-19 who took AAD had a considerably lower probability of mortality. Data from 34 studies with 345,091 individuals with COVID-19 demonstrated that the frequency of concomitant AAD in patients with COVID-19 did not vary among nations or regions. AAD population size would also be a factor in the reliability of the findings.

There has to be more research done to determine the underlying pathogenic processes and potential therapeutic applications (Ming et al. 2022).

In the setting of airway allergies, immunological responses develop upon initial encounters with allergens in a manner associated with type 2 helper T-cell (Th2) functioning. The aforementioned procedure leads to the expression of interleukin (IL)-4, IL-5, and IL-13 inflammatory proteins, as well as the production of allergen-specific antibodies known as immunoglobulin E (IgE) at heightened levels. These substances then facilitate the recruitment and activation of effector cells, such as eosinophils, basophils, and mast cells. The majority of IgE antibodies have affinity for FcεR1 receptors found on mast cells, basophils, and eosinophils. Many mediators of inflammatory processes, including prostaglandins, leukotrienes, and histamine, are released into the bloodstream once allergens and IgE come into interaction. These mediators contribute to the manifestation of early-phase symptoms, including itching, swelling, and bronchoconstriction. The late-phase responses are facilitated by the infiltration of T cells and the release of their produced cytokines, which contribute to the exacerbation of tissue inflammation. Upon rechallenge, the durable specific memory B and T cells exhibit fast proliferation, therefore establishing an immunologic memory and serving as a prompt reservoir for further immune responses.

Efficient treatment techniques are required for the control and management of allergic disorders. Medicinal interventions for respiratory conditions include the use of bronchodilators and inhaled corticosteroids, which have demonstrated efficacy in mitigating inflammation and alleviating associated symptoms. However, it is important to note that these medications are unable to alter the inherent nature and course of the disease. Therefore, the need is crucial to clarify the shared major physiological variables and biological mechanisms driving AADs, because any therapy that targets these "master switches" is better than the standard approach.

1.1.1 Immunopathology of AAD

As a way to highlight the common vulnerability and pathophysiology across allergic illnesses, the term "atopic disease" is frequently used. An individual with atopy has a genetic or inherited predisposition to develop environmental sensitization and an IgE antibody response (Johansson et al. 2004). Atopic diseases have a complicated and as-yet-ununderstood etiology. There is considerable evidence that both hereditary and environment-related elements contribute to AAD pathogenesis. Many cases of allergic illness begin in infancy or early childhood, and the onset of symptoms typically follows a predictable pattern known as the "atopic march" (Spergel and Paller 2003). Environmental allergen exposure in conjunction with adjuvant variables throughout infancy and early childhood influences the atopic status of genetically sensitive patients (Björkstén 1999; Peden 2000). Atopic dermatitis (AD) can be diagnosed in children as early as 1 year old, and it frequently occurs alongside an allergy to cow milk in children with a propensity for allergic illnesses. Diagnosis of AR and, subsequently, AA in these kids is possible as early as age 6–7.

Children that possess a predisposition for allergic illnesses are frequently identified as having atopic dermatitis (AD) at a tender age of 1 year, commonly accompanied by an allergic reaction to cow milk. These children frequently exhibit a pattern of advancement leading to the development of AR and subsequently AA, which may typically be identified by diagnosis at about 6–7 years of age (Čelakovská et al. 2020). Several children will carry their asthma attacks into adulthood, while others may see a remission of their symptoms or a total disappearance of their condition as they become grow older. Although hereditary variables influence vulnerability, environmental triggers can cause the onset of the illness, as shown by the greater prevalence of asthma in infants who suffered from virally (respiratory syncytial virus, RSV) produced bronchiolitis in infancy (Mackay 1985). There are extensive environmental and host-based biological networks involved in the pathophysiology of inflammatory airway disorders. The airways' epithelium serves as a main line of defense against aeroallergens inside the individual's respiratory system in the context of AR. Nonetheless, allergens' intrinsic proteolytic activity may lead the epithelium to lose its passive structural barrier role, enabling allergens to enter local tissues and set off ongoing chronic inflammatory processes (Bashir et al. 2013; Stamataki et al. 2021). The genesis of immune-mediated illness including AR, CRS, and AA has been further investigated, and it has been demonstrated that intrinsic barrier failure contributes to the disease (Steelant et al. 2016; Hellings and Steelant 2020). However, the relative contributions of basic (genetic) and secondary (inflammatory) pathways in causing barrier dysfunction remain unclear. Although epithelial cells certainly have a role in providing a barrier of tissue against inhaled allergens, they also participate in airway inflammation by sensing and reacting to external triggers. In response to allergens or infections, the airway epithelium induces toll-like receptors, which are proteins that recognize patterns (Radman et al. 2015; van Tongeren et al. 2015a), which then trigger the synthesis of various mediators. These mediators provide an environment that promotes local inflammation by altering the activity of immune cells and influencing the migration of inflammatory cells to affected tissues (Van Tongeren et al. 2015b). An inability to turn off the activation response may contribute to the persistent activation of the nasal epithelium seen in allergic illness (Golebski et al. 2015).

Airborne allergens such dust mite feces, cockroach dander, pet dander, mold spores, and pollens trigger AR by invading the nasal cavity lining with inflammatory cells like macrophages, mast cells, B lymphocytes, CD4+ T lymphocytes, and eosinophils. IgE is generated by plasma cells, while IL-3, IL-4, IL-5, and IL-13 are induced in the circulation by Th2 cells. However, the mechanism that generates IgE remains a mystery to many. It has been discovered in recent years that follicular helper T (Tfh) cells, a subgroup of CD4+ T-effector cells, rather than Th2 cells, are the key regulators of IgE production. The interaction between allergen cross-linked IgE and mast cells results in the release of a number of mediators (such as leukotrienes and histamine) which promote vascular permeability and arteriole dilatation, the contraction of pulmonary smooth muscle, an accumulation of mucus, chronic runny nose, and itchy skin (Small et al. 2007; Dykewicz and Hamilos 2010). Cellular late inflammatory responses are triggered by the released mediators and cytokines

4–8 h later, causing a return of symptoms, most often nasal congestion, that typically lasts for days. Both AR and AA have comparable immunopathological characteristics, including the invasion of eosinophils, mast cells, and Th2 cells. Airway remodeling structural alterations are most understood in AA, although they may also occur in AR. AR and AA also differ on a pathophysiological level. Epithelial hyperplasia, increased mucus production, and goblet cell metaplasia are all examples of mucosal pathological abnormalities seen with AA illness (Fahy 2015). Airway constriction and increased mucus formation characterize an asthma episode, accompanied by symptoms including breathlessness, wheezing, chest discomfort, and coughing, all of which originate in the submucosal layer, where significant mucus glands, collagen adhesion, and smooth muscle hypertrophy predominate (Mims 2015) (Fig. 1.1).

The imbalance of Th cells has been hypothesized to be at the root of AA, an adaptive immune illness (Busse and Rosenwasser 2003). The Th hypothesis has been called into question due to the lack of a consistent response to Th2-directed therapies for AA, the presence of innate signatures in subtypes of asthmatics, and the association between respiratory viral infections and the onset and exacerbation of disease (Holtzman 2012; Brasier 2013). There is mounting evidence that antigen-independent mucosal responses, known as "innate" responses, play crucial roles in both the onset and development of the illness. Both viral infections and exposure to common aeroallergens can set off innate immune responses (IIRs) in the normal and asthmatic airway. In addition, atopy modifies the IIR to viral pathogens, and viral

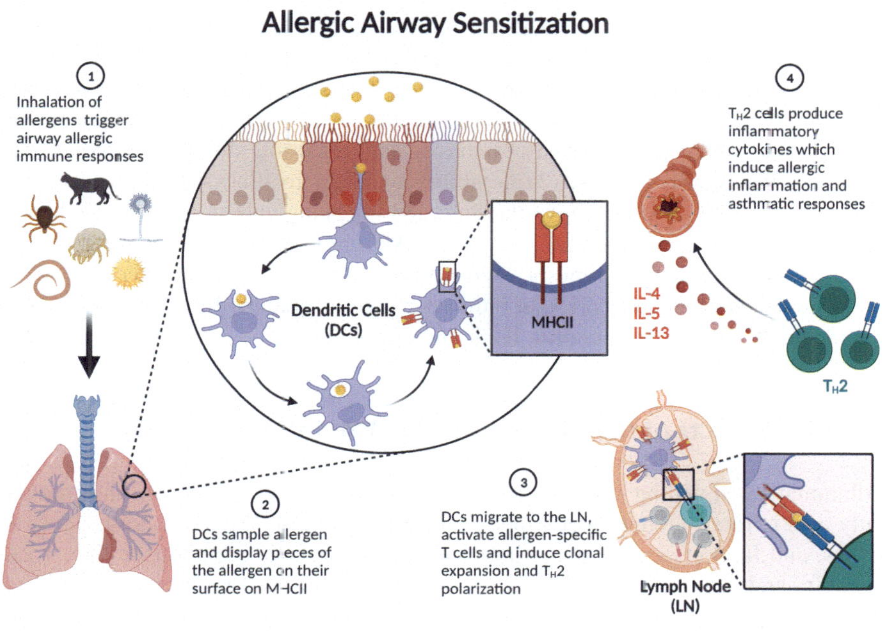

Fig. 1.1 Exhibits aeroallergens-mediated allergic immune response in AAD

infections themselves predispose to the development of atopy (Message et al. 2008; Sigurs et al. 2010). This suggests a more holistic understanding of mucosal innate immune dysregulation in AA. New evidence supports the treatment development route by showing that the NFB/RelA-BRD4 complex facilitates epithelial barrier breakdown and remodeling in response to viruses and allergens (Brasier 2019). Emerging scientific investigations suggest that modifications to the epigenome have a role in changes to the morphology of epithelial cells (Seaton et al. 1994; Hewitt and Lloyd 2021; Wang et al. 2023). Epigenomic moderators may also be a way to restore normalcy to the epithelium injury-repair process by inhibiting the antiviral IFN signal and modification of innate communication pathway. Reversing structural remodeling and restoring normal IIR in AA may be possible with the use of therapies that target epigenetic alterations in the mucosa (Brasier 2019).

1.2 Biomarkers of AAD

New "biomarkers" for illness existence, vulnerability, and even varied response to therapy have emerged as the direct outcome of the fast expansion and utilization of molecular methods in the area of AAD. These biomarkers are helping to pave the way for "personalized" medicine, in which patients receive care that is more precisely customized to their unique needs and preferences and less in accordance with the current "one-size-fits-all" clinical management recommendations. Existing and new biomarkers are expected to influence daily clinical decision-making by physicians; this will allow them to focus medicines to patients who will benefit while sparing expense and possible negative consequences for individuals who will not.

Although there is no such thing as a perfect biomarker, an ideal biomarker for AA and other AAD would have the following features: (1) the ability to reliably differentiate among health conditions and diseases with high positive and negative anticipatory values; (2) the ability to convey data regarding disease prognosis and clinical efficacy conclusions; (3) flexibility to adapt to illness development and return to "normal" after effective therapy; and (4) being able to be replicated successfully in a healthcare setting. AA and AAD biomarkers may potentially have several potential points of genesis, each with its own set of benefits and drawbacks.

Notable and potentially relevant biomarkers are outside the focus of this chapter but include CRP, TNF-α, pendrin, IL-2, and osteopontin.

1.2.1 Products of the Arachidonic Acid Metabolic Pathway

Inflammatory signaling involves the release of arachidonic acid, an unsaturated fatty acid normally found in cell membrane phospholipids. Prostaglandins, leukotrienes, and lipoxins are all byproducts of arachidonic acid metabolism (Fig. 1.2). Lipoxins and leukotriene E4 (LTE4), a cysteinyl leukotriene, are the primary targets of ongoing biomarker research.

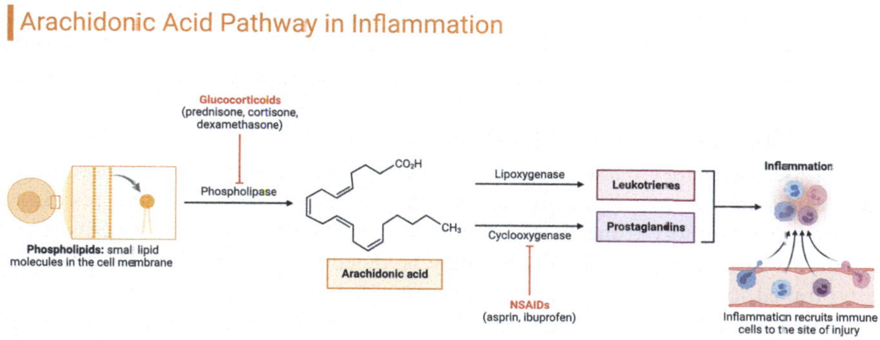

Fig. 1.2 Illustrates the role of arachidonic acid pathway in inflammatory consequences of AAD

1.2.1.1 Lipoxins

Furthermore, to their role in chemotaxis and the transduction of signals that goes along with it, lipoxins also suppress inflammation in a number of different ways. Lipoxin A4 values are lesser in critically sufferers of AA than in individuals with milder forms of the disease (Planagumà et al. 2008), and this is accompanied by extended innate lymphoid cell activity and eosinophilia, in addition to missing of activity in essential enzymes and receptors necessary for lipoxin synthesis (Barnig et al. 2013). In serious asthmatics, decreased airway macrophage lipopolysaccharide-stimulated lipoxin A4 synthesis is strongly linked with airflow limitation (Bhavsar et al. 2010). Lipoxin A4 production has been shown to be increased in airway specimens from individuals with AA, suggesting that blockers of soluble epoxide hydrolase ability may be useful for the therapeutic management of life-threatening asthma.

1.2.1.2 The E4 Leukotriene

Leukotriene (LT) E4, a stable consequence of cysteinyl leukotriene breakdown, is readily identifiable in specimens of urine and does not need any invasive method for detection. Multiple studies have shown that both teens suffering from AA and adults suffering aspirin-induced airway disease had increased amounts of LTE4 in their urine (Rabinovitch et al. 2006; Bochenek et al. 2014). Urinary LTE4 concentrations have also been linked to asthma exacerbations and airflow restriction severity, albeit this biomarker may be muddled by other factors (Lang et al. 2013). However, in children with AA who have been impacted by passive smoking, elevated LTE4 levels are no longer a good predictor of future exacerbations. Other prospective clinical studies using urine LTE4 as a biomarker have shown promising findings (Rabinovitch et al. 2011; Kolmert et al. 2021). In research, a leukotriene receptor antagonist outperformed inhaled corticosteroids in improving pulmonary function and asthma management in children having a greater baseline ratio of urine LTE4 to FeNO. Add-on treatment with the combination of an inhaled corticosteroid and an antagonist of the LT receptor was shown to be more effective than long-acting agonists, regardless of baseline urine LTE4 concentrations, according to a more recent investigation (Rabinovitch et al. 2010; Hu et al. 2023). This correlation was small and may have been underpowered, but the results nonetheless support future investigation into the use of LTE4 as a biomarker in the selection of asthma medication.

1.2.1.3 3-Bromotyrosine in the Urine

During the respiratory burst process, activated eosinophils produce hypobromous acid, which may be used to posttranslationally modify tyrosine protein residues, resulting in 3-bromotyrosine. 3-Bromotyrosine's potential biomarker benefits from its stability and the ease with which it may be detected in the urine. In addition to being linked to airway limitation, asthma management, and future exacerbations, recent investigations have revealed that 3-bromotyrosine concentrations are greater in individuals with AA (Wedes et al. 2009; Wang et al. 2022). Other research has indicated that the use of inhaled corticosteroids leads to a reduction in urine 3-bromotyrosine concentrations.

1.2.1.4 Chitinase Homologous Proteins

Extracellular matrix remodeling and control are processes that may include chitinases, a class of hydrolases distinguished by their capacity to split chitin. Airflow restriction and other clinical markers of illness severity are linked to the chitinase-like protein YKL-40 (also known as human cartilage glycoprotein-39), which is present in the blood and airways of individuals with AA (Chupp et al. 2007). Additionally, YKL-40 expression is elevated during acute asthma exacerbations, suggesting that it may be used as a predictor of the progressive loss of lung function that occurs in response to cigarette smoking. In children, researchers have shown that thicker bronchial walls are linked to higher levels of YKL-40 expression in both bronchoalveolar lavage fluid and serum (Tang et al. 2010). Further research is required; however, YKL-40 shows promise as a tool for predicting asthma outcomes and risks in the future.

1.2.1.5 Periostin

Periostin is secreted by airway epithelial cells in anticipation of IL-13, a cytokine associated with inflammatory responses that activates the TGF-beta regulatory pathway and affects epithelial-mesenchymal communications (Corren et al. 2011; Lopez-Guisa et al. 2012). Serum testing is a noninvasive technique to check for elevated periostin levels, which have been proven to be present in the airways of people with AA. However, there is little evidence that these two measures correspond (Wagener et al. 2015). Serum periostin values have been demonstrated to have no or little correlation with bronchial eosinophils according to recent studies (Kim et al. 2014). However, higher blood periostin concentrations in people with asthma have been linked to more severe airflow restriction, a steeper loss in lung function, and other markers of asthma severity (Kanemitsu et al. 2013).

1.3 Treatment Approaches for AAD

1.3.1 Mast Cell-Specific Therapy

Histamine, produced mostly by mast cells, stimulates the H1 receptor to generate a cascade of reactions that manifest as allergy and inflammatory illness symptoms (such as urticaria) in susceptible individuals (Thangam et al. 2018). Treatment for numerous mast cell-driven disorders has relied heavily on H1 antihistamines since their discovery.

There has been mixed success treating AAD with leukotriene inhibitors including montelukast, zafirlukast, pranluton, and zileuton. For the management of AA and AR, the US FDA has approved the therapeutic application of montelukast. It is believed to be less effective than inhaled or intranasal corticosteroids in treating both diseases. The Food and Medicine Administration warned of significant neuropsychiatric events in 2020 among montelukast users and that the medicine be used only as a last resort in the treatment of AR.

1.3.2 Target Lipid Messenger: PGs and LTs

PGD2, a lipid messenger produced mostly by mast cells, has been linked to urticaria and allergies through its impact on chemoattractant receptor homologue substance presented on T helper 2 cells (TH2 cells). AZD1981 (a selective CRTh2 antagonist) was well tolerated, lowered weekly itch ratings more than wheals, and blocked PGD2-mediated eosinophil morphology change in individuals with antihistamine-resistant chronic sclerodactylic urticaria (CSU) (Oliver et al. 2019). In the treatment of people with AA and AR, various studies have found little or poor clinical success for many CRTh2 antagonists, such as setipiprant, AZD1981, timapiprant (OC-459), fevipiprant, and BI671800 (Brightling et al. 2021). In individuals with severe asthma, for instance, fevipiprant did not demonstrate substantial improvement in phase III studies, but another CRTh2 antagonist, GB001, did show modest efficacy in reducing asthma symptoms (Asano et al. 2020).

1.3.3 Immunotherapy

Allergen immunotherapy is an alternative to pharmacological treatment that modifies the immune system's response to allergens by means of prolonged, carefully monitored exposure. Treatment with this method usually takes between 3 and 5 years. Subcutaneous (SCIT) and sublingual (SLIT) immunotherapy are two forms of this treatment. While anaphylaxis is a worry with SCIT, it is considerably less likely to occur with SLIT (Solelhac and Charpin 2014).

Currently, the treatment of AADs with the biologic agents omalizumab, benralizumab, mepolizumab, reslizumab, and dupilumab is approved. Other anti-inflammatory medicines used to treat asthma include bronchodilating agents, chromones, and theophylline, and in recent years, researchers have developed antibodies against IL-4/IL-13 receptors and IL-5/IL-5 receptors.

1.4 Conclusion

Over the course of many decades of study, scientists have come to a much better knowledge of the immunological and environmental underpinnings of AAD. Unfortunately, the state-of-the-art treatments for these disorders have yet to catch up. Allergies are immunologically complicated, and they might entail many different types of hypersensitivity that all work together. The immunological response of the airways to particles and allergens in the air is complex and multifaceted, as shown by recent investigations. Asthma and rhinosinusitis have been associated to infections that are of the airway tract caused by viruses, particularly HRV and fungi, according to recent research. Therefore, it is anticipated that future therapies will aim to both reduce environmental factors that promote the development of allergic responses and reduce endogenous chemicals (such as cytokines) that coordinate allergic inflammation in AAD.

References

Asano K, Sagara H, Ichinose M, Hirata M, Nakajima A, Ortega H, Tohda Y (2020) A phase 2a study of DP(2) antagonist GB001 for asthma. J Allergy Clin Immunol Pract 8(4):1275–1283.e1271

Barnig C, Cernadas M, Dutile S, Liu X, Perrella MA, Kazani S, Wechsler ME, Israel E, Levy BD (2013) Lipoxin A4 regulates natural killer cell and type 2 innate lymphoid cell activation in asthma. Sci Transl Med 5(174):174ra126-174ra126

Bashir MEH, Ward JM, Cummings M, Karrar EE, Root M, Mohamed ABA, Naclerio RM, Preuss D (2013) Dual function of novel pollen coat (surface) proteins: IgE-binding capacity and proteolytic activity disrupting the airway epithelial barrier. PLoS One 8(1):e53337

Bhavsar PK, Levy BD, Hew MJ, Pfeffer MA, Kazani S, Israel E, Fan Chung K (2010) Corticosteroid suppression of lipoxin A4 and leukotriene B4from alveolar macrophages in severe asthma. Respir Res 11(1):1–9

Björkstén B (1999) The environmental influence on childhood asthma. Allergy 54(Suppl 49):17–23

Bochenek G, Kuschill-Dziurda J, Szafraniec K, Plutecka H, Szczeklik A, Nizankowska-Mogilnicka E (2014) Certain subphenotypes of aspirin-exacerbated respiratory disease distinguished by latent class analysis. J Allergy Clin Immunol 133(1):98–103. e106

Brasier AR (2013) Identification of innate immune response endotypes in asthma: implications for personalized medicine. Curr Allergy Asthma Rep 13(5):462–468

Brasier AR (2019) Mechanisms how mucosal innate immunity affects progression of allergic airway disease. Expert Rev Respir Med 13(4):349–356

Brightling CE, Gaga M, Inoue H, Li J, Maspero J, Wenzel S, Maitra S, Lawrence D, Brockhaus F, Lehmann T, Brindicci C, Knorr B, Bleecker ER (2021) Effectiveness of fevipiprant in reducing exacerbations in patients with severe asthma (LUSTER-1 and LUSTER-2): two phase 3 randomised controlled trials. Lancet Respir Med 9(1):43–56

Busse WW, Rosenwasser LJ (2003) Mechanisms of asthma. J Allergy Clin Immunol 111(3):S799–S804

Čelakovská J, Bukač J, Vaňková R, Krcmova I, Krejsek J, Andrýs C (2020) ISAC multiplex testing–results of examination in 100 patients suffering from atopic dermatitis. Food Agric Immunol 31(1):1014–1035
Chupp GL, Lee CG, Jarjour N, Shim YM, Holm CT, He S, Dziura JD, Reed J, Coyle AJ, Kiener P (2007) A chitinase-like protein in the lung and circulation of patients with severe asthma. N Engl J Med 357(20):2016–2027
Corren J, Lemanske RF Jr, Hanania NA, Korenblat PE, Parsey MV, Arron JR, Harris JM, Scheerens H, Wu LC, Su Z (2011) Lebrikizumab treatment in adults with asthma. N Engl J Med 365(12):1088–1098
Dykewicz MS, Hamilos DL (2010) Rhinitis and sinusitis. J Allergy Clin Immunol 125(2 Suppl 2):S103–S115
Fahy JV (2015) Type 2 inflammation in asthma—present in most, absent in many. Nat Rev Immunol 15(1):57–65
Golebski K, van Egmond D, de Groot EJ, Roschmann KI, Fokkens WJ, van Drunen CM (2015) EGR-1 and DUSP-1 are important negative regulators of pro-allergic responses in airway epithelium. Mol Immunol 65(1):43–50
Hellings PW, Steelant B (2020) Epithelial barriers in allergy and asthma. J Allergy Clin Immunol 145(6):1499–1509
Hewitt RJ, Lloyd CM (2021) Regulation of immune responses by the airway epithelial cell landscape. Nat Rev Immunol 21(6):347–362
Holtzman MJ (2012) Asthma as a chronic disease of the innate and adaptive immune systems responding to viruses and allergens. J Clin Invest 122(8):2741–2748
Hu Y, Zheng S, Chen Z, Yu D, Lai T, Chen Y, Liao W (2023) Validity of fractional exhaled nitric oxide and small airway lung function measured by IOS in the diagnosis of cough variant asthma in preschool children with chronic cough. Allergy, Asthma Clin Immunol 19(1):83
Johansson SGO, Bieber T, Dahl R, Friedmann PS, Lanier BQ, Lockey RF, Motala C, Ortega Martell JA, Platts-Mills TAE, Ring J, Thien F, Van Cauwenberge P, Williams HC (2004) Revised nomenclature for allergy for global use: report of the nomenclature review Committee of the World Allergy Organization, October 2003. J Allergy Clin Immunol 113(5):832–836
Kanemitsu Y, Matsumoto H, Izuhara K, Tohda Y, Kita H, Horiguchi T, Kuwabara K, Tomii K, Otsuka K, Fujimura M (2013) Increased periostin associates with greater airflow limitation in patients receiving inhaled corticosteroids. J Allergy Clin Immunol 132(2):305–312. e303
Kim M-A, Izuhara K, Ohta S, Ono J, Yoon MK, Ban GY, Yoo H-S, Shin YS, Ye Y-M, Nahm D-H (2014) Association of serum periostin with aspirin-exacerbated respiratory disease. Ann Allergy Asthma Immunol 113(3):314–320
Kolmert J, Gómez C, Balgoma D, Sjödin M, Bood J, Konradsen JR, Ericsson M, Thörngren JO, James A, Mikus M, Sousa AR, Riley JH, Bates S, Bakke PS, Pandis I, Caruso M, Chanez P, Fowler SJ, Geiser T, Howarth P, Horváth I, Krug N, Montuschi P, Sanak M, Behndig A, Shaw DE, Knowles RG, Holweg CTJ, Wheelock MÅ, Dahlén B, Nordlund B, Alving K, Hedlin G, Chung KF, Adcock IM, Sterk PJ, Djukanovic R, Dahlén SE, Wheelock CE (2021) Urinary leukotriene E(4) and prostaglandin D(2) metabolites increase in adult and childhood severe asthma characterized by type 2 inflammation. A clinical observational study. Am J Respir Crit Care Med 203(1):37–53
Lang JE, Dozor AJ, Holbrook JT, Mougey E, Krishnan S, Sweeten S, Wise RA, Teague WG, Wei CY, Shade D (2013) Biologic mechanisms of environmental tobacco smoke in children with poorly controlled asthma: results from a multicenter clinical trial. J Allergy Clin Immunol Pract 1(2):172–180. e172
Lopez-Guisa JM, Powers C, File D, Cochrane E, Jimenez N, Debley JS (2012) Airway epithelial cells from asthmatic children differentially express proremodeling factors. J Allergy Clin Immunol 129(4):990–997. e996
Mackay M (1985) Asthma and viral infections. N Z Med J 98(778):352–353
Message SD, Laza-Stanca V, Mallia P, Parker HL, Zhu J, Kebadze T, Contoli M, Sanderson G, Kon OM, Papi A, Jeffery PK, Stanciu LA, Johnston SL (2008) Rhinovirus-induced lower respira-

tory illness is increased in asthma and related to virus load and Th1/2 cytokine and IL-10 production. Proc Natl Acad Sci USA 105(36):13562–13567
Mims JW (2015) Asthma: definitions and pathophysiology. Int Forum Allergy Rhinol 5(Suppl 1):S2–S6
Ming W, Zuo J, Han J, Chen J (2022) The impact of comorbid allergic airway disease on the severity and mortality of COVID-19: a systematic review and meta-analysis. Eur Arch Otorrinolaringol 279(4):1675–1690
Oliver ET, Chichester K, Devine K, Sterba PM, Wegner C, Vonakis BM, Saini SS (2019) Effects of an Oral CRTh2 antagonist (AZD1981) on eosinophil activity and symptoms in chronic spontaneous urticaria. Int Arch Allergy Immunol 179(1):21–30
Peden DB (2000) Development of atopy and asthma: candidate environmental influences and important periods of exposure. Environ Health Perspect 108(Suppl 3):475–482
Planagumà A, Kazani S, Marigowda G, Haworth O, Mariani TJ, Israel E, Bleecker ER, Curran-Everett D, Erzurum SC, Calhoun WJ (2008) Airway lipoxin A4 generation and lipoxin A4 receptor expression are decreased in severe asthma. Am J Respir Crit Care Med 178(6):574–582
Rabinovitch N, Graber NJ, Chinchilli VM, Sorkness CA, Zeiger RS, Strunk RC, Bacharier LB, Martinez FD, Szefler SJ, Research CA (2010) Urinary leukotriene E4/exhaled nitric oxide ratio and montelukast response in childhood asthma. J Allergy Clin Immunol 126(3):545–551. e544
Rabinovitch N, Reisdorph N, Silveira L, Gelfand EW (2011) Urinary leukotriene E4 levels identify children with tobacco smoke exposure at risk for asthma exacerbation. J Allergy Clin Immunol 128(2):323–327
Rabinovitch N, Zhang L, Gelfand EW (2006) Urine leukotriene E4 levels are associated with decreased pulmonary function in children with persistent airway obstruction. J Allergy Clin Immunol 118(3):635–640
Radman M, Golshiri A, Shamsizadeh A, Zainodini N, Bagheri V, Arababadi M, Kennedy D (2015) Toll-like receptor 4 plays significant roles during allergic rhinitis. Allergol Immunopathol 43(4):416–420
Seaton A, Godden DJ, Brown K (1994) Increase in asthma: a more toxic environment or a more susceptible population? Thorax 49(2):171
Sigurs N, Aljassim F, Kjellman B, Robinson PD, Sigurbergsson F, Bjarnason R, Gustafsson PM (2010) Asthma and allergy patterns over 18 years after severe RSV bronchiolitis in the first year of life. Thorax 65(12):1045–1052
Small P, Frenkiel S, Becker A, Boisvert P, Bouchard J, Carr S, Cockcroft D, Denburg J, Desrosiers M, Gall R (2007) Rhinitis: a practical and comprehensive approach to assessment and therapy. J Otolaryngol 36
Solelhac G, Charpin D (2014) Management of allergic rhinitis. F1000Prime Rep 6
Spergel JM, Paller AS (2003) Atopic dermatitis and the atopic march. J Allergy Clin Immunol 112(6):S118–S127
Stamataki S, Papadopoulos N, Lakoumentas J, Georgountzou A, Maggina P, Xepapadaki P, Andreakos E, Prokopakis E, Legaki E, Taka S (2021) Nasal epithelium: new insights and differences of the cytokine profile between normal subjects and subjects with allergic rhinitis. Rhinology Online 4:223–232
Steelant B, Farré R, Wawrzyniak P, Belmans J, Dekimpe E, Vanheel H, Van Gerven L, Krohn IK, Bullens DM, Ceuppens JL (2016) Impaired barrier function in patients with house dust mite–induced allergic rhinitis is accompanied by decreased occludin and zonula occludens-1 expression. J Allergy Clin Immunol 137(4):1043–1053. e1045
Tang H, Fang Z, Sun Y, Li B, Shi Z, Chen J, Zhang T, Xiu Q (2010) YKL-40 in asthmatic patients, and its correlations with exacerbation, eosinophils and immunoglobulin E. Eur Respir J 35(4):757–760
Thangam EB, Jemima EA, Singh H, Baig MS, Khan M, Mathias CB, Church MK, Saluja R (2018) The role of histamine and histamine receptors in mast cell-mediated allergy and inflammation: the hunt for new therapeutic targets. Front Immunol 9:1873

van Tongeren J, Golebski K, Van Egmond D, de Groot EJ, Fokkens WJ, van Drunen CM (2015a) Synergy between TLR-2 and TLR-3 signaling in primary human nasal epithelial cells. Immunobiology 220(4):445–451

Van Tongeren J, Röschmann K, Reinartz S, Luiten S, Fokkens W, De Jong E, Van Drunen C (2015b) Expression profiling and functional analysis of toll-like receptors in primary healthy human nasal epithelial cells shows no correlation and a refractory LPS response. Clin Transl Allergy 5(1):1–9

Wagener AH, de Nijs SB, Lutter R, Sousa AR, Weersink EJ, Bel EH, Sterk PJ (2015) External validation of blood eosinophils, FENO and serum periostin as surrogates for sputum eosinophils in asthma. Thorax 70(2):115–120

Wang J, Zhou Y, Zhang H, Hu L, Liu J, Wang L, Wang T, Zhang H, Cong L, Wang Q (2023) Pathogenesis of allergic diseases and implications for therapeutic interventions. Signal Transduct Target Ther 8(1):138

Wang Z, Xu W, Comhair SAA, Fu X, Shao Z, Bearden R, Zein JG, Bleecker ER, Castro M, Denlinger LC, Fahy JV, Israel E, Levy BD, Jarjour NN, Moore WC, Wenzel SE, Mauger DT, Gaston B, Hazen SL, Erzurum SC (2022) Urinary total conjugated 3-bromotyrosine, asthma severity, and exacerbation risk. Am J Phys Lung Cell Mol Phys 323(5):L548–L557

Wedes SH, Khatri SB, Zhang R, Wu W, Comhair SA, Wenzel S, Teague WG, Israel E, Erzurum SC, Hazen SL (2009) Noninvasive markers of airway inflammation in asthma. Clin Transl Sci 2(2):112–117

Pathophysiology of Allergic Airways Disease

2

Gaurav Gupta, Riya Thapa, Asif Ahmad Bhat, Sarita Rawat, Karuna Dhaundhiyal, Ishwar Singh Dhramshaktu, and Abhijeet Ojha

Abstract

Allergic airways disease, encompassing conditions such as asthma and allergic rhinitis, presents a significant global health burden. This chapter provides a comprehensive exploration of its pathophysiology, including immune responses, inflammatory mediators, airway remodeling, and the role of allergens. Current diagnostic methodologies and biomarkers crucial for accurate diagnosis and monitoring are discussed, emphasizing the need for a multidimensional diagnostic approach. Furthermore, this chapter sheds light on the implications of these findings for clinical practice and research, advocating for personalized therapeutic strategies and further interdisciplinary studies to propel advancements in the field, ultimately enhancing patient care and outcomes.

Keywords

Allergens · Airway · Pathophysiology · Cytokines · Inflammatory mediators

---

G. Gupta (✉) · R. Thapa · A. A. Bhat
School of Pharmacy, Suresh Gyan Vihar University, Jaipur, Rajasthan, India

S. Rawat
School of Pharmacy, Suresh Gyan Vihar University, Jaipur, Rajasthan, India

Faculty of Pharmacy, Amrapali Group of Institute, Haldwani, Nainital, Uttarakhand, India

K. Dhaundhiyal
Faculty of Pharmacy, Amrapali Group of Institute, Haldwani, Nainital, Uttarakhand, India

Graphic Era Hill University, Dehradun, Uttarakhand, India

I. S. Dhramshaktu
Dr. Sushila Tiwari Medical college and Hospital, Haldwani, Nainital, Uttarakhand, India

A. Ojha
Faculty of Pharmacy, Amrapali Group of Institute, Haldwani, Nainital, Uttarakhand, India

P. Prasher et al. (eds.), *Understanding Allergic Airway Diseases*,
https://doi.org/10.1007/978-981-97-1953-2_2

2.1 Introduction

Allergic airways disease, also known as allergic respiratory disease or allergic airway inflammation, encompasses a group of chronic respiratory disorders characterized by hypersensitive airways and immune system responses to specific allergens. This condition primarily involves the respiratory tract, including the nasal passages, throat, bronchi, and lungs. It significantly affects the quality of life for those afflicted and represents a substantial global health concern. Allergic airways diseases can manifest in various forms, including allergic rhinitis (hay fever), allergic asthma, allergic bronchitis, and allergic sinusitis (Ackerman et al. 2016; Adams et al. 2019). The common thread among these conditions is an exaggerated immune response triggered by exposure to allergens. Allergens are substances from the environment that, for most individuals, are harmless but induce an immune response in allergic individuals. The pathophysiology of allergic airways disease centers on an abnormal immune response, primarily involving two key immune components: immunoglobulin E (IgE) and various immune cells, such as mast cells, eosinophils, and T lymphocytes (Agresta et al. 2014; Ahmad et al. 2023).

When an allergic individual is exposed to an allergen, the immune system perceives it as a threat, leading to the production of allergen-specific IgE antibodies. These antibodies bind to specific receptors on mast cells and basophils, priming these cells for subsequent allergen encounters. Upon re-exposure to the allergen, it cross-links with the IgE antibodies on these cells, triggering the release of inflammatory mediators like histamine, leukotrienes, and cytokines. Histamine and other mediators induce a cascade of reactions, resulting in typical allergic symptoms. In the upper respiratory tract, this can manifest as sneezing, nasal congestion, and itching, while in the lower respiratory tract, it may lead to coughing, wheezing, shortness of breath, and chest tightness, which are characteristics of asthma. The inflammatory response in allergic airways disease involves the recruitment and activation of eosinophils, which release pro-inflammatory substances, contributing to airway inflammation and tissue damage (Al Heialy et al. 2011; Allen et al. 2009). This chronic inflammation leads to airway remodeling, characterized by structural changes in the airways, such as increased smooth muscle mass, collagen deposition, and mucus gland hyperplasia. These alterations further exacerbate the symptoms and complicate the management of the disease.

Environmental factors play a significant role in the development and exacerbation of allergic airways disease. Common allergens include pollen, pet dander, mold spores, dust mites, cockroach droppings, and certain foods. Additionally, respiratory infections, cigarette smoke, air pollution, and occupational exposures can worsen the condition. Management of allergic airways disease typically involves allergen avoidance, pharmacotherapy (e.g., antihistamines, corticosteroids, bronchodilators), allergen immunotherapy (desensitization), and lifestyle modifications. However, there is ongoing research and development of new treatment modalities to improve outcomes and provide more targeted and personalized approaches (Arm and Lee 1992; Asano et al. 2020; Athari 2019; Athari et al. 2017).

Understanding the pathophysiology of allergic airways disease is crucial for tailoring effective treatments and interventions. It provides insights into the mechanisms underlying the disease, aiding in the development of targeted therapies. Furthermore, comprehension of the disease's molecular and cellular pathways enables early diagnosis, accurate prognosis, and informed patient education. It forms the foundation for ongoing research, fostering innovation and advancements in medical approaches, ultimately striving for improved management and, potentially, prevention of this prevalent and burdensome respiratory condition (Audrit et al. 2017; Bachert et al. 2006, 2004; Backaert et al. 2021).

The primary objectives of this chapter on the pathophysiology of allergic airways disease are to consolidate existing knowledge and research, offering a comprehensive understanding of the disease's mechanisms. It aims to synthesize diverse findings, highlighting key immune processes, allergens, and triggers. Additionally, the chapter seeks to identify gaps in current knowledge, providing a basis for future research directions. Ultimately, by presenting a structured and informed overview, this chapter aspires to enhance clinical management and therapeutic strategies, ultimately improving the quality of life for individuals afflicted by allergic airways disease (Baraniuk 1998; Bates 2016; Bates et al. 2009; Blanchet et al. 2012; Bochner et al. 1994).

The scope of this chapter encompasses an in-depth analysis of the pathophysiology of allergic airways disease, focusing on immune mechanisms, airway inflammation, allergens, clinical manifestations, diagnostics, and therapeutic approaches (Bonay and Aubier 2007; Bonekat and Hardin 2003; Bonini and Silvers 2018). The chapter integrates historical perspectives, current research, and future prospects to provide a comprehensive understanding. Structurally, it follows a logical flow, commencing with an introduction and historical context, followed by sections delving into specific aspects of the disease. The chapter culminates in a conclusion summarizing key findings and implications, ensuring a thorough and organized presentation of the complex pathophysiology of allergic airways disease.

2.2 Historical Perspective

2.2.1 Early Discoveries and Descriptions of Allergic Airways Disease

Early discoveries and descriptions of allergic airways disease trace back to antiquity, although a comprehensive understanding has evolved significantly over time. Historically, the first recorded observations can be attributed to ancient civilizations like the Egyptians, Greeks, and Chinese, who documented symptoms resembling those of modern allergic airways disease. In ancient Egypt, writings dating back to 1500 BCE mention symptoms akin to hay fever, describing nasal congestion, sneezing, and eye irritation in response to certain plant exposures. Similarly, ancient Greek texts, such as those by Hippocrates (460–370 BCE), documented respiratory symptoms related to environmental exposures, emphasizing the influence of climate and seasons on respiratory health (Bourke et al. 1989; Braido et al. 2011). Moving forward in history, the Renaissance period witnessed further documentation of allergic-like symptoms. In the sixteenth century, Italian physician Girolamo Fracastoro described respiratory reactions to dust and various irritants, offering early insights into the triggers of respiratory ailments. Later, in the seventeenth and eighteenth centuries, physicians like Bernardino Ramazzini and John Floyer made notable contributions by linking occupational exposures to respiratory symptoms, highlighting the role of specific environmental factors. The term "hay fever" was coined by Dr. John Bostock in 1819, acknowledging the seasonal nature of allergic symptoms related to exposure to plant pollens. This marked a significant milestone, setting the stage for the identification of allergic airways disease as a distinct clinical entity (Braunstahl and Hellings 2003, 2006; Brown et al. 2007).

However, a comprehensive understanding of the underlying immunological mechanisms was not realized until much later. The late nineteenth and early twentieth centuries witnessed breakthroughs in immunology, especially the discovery of immunoglobulins and the elucidation of the antibody-mediated immune response by scientists like Paul Ehrlich and Emil von Behring. These foundational discoveries laid the groundwork for linking allergic reactions to immune responses (Calhoun et al. 1991; Calzetta et al. 2014).

2.2.2 Milestones in Understanding the Pathophysiology

Milestones in understanding the pathophysiology of allergic airways disease represent a continuum of discoveries, each adding a layer of understanding to the complex interplay of immune responses and environmental triggers. In the mid-twentieth century, advances in immunology brought the concept of atopy to the forefront. Researchers like Coca and Cooke introduced the idea of an allergic diathesis, emphasizing inherited predispositions to allergic reactions. This laid the foundation for recognizing genetic factors in allergic airways disease (Campanella and Asher 2001; Cao et al. 2016).

The 1960s and 1970s saw significant progress in identifying specific immunoglobulins associated with allergic reactions, particularly IgE. In 1967, Ishizaka and colleagues isolated and characterized IgE, linking it to immediate hypersensitivity reactions. This discovery revolutionized our understanding of allergic diseases and paved the way for targeted therapeutic interventions. Further understanding of the inflammatory processes came with the discovery of the role of eosinophils and their mediators in allergic airways disease during the 1980s and 1990s. Eosinophils were identified as major contributors to tissue damage and inflammation in the airways. The advent of advanced imaging techniques and molecular biology in the late twentieth century allowed for a more detailed exploration of airway remodeling, providing insights into the structural changes that characterize chronic allergic airways disease (Cavaliere et al. 2018; Chapman et al. 2019, 2014).

Recent decades have witnessed the discovery of intricate immune signaling pathways and cytokines, shedding light on the complex interactions driving the inflammatory response in allergic airways disease. The identification of Th2 cell activation and their role in orchestrating allergic inflammation has notably advanced our understanding and facilitated the development of targeted therapies (Chawes 2011; Chiba et al. 2007).

2.3 Immune Mechanisms in Allergic Airways Disease

2.3.1 Overview of the Immune System Involvement

The immune system's role in allergic airways disease is intricate and pivotal. It involves a complex interplay of immune cells, antibodies, and signaling molecules in response to allergens, which are typically harmless substances triggering an exaggerated immune response in allergic individuals. Sensitization marks the initial encounter with allergens, where antigen-presenting cells process and present allergens to T-helper cells, particularly Th2 cells. Th2 cells play a central role in activating B cells to produce allergen-specific immunoglobulin E (IgE) antibodies. Upon subsequent exposure to the allergen, IgE antibodies bind to receptors on mast cells and basophils, priming them. When the allergen reencounters these primed cells, it triggers a cascade of events leading to the release of inflammatory mediators such as histamine, leukotrienes, and cytokines (Chitano 2011; Cho et al. 2021). These mediators induce the characteristic allergic symptoms, including bronchoconstriction, mucus secretion, itching, and tissue inflammation. Concurrently, the immune response recruits inflammatory cells, notably eosinophils, amplifying the inflammation. Prolonged exposure to allergens results in chronic airway inflammation, contributing to the characteristic features of allergic airways disease. Eosinophils and other immune cells play crucial roles in this sustained inflammation, leading to airway remodeling, characterized by structural changes in the airways. These alterations include smooth muscle hypertrophy, mucus gland hyperplasia, and thickening of the airway walls, further exacerbating airway obstruction and dysfunction. Understanding this intricate immune response is paramount for the development of targeted therapeutic interventions that modulate specific components of the immune system, providing effective relief and improved quality of life for individuals affected by allergic airways disease. Ongoing research and advancements in immunology continue to unveil new insights, fueling the quest for more precise and efficient treatments (Chung 2017; Cockcroft et al. 2007; Cohn et al. 2004).

2.3.2 Role of Immunoglobulin E (IgE) in Allergic Reactions

Immunoglobulin E (IgE) stands at the forefront of allergic reactions, playing a pivotal role in the immune system's response to allergens. IgE is a specialized class of antibody produced by B cells in response to exposure to specific allergens like pollen, pet dander, or certain foods. Upon initial contact with an allergen, the immune system recognizes it as a threat and initiates a sensitization process.

Antigen-presenting cells present the allergen to T-helper cells, triggering the production of Th2 cytokines that instruct B cells to differentiate into plasma cells, ultimately generating allergen-specific IgE antibodies. These IgE antibodies have a high affinity for FcεRI receptors found on the surface of mast cells and basophils. Subsequent encounters with the same allergen lead to cross-linking of IgE antibodies on mast cells and basophils. This cross-linking triggers the release of a plethora of inflammatory mediators, most notably histamine, which is central in initiating allergic responses. Histamine induces vasodilation, increased vascular permeability, and smooth muscle contraction, culminating in the characteristic allergic symptoms such as itching, sneezing, bronchoconstriction, and tissue inflammation (Coleman and Shaw 2017; Constantino Gde and Mello Jr. 2009; Corren 1997).

The high specificity and affinity of IgE antibodies for allergens make them potent mediators of allergic reactions. Understanding the role of IgE in allergic responses is vital for the development of targeted therapies, including anti-IgE treatments, that aim to block the allergic cascade by neutralizing IgE or preventing its binding to mast cells and basophils. These therapies have revolutionized the management of allergic airways disease and other allergic conditions, significantly improving the quality of life for affected individuals (Corren 2007; D'Agostino et al. 2019).

2.3.3 T-Cell-Mediated Responses in Allergic Airways Disease

T-cell-mediated responses are significant contributors to the pathophysiology of allergic airways disease, representing a crucial facet of the immune system's involvement. In allergic individuals, T-helper cells, specifically Th2 cells, play a central role in orchestrating immune responses to allergens. Upon allergen exposure, antigen-presenting cells process and present allergenic peptides to naïve T cells, leading to the activation and differentiation of Th2 cells. Th2 cells secrete a range of cytokines such as interleukin-4 (IL-4), interleukin-5 (IL-5), and interleukin-13 (IL-13), among others, which drive and perpetuate allergic inflammation. IL-4 facilitates the production of allergen-specific immunoglobulin E (IgE) antibodies by B cells, contributing to sensitization. IL-5 promotes the recruitment, activation, and survival of eosinophils, a characteristic cell type involved in allergic airway inflammation. Eosinophils release inflammatory mediators that perpetuate tissue damage and amplify the allergic response. IL-13 plays a role in airway hyperreactivity, mucus hypersecretion, and airway remodeling, further contributing to the chronicity of the disease (Davis and Cockcroft 2012; De et al. 2018; de Benedictis and Bush 2020).

The delicate balance between Th1 and Th2 responses is disrupted in allergic airways disease, with an exaggerated Th2 response prevailing. Understanding the intricacies of T-cell-mediated responses provides valuable insights into the immune dysregulation that underlies allergic airways disease. Targeted therapies aimed at modulating these responses, such as monoclonal antibodies against specific cytokines or receptors, hold promise in effectively managing and mitigating the impact of allergic airways disease, providing avenues for improved treatments and enhanced quality of life for affected individuals (Dembele et al. 2021; Denburg et al. 2000).

2.3.4 Mast Cell Activation and its Effects on Airways

Mast cell activation stands as a pivotal mechanism in allergic airways disease, exerting profound effects on the respiratory system. Mast cells, abundantly present in tissues throughout the body, especially in proximity to surfaces exposed to the external environment such as the skin and mucosal linings of the airways and gastrointestinal tract, are armed with an array of preformed granules containing histamine, proteases, leukotrienes, and cytokines. Upon encountering allergens to which the individual is sensitized, cross-linking of allergen-specific immunoglobulin E (IgE) antibodies on the surface of mast cells occurs, initiating their degranulation. This degranulation is swift and results in the release of inflammatory mediators into the surrounding tissues and bloodstream (Dixon 2009; Djukanović 2000). Histamine, a primary mediator, induces vasodilation, increased vascular permeability, and smooth muscle contraction in the airways. This cascade of events leads to the characteristic symptoms of allergic airways disease, including bronchoconstriction, shortness of breath, wheezing, coughing, and nasal congestion. Moreover, the released leukotrienes, prostaglandins, and cytokines further enhance the inflammatory response, attracting and activating other immune cells like eosinophils, amplifying the allergic inflammation and tissue damage. Chronic mast cell activation contributes to airway remodeling, characterized by structural alterations like smooth muscle hypertrophy, subepithelial fibrosis, and increased mucus production, further exacerbating airway obstruction and dysfunction in allergic individuals (Donovan et al. 2015; Drolet et al. 2010; Elieh Ali Komi and Bjermer 2019) (Fig. 2.1).

Understanding mast cell activation and its effects on the airways is pivotal in developing targeted therapies that aim to mitigate the impact of allergic airways disease. Antihistamines and mast cell stabilizers are common medications used to alleviate symptoms by modulating mast cell activity and reducing the release of inflammatory mediators, highlighting the significance of mast cells in the pathophysiology of allergic airways disease and the potential for therapeutic interventions targeting this immune mechanism (Erjefält 2014; Fanucchi et al. 2004).

2.4 Airway Inflammation and Remodeling

2.4.1 Inflammatory Mediators and their Role

Inflammatory mediators are key molecular players in the pathophysiology of allergic airways disease, orchestrating a cascade of events that drive and sustain the inflammatory response. These mediators, including histamine, leukotrienes, prostaglandins, and cytokines, are released by various immune cells, particularly mast

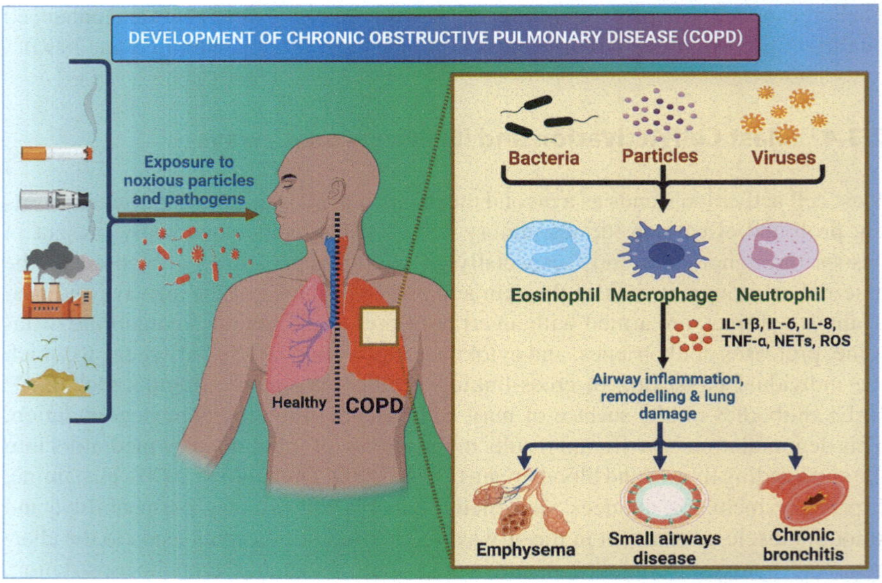

Fig. 2.1 Development of chronic obstructive pulmonary disease

cells, eosinophils, and basophils, in response to allergen exposure (Fasano 2010; Fear et al. 2016).

Histamine, a potent vasodilator, increases vascular permeability and smooth muscle contraction upon binding to its receptors, initiating the classic allergic symptoms such as itching, sneezing, bronchoconstriction, and tissue swelling. Leukotrienes, derived from arachidonic acid, induce bronchoconstriction and promote mucus production, contributing to airway obstruction and coughing. Prostaglandins also play a role in bronchoconstriction and mucus secretion, exacerbating airway inflammation and hyperreactivity. Cytokines, another class of inflammatory mediators, are crucial in immune cell communication and regulation of the immune response. Interleukins, such as IL-4, IL-5, and IL-13, are central in allergic airways disease. IL-4 facilitates IgE production, further sensitizing the individual to allergens, while IL-5 promotes eosinophil activation and recruitment, intensifying inflammation and tissue damage. IL-13 contributes to airway hyperreactivity and mucus hypersecretion, contributing to the chronicity and severity of allergic airways disease (Fireman 1993; Fokkens and Scheeren 2000; Fonseca and Arrobas 2006).

Understanding the role of these inflammatory mediators is essential for developing targeted therapeutic approaches. Current treatments, including antihistamines, leukotriene inhibitors, and monoclonal antibodies targeting specific cytokines, aim to modulate these mediators to alleviate symptoms and reduce inflammation. Advances in research targeting these mediators hold promise for more effective and tailored treatments, ultimately improving the management and outcomes of allergic airways disease (Foong et al. 2016; Gelfand 2004).

2.4.2 Cellular Components of Airway Inflammation

Cellular components of airway inflammation in allergic airways disease encompass a diverse array of immune cells, each playing a distinct yet interconnected role in the inflammatory cascade. Among these, eosinophils are prominent and pivotal. Eosinophils are heavily involved in allergic responses, releasing toxic granules and pro-inflammatory cytokines that contribute to tissue damage and exacerbate inflammation. Their infiltration into the airways is a hallmark of allergic airways disease, contributing to bronchial hyperreactivity and airway remodeling (Gern 2003; Gillissen et al. 2005).

Neutrophils also play a role, particularly in severe or chronic cases. They release inflammatory mediators and enzymes, contributing to tissue destruction and amplifying the inflammatory response. Mast cells, as primary mediators of allergic reactions, release histamine and other inflammatory molecules upon activation, triggering immediate allergic responses (Godard et al. 1989; Goldstein et al. 1994).

Furthermore, T lymphocytes, particularly Th2 cells, orchestrate and perpetuate allergic inflammation by releasing cytokines like IL-4, IL-5, and IL-13, which stimulate B cells to produce IgE and activate eosinophils. B lymphocytes, in turn, produce allergen-specific IgE antibodies that bind to mast cells and basophils, priming them for subsequent allergic reactions (Greenberger and Grammer 2010; Groneberg and Chung 2004).

Monocytes and macrophages contribute by producing inflammatory cytokines and engulfing foreign particles, further intensifying the inflammatory response. These cellular components collectively drive the chronic inflammatory state characteristic of allergic airways disease, resulting in airway hyperreactivity, mucus hypersecretion, and structural changes in the airways. Understanding and targeting these cellular components is critical for developing therapies that effectively modulate the inflammatory cascade and mitigate the impact of allergic airways disease (Groneberg et al. 2004; Guan et al. 2018).

2.4.3 Inflammatory Signaling Pathways

Inflammatory signaling pathways are intricate networks of molecular interactions that govern the initiation, progression, and resolution of inflammation, a critical component of the immune response in allergic airways disease. Several pathways are implicated, often interlinked, playing a significant role in orchestrating the inflammatory cascade (Haccuria et al. 2018; Hall and Agrawal 2016).

One of the key pathways is the NF-κB (nuclear factor-kappa B) pathway, a central regulator of immune and inflammatory responses. Activation of NF-κB leads to the production of pro-inflammatory cytokines, chemokines, and adhesion molecules, promoting immune cell recruitment and amplifying inflammation. Another vital pathway is the JAK-STAT (Janus kinase-Signal Transducer and Activator of Transcription) pathway, particularly associated with cytokine signaling. Cytokines like IL-4, IL-5, and IL-13 activate JAK-STAT signaling, amplifying Th2 responses and eosinophil activation. Phosphoinositide 3-kinase (PI3K) and MAPK (Mitogen-Activated Protein Kinase) pathways are also crucial in the inflammatory process. These pathways are involved in cell survival, activation, and proliferation, contributing to the perpetuation of inflammation. Additionally, the arachidonic acid pathway, involving enzymes like cyclooxygenase and lipoxygenase, produces eicosanoids such as prostaglandins and leukotrienes, potent mediators of inflammation and bronchoconstriction (Halpern 2012; Hamelmann and Gelfand 2001; Hamid and Tulic 2009). Understanding these signaling pathways is pivotal for targeted therapeutic interventions aimed at modulating the inflammatory response in allergic airways disease. Current research focuses on developing inhibitors and biologics that selectively target specific molecules within these pathways, providing opportunities for more precise and effective anti-inflammatory treatments to alleviate symptoms and improve the management of allergic airways disease (Hamid and Tulic 2007; Hellings and Prokopakis 2010).

2.4.4 Airway Remodeling Processes and their Consequences

Airway remodeling is a complex and chronic process that occurs in allergic airways disease, leading to structural changes in the airways. This phenomenon involves alterations in the airway's cellular and structural components, including the epithelium, smooth muscle, blood vessels, and extracellular matrix. Over time, persistent inflammation and repeated exposure to allergens cause the airways to undergo remodeling, significantly impacting lung function and contributing to disease progression (Hellings and Steelant 2020; Henry et al. 2005). One of the primary consequences of airway remodeling is increased smooth muscle mass or hypertrophy. Smooth muscle cells undergo proliferation and hypertrophy, leading to airway narrowing and increased airway responsiveness, a hallmark of asthma. Additionally, the mucus-producing cells in the airways, known as goblet cells, proliferate and hypersecrete mucus, resulting in increased mucus production and airway obstruction. Moreover, there is thickening of the airway walls due to subepithelial fibrosis, involving the accumulation of collagen and other extracellular matrix proteins. This structural alteration further compromises airway function, exacerbating airflow limitation. Blood vessels in the airway may also undergo angiogenesis, leading to increased vascularity and enhanced inflammatory responses (Hens and Hellings 2006; Hogg et al. 1979).

Collectively, these remodeling processes result in reduced airflow, persistent airway inflammation, and heightened bronchial hyperresponsiveness. The consequences of airway remodeling are profound, affecting the individual's ability to breathe freely, worsened disease severity, and increased frequency and severity of asthma exacerbations. Understanding these remodeling processes and their implications is crucial for developing therapeutic strategies that can target and potentially reverse the structural changes, providing better management and improving the quality of life for individuals with allergic airways disease (Hohlfeld 2002; Holgate 2008) (Fig. 2.2).

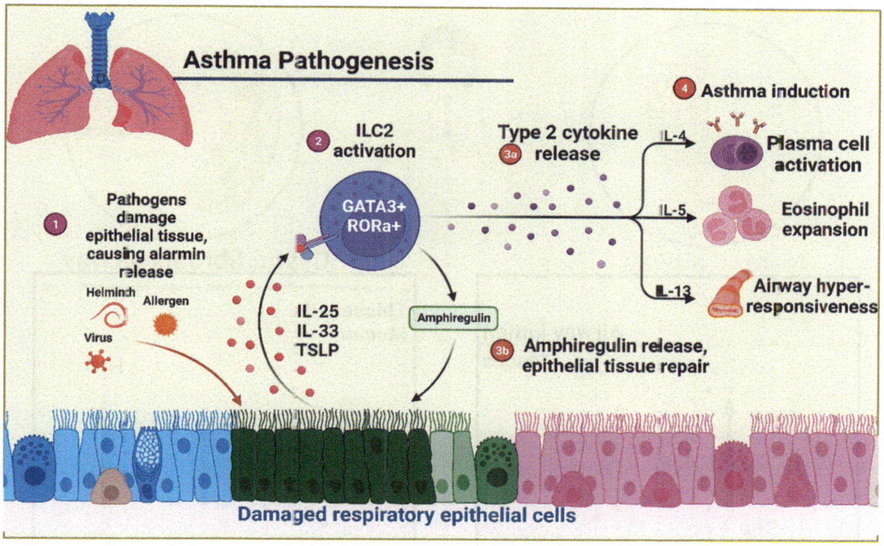

Fig. 2.2 Pathogenesis of asthma

2.5 Allergens and Triggers

2.5.1 Common Allergens in Allergic Airways Disease

Common allergens in allergic airways disease encompass a spectrum of substances that provoke immune hypersensitivity in susceptible individuals, triggering allergic reactions within the respiratory system. These allergens primarily consist of proteins or glycoproteins derived from sources such as pollens, dust mites, animal dander, mold spores, cockroach droppings, and certain foods. Pollens, prevalent during specific seasons, are significant contributors to allergic rhinitis and often induce hay fever symptoms. House dust mites, predominantly *Dermatophagoides pteronyssinus* and *Dermatophagoides farinae*, are common indoor allergens found in bedding, carpets, and upholstery, triggering allergic reactions (Holgate 2010, 2011). Animal allergens, arising from pet skin flakes, urine, and saliva, are notable culprits in allergic asthma, particularly from cats, dogs, and rodents. Mold spores from damp and humid environments, such as Aspergillus and Penicillium, are additional allergenic sources. Cockroach droppings, primarily from German cockroaches, contain allergenic proteins and are prevalent in urban dwellings. Certain foods like peanuts, tree nuts, milk, eggs, wheat, soy, and shellfish can induce allergic reactions, not limited to the gastrointestinal tract but often manifesting in respiratory symptoms. Occupational allergens, including chemicals, dust, and fumes in various workplaces, also play a role in allergic airways disease. Understanding and identifying these common allergens are fundamental for accurate diagnosis, effective allergen

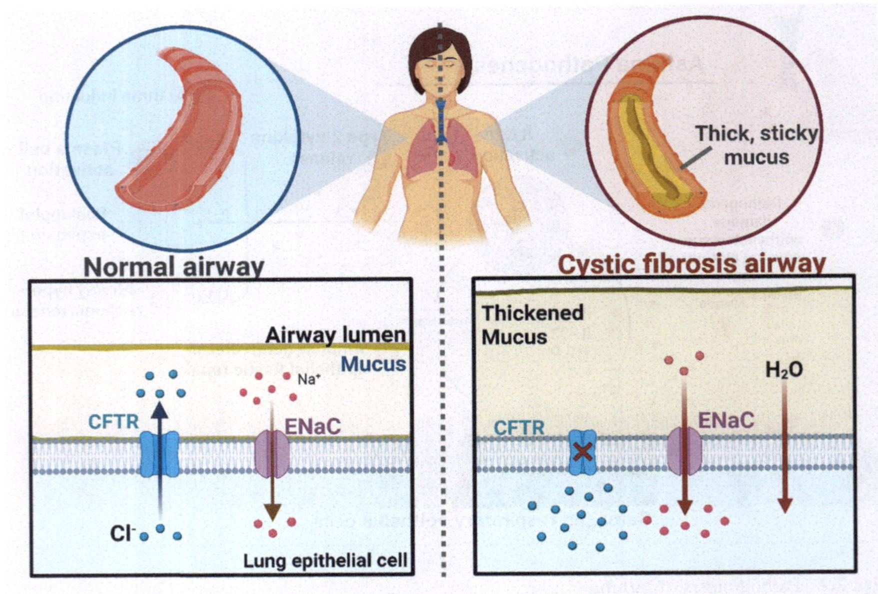

Fig. 2.3 A difference among normal airway and cystic fibrosis airway

avoidance strategies, and targeted therapeutic interventions in managing allergic airways disease (Holloway et al. 2010; Humbles et al. 1997) (Fig. 2.3).

2.5.2 Mechanisms of Allergen-Induced Airway Responses

The mechanisms of allergen-induced airway responses are intricate and central to the pathophysiology of allergic airways disease. Upon exposure to specific allergens, the immune system in susceptible individuals mounts a hypersensitive response characterized by distinct phases. Sensitization occurs during the initial exposure, wherein the allergen is recognized by antigen-presenting cells like dendritic cells. These cells process the allergen and present it to T-helper cells (primarily Th2 cells). Th2 cells, upon activation, produce cytokines such as interleukin-4 (IL-4), which stimulate B cells to differentiate into plasma cells and produce allergen-specific immunoglobulin E (IgE) antibodies (Idzko et al. 2015; Inoue 2011). IgE antibodies bind to receptors on mast cells, particularly FcεRI receptors. Upon subsequent exposure to the same allergen, the allergen cross-links with IgE antibodies on mast cells, leading to their degranulation. Degranulation results in the release of various inflammatory mediators, including histamine, leukotrienes, and prostaglandins, initiating the early-phase allergic response. Histamine induces smooth muscle contraction, increased vascular permeability, and mucus secretion, culminating in bronchoconstriction, edema, and mucus production. This leads to acute symptoms such as wheezing, coughing, and shortness of breath (Jakwerth et al. 2022; Kallinich et al. 2005).

Following the early-phase response, a late-phase response occurs, characterized by the recruitment of inflammatory cells, primarily eosinophils, and the release of additional inflammatory cytokines. Eosinophils play a central role in allergic airway inflammation, releasing toxic granules and further amplifying tissue damage and inflammation. These complex mechanisms of allergen-induced airway responses underline the chronic and recurring nature of allergic airways disease, necessitating targeted therapeutic interventions to mitigate symptoms and prevent long-term airway damage (Karta et al. 2016; Kaur and Chupp 2019).

2.5.3 Interaction between Genetic and Environmental Factors

The interaction between genetic and environmental factors is a fundamental determinant in the onset, progression, and manifestation of various diseases, including allergic airways disease. Genetic predisposition, often involving multiple genes, influences an individual's susceptibility to the development of allergies. Genetic variations can affect immune responses, alter the production of specific antibodies (e.g., IgE), and modulate inflammatory pathways, thereby impacting an individual's response to allergens. Additionally, genetic factors may influence the likelihood of developing conditions such as asthma, allergic rhinitis, or atopic dermatitis (Keir and Page 2008; Khan 2014).

On the other hand, environmental factors play a pivotal role in triggering and exacerbating allergic reactions. Exposure to allergens, such as pollen, dust mites, pet dander, mold spores, and certain foods, can induce allergic responses in genetically susceptible individuals. Furthermore, environmental pollutants like tobacco smoke, air pollutants, and occupational hazards can aggravate allergic airway symptoms and worsen disease severity (KleinJan et al. 2010; Koterba and Saltoun 2012).

The interplay between these genetic and environmental factors is complex and dynamic. Genetic predisposition sets the stage, influencing an individual's susceptibility, while environmental exposures act as triggers and modifiers. Understanding these interactions is crucial for personalized medicine approaches, enabling tailored strategies for prevention, early intervention, and targeted treatments, ultimately improving the management and outcomes of allergic airways disease. Furthermore, ongoing research aims to unravel the intricate relationship between genetics and the environment, paving the way for advancements in precision medicine and more effective therapeutic interventions (Kraneveld et al. 1997; Krouse and Altman 2010).

2.6 Clinical Manifestations and Symptomatology

2.6.1 Common Symptoms Experienced by Patients

Patients with allergic airways disease often experience a range of common symptoms that significantly impact their respiratory health and overall quality of life. Wheezing, a characteristic high-pitched whistling sound during breathing, is a hallmark symptom and often accompanies shortness of breath or difficulty in breathing, particularly during physical exertion or at night. Persistent coughing, especially at night or in the early morning, is another prevalent symptom, sometimes accompanied by sputum production or a sensation of tightness in the chest. Nasal symptoms include frequent sneezing, a runny or congested nose, and itching or irritation in the nasal passages, often resembling symptoms of allergic rhinitis. Itchy, watery, or red eyes, known as allergic conjunctivitis, are also common. In some cases, patients may experience an exacerbation of symptoms, known as asthma attacks or flare-ups, characterized by a sudden intensification of wheezing, coughing, and shortness of breath, often requiring immediate medical attention (Lang 2019; Lee et al. 2015). These symptoms can be triggered or worsened by exposure to specific allergens, such as pollen, dust mites, pet dander, or certain environmental factors like smoke or pollutants. The presence and severity of these symptoms may vary from person to person and can fluctuate over time. Understanding and effectively managing these common symptoms are essential for patients, enabling them to proactively address their condition, adhere to prescribed treatment plans, and seek timely medical intervention to alleviate symptoms and improve their day-to-day functioning and well-being (Li et al. 2019; Lindén 2006).

2.6.2 Variability of Symptoms and their Correlation with Pathophysiological Changes

The variability of symptoms in allergic airways disease often correlates with underlying pathophysiological changes within the respiratory system. This variability is a hallmark of the disease, with symptoms ranging from mild, intermittent episodes to severe and persistent manifestations. Airway inflammation plays a central role; its intensity, persistence, and response to treatment directly influence symptom severity. Inflammation in the airways, marked by the presence of immune cells like eosinophils, mast cells, and T lymphocytes, leads to bronchial hyperreactivity and increased mucus production, contributing to symptoms like wheezing, coughing, and shortness of breath. The cyclical and unpredictable nature of allergic airways disease stems from the interplay of various factors. Allergen exposure, environmental pollutants, respiratory infections, and even emotional stress can trigger or exacerbate symptoms. Additionally, individual genetic predispositions and immune responses contribute to symptom variability (Liu et al. 2020, 2017, 2022).

Long-term structural changes in the airways, known as airway remodeling, further exacerbate symptoms and their variability. Remodeling involves alterations in airway smooth muscle, epithelial cells, and the extracellular matrix, leading to increased airway hyperreactivity and reduced lung function. Understanding the relationship between symptom variability and pathophysiological changes is critical for tailoring treatment plans. Personalized approaches, including identifying specific triggers, targeting inflammation, and addressing airway remodeling, are key strategies to effectively manage the variable nature of symptoms and optimize the quality of life for individuals with allergic airways disease. Further research in this area holds promise for better understanding these complex relationships and improving targeted therapeutic interventions (Liu et al. 2015; Lo et al. 2018; Locke et al. 2007).

2.6.3 Impact of Allergic Airways Disease on Quality of Life

Allergic airways disease exerts a profound and multifaceted impact on an individual's quality of life, encompassing physical, emotional, and social dimensions. Physically, the persistent and recurrent symptoms such as wheezing, coughing, shortness of breath, and nasal congestion can significantly impair daily activities, exercise, and sleep patterns. Patients often experience fatigue and reduced stamina due to the effort required to breathe, hindering overall productivity. Frequent asthma exacerbations can necessitate hospitalizations or emergency visits, disrupting normal routines and causing distress (Löser et al. 2017; Maarsingh et al. 2008). Emotionally, the chronic nature of allergic airways disease can lead to heightened levels of anxiety, frustration, and fear, stemming from the unpredictability and variability of symptoms. The constant need for vigilance and the potential for sudden exacerbations can negatively impact mental well-being and self-confidence. Socially, the limitations imposed by the disease can result in social isolation and decreased participation in social and recreational activities. Missed school or work days due to illness can also affect academic and professional performance, potentially limiting career growth and opportunities (Maarsingh et al. 2009; Macdowell and Peters 2007).

Moreover, the financial burden of medical expenses, medications, and healthcare visits adds an additional layer of stress. The collective impact of these physical, emotional, and social factors diminishes the overall quality of life for individuals with allergic airways disease. Comprehensive management strategies focusing on symptom control, patient education, psychosocial support, and targeted interventions are essential to mitigate this impact and enhance the well-being and daily functioning of affected individuals (Marple 2010; Marseglia et al. 2007).

2.7 Diagnostic and Monitoring Approaches

2.7.1 Imaging Techniques for Assessing Airway Structure and Function

Imaging techniques are pivotal in evaluating both airway structure and function, providing critical insights into respiratory health. High-resolution computed tomography (HRCT) stands out for its ability to capture detailed airway morphology, showcasing structural abnormalities, airway wall thickness, and bronchial dimensions. HRCT can noninvasively detect bronchial wall thickening, a hallmark of airway remodeling in allergic airways disease. Complementary techniques like magnetic resonance imaging (MRI) offer advantages in assessing dynamic changes in airway dimensions during the respiratory cycle, allowing for a better understanding of airway mechanics and function (Marseglia et al. 2011; Meyer and Akdis 2013). Functional imaging modalities, including positron emission tomography (PET) and single-photon emission computed tomography (SPECT), can provide valuable data on regional lung ventilation, perfusion, and inflammation, aiding in diagnosing and monitoring airway diseases. Lung scintigraphy, utilizing inhaled radioactive gases or aerosols, offers valuable functional data on regional airflow, making it useful in assessing regional lung function and identifying ventilation abnormalities (Millien et al. 2014; Montuschi et al. 2007).

Furthermore, emerging techniques like optical coherence tomography (OCT) provide high-resolution, real-time imaging of airway microstructure, enabling visualization of cellular and subcellular details within the airway epithelium. These advanced imaging technologies collectively offer a comprehensive understanding of airway structure and function, aiding in disease diagnosis, progression monitoring, and treatment planning. Integrating these imaging tools in research and clinical practice can significantly contribute to personalized medicine approaches, guiding precise therapeutic strategies and ultimately improving the management and outcomes of allergic airways disease (Morrissey and Evans 2003; Movassagh et al. 2017).

2.7.2 Biomarkers of Allergic Airways Disease

Biomarkers in allergic airways disease are crucial indicators that aid in diagnosing, predicting disease progression, and monitoring treatment response. Immunoglobulin E (IgE) levels, central to allergic reactions, are a classic biomarker. Elevated total and allergen-specific IgE levels in blood indicate sensitization to specific allergens. Eosinophils, a type of white blood cell, are key players in allergic inflammation. Elevated eosinophil counts in blood, sputum, or bronchoalveolar lavage fluid serve as biomarkers for airway inflammation and disease severity. Fractional exhaled nitric oxide (FeNO), a noninvasive marker, indicates eosinophilic airway inflammation and helps in asthma diagnosis and management (Mur et al. 2020; Nassenstein et al. 2018; Nelson 2007). Additionally, periostin, a protein associated with airway remodeling, is a promising biomarker, reflecting disease severity and response to specific therapies. Inflammatory cytokines like interleukins (e.g., IL-4, IL-5, IL-13), measured in blood, sputum, or bronchoalveolar lavage, provide insights into the immune response and can guide treatment decisions. Genetic markers, such as specific polymorphisms in genes related to asthma or allergic rhinitis, are being increasingly recognized as potential biomarkers, aiding in identifying individuals at higher risk. Advancements in omics technologies, including genomics, proteomics, and metabolomics, are expanding the landscape of biomarker discovery, enabling the identification of novel molecules that could further enhance disease understanding and management. Integrating these biomarkers into clinical practice allows for personalized treatment strategies, optimizing therapeutic interventions, and improving outcomes for individuals with allergic airways disease (Nockher and Renz 2006; Nur Husna et al. 2022; Ogawa et al. 2014).

2.7.3 Current Diagnostic Methodologies and their Limitations

Current diagnostic methodologies for allergic airways disease primarily rely on a combination of clinical evaluation, patient history, lung function tests, and specific biomarker measurements. Physical examination and patient-reported symptoms are often the initial steps, followed by spirometry and peak expiratory flow (PEF) measurements to assess lung function and detect airflow limitation. Skin prick tests or specific IgE blood tests are commonly used to identify allergens triggering allergic reactions. Fractional exhaled nitric oxide (FeNO) measurement aids in assessing airway inflammation, while chest X-rays or computed tomography (CT) scans provide structural information about the airways. Bronchial challenge tests, including methacholine or mannitol challenge, help confirm airway hyperresponsiveness (Oguma et al. 2008; Okano et al. 2015; Olaguibel et al. 2021; Padem and Saltoun 2019).

However, these methodologies have limitations. Clinical evaluation and symptoms can overlap with other respiratory conditions, making accurate diagnosis challenging. Lung function tests may not always reflect early-stage disease or adequately capture variations in symptoms. Allergy tests may yield false positives or negatives, and FeNO levels can be influenced by factors other than airway inflammation. Radiation exposure and cost-effectiveness are concerns with imaging techniques. Additionally, bronchial challenge tests are resource-intensive and may not always be feasible. Thus, a multimodal approach, taking into account patient history, clinical examination, lung function tests, specific biomarkers, and imaging, is essential to enhance diagnostic accuracy and guide appropriate management strategies for allergic airways disease. Ongoing research aims to improve diagnostic methodologies, seeking more accurate, accessible, and noninvasive techniques (Paiva Ferreira et al. 2019; Pakdaman et al. 2011; Pandya et al. 2002; Passalacqua and Canonica 2001).

2.8 Conclusion

The literature on allergic airways disease illuminates a complex interplay of immunological, environmental, and genetic factors contributing to disease onset, progression, and symptom variability. Immunoglobulin E (IgE)-mediated responses, T-cell involvement, inflammatory signaling pathways, and airway remodeling are fundamental mechanisms driving the disease. Common allergens, such as pollen, dust mites, and pet dander, play pivotal roles in exacerbating symptoms. The variability of symptoms corresponds with distinct pathophysiological changes, emphasizing the need for personalized therapeutic approaches. Imaging techniques like HRCT and functional modalities offer vital insights into airway structure and function. Biomarkers such as eosinophil count, FeNO levels, and inflammatory cytokines serve as crucial diagnostic and prognostic indicators. Understanding these facets of allergic airways disease is vital for improving diagnosis, tailoring treatment strategies, and ultimately enhancing patient outcomes.

In clinical practice, these insights underscore the importance of a multidimensional diagnostic approach, incorporating biomarkers, imaging, and clinical assessment. Personalized medicine, targeting specific pathways and utilizing advanced imaging technologies, holds promise for optimal treatment outcomes. Furthermore, research should focus on refining existing biomarkers, exploring novel ones, and investigating genetic associations to propel the field forward. Therapeutic strategies aiming at modulating immune responses and airway remodeling could revolutionize disease management.

Encouragement for further studies in allergic airways disease is warranted to unravel its intricate pathogenesis fully. Longitudinal studies exploring the role of emerging biomarkers, the impact of environmental changes, and the effectiveness of targeted therapies are essential. Comparative research to elucidate different phenotypes and endotypes within allergic airways disease will aid in tailored treatments. Additionally, fostering interdisciplinary collaborations integrating immunology, genetics, and technology is key to advancing our understanding and paving the way for innovative interventions. Ultimately, continuous research endeavors will enhance our ability to manage allergic airways disease comprehensively and improve the lives of affected individuals.

References

Ackerman SJ, Park GY, Christman JW, Nyenhuis S, Berdyshev E, Natarajan V (2016) Polyunsaturated lysophosphatidic acid as a potential asthma biomarker. Biomark Med 10(2):123–135

Adams DC, Miller AJ, Applegate MB, Cho JL, Hamilos DL, Chee A et al (2019) Quantitative assessment of airway remodelling and response to allergen in asthma. Respirology (Carlton, Vic) 24(11):1073–1080

Agresta MF, Saranz RJ, Lozano NA, Lozano A (2014) The relationship between rhinitis and asthma: is that all? Revista de la Facultad de Ciencias Medicas (Cordoba, Argentina) 71(2):111–121

Ahmad JG, Marino MJ, Luong AU (2023) Unified airway disease: future directions. Otolaryngol Clin N Am 56(1):181–195

Al Heialy S, McGovern TK, Martin JG (2011) Insights into asthmatic airway remodelling through murine models. Respirology (Carlton, Vic). 16(4):589–597

Allen JE, Bischof RJ, Sucie Chang HY, Hirota JA, Hirst SJ, Inman MD et al (2009) Animal models of airway inflammation and airway smooth muscle remodelling in asthma. Pulm Pharmacol Ther 22(5):455–465

Arm JP, Lee TH (1992) The pathobiology of bronchial asthma. Adv Immunol 51:323–382

Asano K, Ueki S, Tamari M, Imoto Y, Fujieda S, Taniguchi M (2020) Adult-onset eosinophilic airway diseases. Allergy 75(12):3087–3099

Athari SS (2019) Targeting cell signaling in allergic asthma. Signal Transduct Target Ther 4:45

Athari SS, Athari SM, Beyzay F, Movassaghi M, Mortaz E, Taghavi M (2017) Critical role of Toll-like receptors in pathophysiology of allergic asthma. Eur J Pharmacol 808:21–27

Audrit KJ, Delventhal L, Aydin Ö, Nassenstein C (2017) The nervous system of airways and its remodeling in inflammatory lung diseases. Cell Tissue Res 367(3):571–590

Bachert C, Patou J, Van Cauwenberge P (2006) The role of sinus disease in asthma. Curr Opin Allergy Clin Immunol 6(1):29–36

Bachert C, Vignola AM, Gevaert P, Leynaert B, Van Cauwenberge P, Bousquet J (2004) Allergic rhinitis, rhinosinusitis, and asthma: one airway disease. Immunol Allergy Clin N Am 24(1):19–43

Backaert W, Steelant B, Hellings PW, Talavera K, Van Gerven L (2021) A TRiP through the roles of transient receptor potential cation channels in type 2 upper airway inflammation. Curr Allergy Asthma Rep 21(3):20

Baraniuk JN (1998) Neuropeptides. Am J Rhinol 12(1):9–16

Bates JH (2016) Physiological mechanisms of airway hyperresponsiveness in obese asthma. Am J Respir Cell Mol Biol 54(5):618–623

Bates JH, Rincon M, Irvin CG (2009) Animal models of asthma. Am J Physiol Lung Cell Mol Physiol 297(3):L401–L410

de Benedictis FM, Bush A (2020) Janus looks both ways: how do the upper and lower airways interact? Paediatr Respir Rev 34:59–66

Blanchet MR, Gold MJ, McNagny KM (2012) Mouse models to evaluate the function of genes associated with allergic airway disease. Curr Opin Allergy Clin Immunol 12(5):467–474

Bochner BS, Undem BJ, Lichtenstein LM (1994) Immunological aspects of allergic asthma. Annu Rev Immunol 12:295–335

Bonay M, Aubier M (2007) Air pollution and allergic airway diseases. Med Sci: M/S 23(2):187–192

Bonekat HW, Hardin KA (2003) Severe upper airway obstruction during sleep. Clin Rev Allergy Immunol 25(2):191–210

Bonini M, Silvers W (2018) Exercise-induced bronchoconstriction: background, prevalence, and sport considerations. Immunol Allergy Clin N Am 38(2):205–214

Bourke SJ, Carter R, Anderson K, Boyd J, King S, Douglas B et al (1989) Obstructive airways disease in non-smoking subjects with pigeon fanciers' lung. Clin Exp Allergy 19(6):629–632

Braido F, Baiardini I, Lagasio C, Sclifò F, Canonica GW (2011) Allergic rhinitis in asthma. Panminerva Med 53(2):97–107

Braunstahl GJ, Hellings PW (2003) Allergic rhinitis and asthma: the link further unraveled. Curr Opin Pulm Med 9(1):46–51

Braunstahl GJ, Hellings PW (2006) Nasobronchial interaction mechanisms in allergic airways disease. Curr Opin Otolaryngol Head Neck Surg 14(3):176–182

Brown JL, Behndig AF, Sekerel BE, Pourazar J, Blomberg A, Kelly FJ et al (2007) Lower airways inflammation in allergic rhinitics: a comparison with asthmatics and normal controls. Clin Exp Allergy 37(5):688–695

Calhoun WJ, Sedgwick J, Busse WW (1991) The role of eosinophils in the pathophysiology of asthma. Ann N Y Acad Sci 629:62–72

Calzetta L, Rogliani P, Cazzola M, Matera MG (2014) Advances in asthma drug discovery: evaluating the potential of nasal cell sampling and beyond. Expert Opin Drug Discov 9(6):595–607

Campanella SG, Asher MI (2001) Current controversies: sinus disease and the lower airways. Pediatr Pulmonol 31(2):165–172
Cao PP, Shi LL, Xu K, Yao Y, Liu Z (2016) Dendritic cells in inflammatory sinonasal diseases. Clin Exp Allergy 46(7):894–906
Cavaliere C, Masieri S, Cavaliere F (2018) Therapeutic applications of capsaicin in upper airways. Curr Drug Targets 19(10):1166–1176
Chapman DG, Casey DT, Ather JL, Aliyeva M, Daphtary N, Lahue KG et al (2019) The effect of flavored E-cigarettes on murine allergic airways disease. Sci Rep 9(1):13671
Chapman DG, Tully JE, Nolin JD, Janssen-Heininger YM, Irvin CG (2014) Animal models of allergic airways disease: where are we and where to next? J Cell Biochem 115(12):2055–2064
Chawes BL (2011) Upper and lower airway pathology in young children with allergic- and non-allergic rhinitis. Dan Med Bull 58(5):B4278
Chiba Y, Yu Y, Sakai H, Misawa M (2007) Increase in the expression of matrix metalloproteinase-12 in the airways of rats with allergic bronchial asthma. Biol Pharm Bull 30(2):318–323
Chitano P (2011) Models to understand contractile function in the airways. Pulm Pharmacol Ther 24(5):444–451
Cho HJ, Ha JG, Lee SN, Kim CH, Wang DY, Yoon JH (2021) Differences and similarities between the upper and lower airway: focusing on innate immunity. Rhinology 59(5):441–450
Chung KF (2017) Airway microbial dysbiosis in asthmatic patients: a target for prevention and treatment? J Allergy Clin Immunol 139(4):1071–1081
Cockcroft DW, Hargreave FE, O'Byrne PM, Boulet LP (2007) Understanding allergic asthma from allergen inhalation tests. Can Respir J 14(7):414–418
Cohn L, Elias JA, Chupp GL (2004) Asthma: mechanisms of disease persistence and progression. Annu Rev Immunol 22:789–815
Coleman SL, Shaw OM (2017) Progress in the understanding of the pathology of allergic asthma and the potential of fruit proanthocyanidins as modulators of airway inflammation. Food Funct 8(12):4315–4324
Constantino Gde T, Mello JF Jr (2009) Remodeling of the lower and upper airways. Braz J Otorhinolaryngol 75(1):151–156
Corren J (1997) Allergic rhinitis and asthma: how important is the link? J Allergy Clin Immunol 99(2):S781–S7S6
Corren J (2007) The connection between allergic rhinitis and bronchial asthma. Curr Opin Pulm Med 13(1):13–18
D'Agostino B, Sgambato M, Esposito R, Spaziano G (2019) N/OFQ-NOP system and airways. Handb Exp Pharmacol 254:313–322
Davis BE, Cockcroft DW (2012) Past, present and future uses of methacholine testing. Expert Rev Respir Med 6(3):321–329
De A, Manwani D, Rastogi D (2018) Airway inflammation in sickle cell disease-a translational perspective. Pediatr Pulmonol 53(4):400–411
Dembele M, Tao S, Massoud AH, Miah SMS, Lelias S, De Groot AS et al (2021) Tregitopes improve asthma by promoting highly suppressive and antigen-specific Tregs. Front Immunol 12:634509
Denburg JA, Sehmi R, Saito H, Pil-Seob J, Inman MD, O'Byrne PM (2000) Systemic aspects of allergic disease: bone marrow responses. J Allergy Clin Immunol 106(5 Suppl):S242–S2S6
Dixon AE (2009) Rhinosinusitis and asthma: the missing link. Curr Opin Pulm Med 15(1):19–24
Djukanović R (2000) Asthma: a disease of inflammation and repair. J Allergy Clin Immunol 105(2 Pt 2):S522–S5S6
Donovan C, Bailey SR, Tran J, Haitsma G, Ibrahim ZA, Foster SR et al (2015) Rosiglitazone elicits in vitro relaxation in airways and precision cut lung slices from a mouse model of chronic allergic airways disease. Am J Physiol Lung Cell Mol Physiol 309(10):L1219–28
Drolet JP, Frangie H, Guay J, Hajoui O, Hamid Q, Mazer BD (2010) B lymphocytes in inflammatory airway diseases. Clin Exp Allergy 40(6):841–849
Elieh Ali Komi D, Bjermer L (2019) Mast cell-mediated orchestration of the immune responses in human allergic asthma: current insights. Clin Rev Allergy Immunol 56(2):234–247

Erjefält JS (2014) Mast cells in human airways: the culprit? Eur Respir Rev 23(133):299–307
Fanucchi MV, Schelegle ES, Baker GL, Evans MJ, McDonald RJ, Gershwin LJ et al (2004) Immunostimulatory oligonucleotides attenuate airways remodeling in allergic monkeys. Am J Respir Crit Care Med 170(11) 1153–1157
Fasano MB (2010) Combined airways: impact of upper airway on lower airway. Curr Opin Otolaryngol Head Neck Surg 18(1):15–20
Fear VS, Lai SP, Zosky GR, Perks KL, Gorman S, Blank F et al (2016) A pathogenic role for the integrin CD103 in experimental allergic airways disease. Physiol Rep 4:21
Fireman P (1993) Pathophysiology and pharmacotherapy of common upper respiratory diseases. Pharmacotherapy 13(6 Pt 2):101S–9S. discussion 43S–46S
Fokkens WJ, Scheeren RA (2000) Upper airway defence mechanisms. Paediatr Respir Rev 1(4):336–341
Fonseca AL, Arrobas AM (2006) Allergic inflammatory diseases of the upper airways and their impact on asthma—following on from a case report. Rev Port Pneumol 12(5):563–579
Foong RE, Bosco A, Troy NM, Gorman S, Hart PH, Kicic A et al (2016) Identification of genes differentially regulated by vitamin D deficiency that alter lung pathophysiology and inflammation in allergic airways disease. Am J Physiol Lung Cell Mol Physiol 311(3):L653–L663
Gelfand EW (2004) Inflammatory mediators in allergic rhinitis. J Allergy Clin Immunol 114(5 Suppl):S135–8
Gern JE (2003) Mechanisms of virus-induced asthma. J Pediatr 142(2 Suppl):S9–13. discussion S-4
Gillissen A, Höffken G, Juergens UR (2005) A connection between allergic rhinitis and allergic asthma? The "one-airway-one-disease"-hypothesis. Pneumologie (Stuttgart, Germany) 59(2):112–119
Godard P, Chanez P, Pujol JL, Clauzel AM, Bousquet J, Michel FB (1989) Physiopathology of asthma. Ann Biol Clin 47(6):357–372
Goldstein RA, Paul WE, Metcalfe DD, Busse WW, Reece ER (1994) NIH conference. Asthma. Ann Intern Med 121(9):698–708
Greenberger PA, Grammer LC (2010) Pulmonary disorders, including vocal cord dysfunction. J Allergy Clin Immunol 125(2 Suppl 2):S248–S254
Groneberg DA, Chung KF (2004) Models of chronic obstructive pulmonary disease. Respir Res 5(1):18
Groneberg DA, Quarcoo D, Frossard N, Fischer A (2004) Neurogenic mechanisms in bronchial inflammatory diseases. Allergy 59(11):1139–1152
Guan WJ, Peng Y, Zi XX, Tan KS, He TT, Zhong NS et al (2018) Motile ciliary disorders in chronic airway inflammatory diseases: critical target for interventions. Curr Allergy Asthma Rep 18(9):48
Haccuria A, Van Muylem A, Malinovschi A, Doan V, Michils A (2018) Small airways dysfunction: the link between allergic rhinitis and allergic asthma. Eur Respir J 51(2):1701749
Hall SC, Agrawal DK (2016) Toll-like receptors, triggering receptor expressed on myeloid cells family members and receptor for advanced glycation end-products in allergic airway inflammation. Expert Rev Respir Med 10(2):171–184
Halpern LR (2012) Allergic rhinitis and the unified airway: a therapeutic dilemma. Oral Maxillofac Surg Clin North Am 24(2):205–217. viii
Hamelmann E, Gelfand EW (2001) IL-5-induced airway eosinophilia—the key to asthma? Immunol Rev 179:182–191
Hamid Q, Tulic M (2009) Immunobiology of asthma. Annu Rev Physiol 71:489–507
Hamid Q, Tulic MK (2007) New insights into the pathophysiology of the small airways in asthma. Ann Thorac Med 2(1):28–33
Hellings PW, Prokopakis EP (2010) Global airway disease beyond allergy. Curr Allergy Asthma Rep 10(2):143–149
Hellings PW, Steelant B (2020) Epithelial barriers in allergy and asthma. J Allergy Clin Immunol 145(6):1499–1509

Henry PJ, Mann TS, Goldie RG (2005) A rho kinase inhibitor, Y-27632 inhibits pulmonary eosinophilia, bronchoconstriction and airways hyperresponsiveness in allergic mice. Pulm Pharmacol Ther 18(1):67–74
Hens G, Hellings PW (2006) The nose: gatekeeper and trigger of bronchial disease. Rhinology 44(3):179–187
Hogg JC, Paré PD, Boucher RC, Michoud MC (1979) The pathophysiology of asthma. Can Med Assoc J 121(4):409–414
Hohlfeld JM (2002) The role of surfactant in asthma. Respir Res 3(1):4
Holgate ST (2008) Pathogenesis of asthma. Clin Exp Allergy 38(6):872–897
Holgate ST (2010) Has the time come to rethink the pathogenesis of asthma? Curr Opin Allergy Clin Immunol 10(1):48–53
Holgate ST (2011) The sentinel role of the airway epithelium in asthma pathogenesis. Immunol Rev 242(1):205–219
Holloway JW, Yang IA, Holgate ST (2010) Genetics of allergic disease. J Allergy Clin Immunol 125(2 Suppl 2):S81–S94
Humbles AA, Conroy DM, Marleau S, Rankin SM, Palframan RT, Proudfoot AE et al (1997) Kinetics of eotaxin generation and its relationship to eosinophil accumulation in allergic airways disease: analysis in a Guinea pig model in vivo. J Exp Med 186(4):601–612
Idzko M, Pitchford S, Page C (2015) Role of platelets in allergic airway inflammation. J Allergy Clin Immunol 135(6):1416–1423
Inoue K (2011) Promoting effects of nanoparticles/materials on sensitive lung inflammatory diseases. Environ Health Prev Med 16(3):139–143
Jakwerth CA, Ordovas-Montanes J, Blank S, Schmidt-Weber CB, Zissler UM (2022) Role of respiratory epithelial cells in allergic diseases. Cells 11:9
Kallinich T, Beier KC, Gelfand EW, Kroczek RA, Hamelmann E (2005) Co-stimulatory molecules as potential targets for therapeutic intervention in allergic airway disease. Clin Exp Allergy 35(12):1521–1534
Karta MR, Broide DH, Doherty TA (2016) Insights into group 2 innate lymphoid cells in human airway disease. Curr Allergy Asthma Rep 16(1):8
Kaur R, Chupp G (2019) Phenotypes and endotypes of adult asthma: moving toward precision medicine. J Allergy Clin Immunol 144(1):1–12
Keir S, Page C (2008) The rabbit as a model to study asthma and other lung diseases. Pulm Pharmacol Ther 21(5):721–730
Khan DA (2014) Allergic rhinitis and asthma: epidemiology and common pathophysiology. Allergy Asthma Proc 35(5):357–361
KleinJan A, Willart M, van Nimwegen M, Leman K, Hoogsteden HC, Hendriks RW et al (2010) United airways: circulating Th2 effector cells in an allergic rhinitis model are responsible for promoting lower airways inflammation. Clin Exp Allergy 40(3):494–504
Koterba AP, Saltoun CA (2012) Chapter 9: asthma classification. Allergy Asthma Proc 33(Suppl 1):28–31
Kraneveld AD, Folkerts G, Van Oosterhout AJ, Nijkamp FP (1997) Airway hyperresponsiveness: first eosinophils and then neuropeptides. Int J Immunopharmacol 19(9–10):517–527
Krouse JH, Altman KW (2010) Rhinogenic laryngitis, cough, and the unified airway. Otolaryngol Clin N Am 43(1):111–121. ix-x
Lang JE (2019) Obesity and childhood asthma. Curr Opin Pulm Med 25(1):34–43
Lee DC, Walker SA, Byrne AJ, Gregory LG, Buckley J, Bush A et al (2015) Perinatal paracetamol exposure in mice does not affect the development of allergic airways disease in early life. Thorax 70(6):528–536
Li E, Knight JM, Wu Y, Luong A, Rodriguez A, Kheradmand F et al (2019) Airway mycosis in allergic airway disease. Adv Immunol 142:85–140
Lindén A (2006) Interleukin-17 and airway remodelling. Pulm Pharmacol Ther 19(1):47–50
Liu D, Wagner JG, Harkema JR, Gerlofs-Nijland ME, Pinelli E, Folkerts G et al (2020) Livestock farm particulate matter enhances airway inflammation in mice with or without allergic airway disease. World Allergy Organ J 13(4):100114

Liu G, Cooley MA, Nair PM, Donovan C, Hsu AC, Jarnicki AG et al (2017) Airway remodelling and inflammation in asthma are dependent on the extracellular matrix protein fibulin-1c. J Pathol 243(4):510–523

Liu Y, Sha J, Meng C, Zhu D (2022) Mechanism of lower airway Hyperresponsiveness induced by allergic rhinitis. J Immunol Res 2022:4351345

Liu Y, Zeng M, Liu Z (2015) Th17 response and its regulation in inflammatory upper airway diseases. Clin Exp Allergy 45(3):602–612

Lo D, Kennedy JL, Kurten RC, Panettieri RA Jr, Koziol-White CJ (2018) Modulation of airway hyperresponsiveness by rhinovirus exposure. Respir Res 19(1):208

Locke NR, Royce SG, Wainewright JS, Samuel CS, Tang ML (2007) Comparison of airway remodeling in acute, subacute, and chronic models of allergic airways disease. Am J Respir Cell Mol Biol 36(5):625–632

Löser S, Gregory LG, Zhang Y, Schaefer K, Walker SA, Buckley J et al (2017) Pulmonary ORMDL3 is critical for induction of Alternaria-induced allergic airways disease. J Allergy Clin Immunol 139(5):1496–1507.e3

Maarsingh H, Zaagsma J, Meurs H (2008) Arginine homeostasis in allergic asthma. Eur J Pharmacol 585(2–3):375–384

Maarsingh H, Zaagsma J, Meurs H (2009) Arginase: a key enzyme in the pathophysiology of allergic asthma opening novel therapeutic perspectives. Br J Pharmacol 158(3):652–664

Macdowell AL, Peters SP (2007) Neutrophils in asthma. Curr Allergy Asthma Rep 7(6):464–468

Marple BF (2010) Allergic rhinitis and inflammatory airway disease: interactions within the unified airspace. Am J Rhinol Allergy 24(4):249–254

Marseglia GL, Cirillo I, Vizzaccaro A, Klersy C, Tosca MA, La Rosa M et al (2007) Role of forced expiratory flow at 25-75% as an early marker of small airways impairment in subjects with allergic rhinitis. Allergy Asthma Proc 28(1):74–78

Marseglia GL, Merli P, Caimmi D, Licari A, Labó E, Marseglia A et al (2011) Nasal disease and asthma. Int J Immunopathol Pharmacol 24(4 Suppl):7–12

Meyer N, Akdis CA (2013) Vascular endothelial growth factor as a key inducer of angiogenesis in the asthmatic airways. Curr Allergy Asthma Rep 13(1):1–9

Millien VO, Lu W, Mak G, Yuan X, Knight JM, Porter P et al (2014) Airway fibrinogenolysis and the initiation of allergic inflammation. Ann Am Thorac Soc 11(Suppl 5):S277–83

Montuschi P, Sala A, Dahlén SE, Folco G (2007) Pharmacological modulation of the leukotriene pathway in allergic airway disease. Drug Discov Today 12(9–10):404–412

Morrissey BM, Evans SJ (2003) Severe bronchiectasis. Clin Rev Allergy Immunol 25(3):233–247

Movassagh H, Shan L, Duke-Cohan JS, Halayko AJ, Uzonna JE, Gounni AS (2017) Semaphorin 3E alleviates hallmarks of house dust mite-induced allergic airway disease. Am J Pathol 187(7):1566–1576

Mur T, Brook C, Platt M (2020) Extranasal manifestations of allergy in the head and neck. Curr Allergy Asthma Rep 20(7):21

Nassenstein C, Krasteva-Christ G, Renz H (2018) New aspects of neuroinflammation and neuroimmune crosstalk in the airways. J Allergy Clin Immunol 142(5):1415–1422

Nelson HS (2007) Advances in upper airway diseases and allergen immunotherapy. J Allergy Clin Immunol 119(4):872–880

Nockher WA, Renz H (2006) Neurotrophins in allergic diseases: from neuronal growth factors to intercellular signaling molecules. J Allergy Clin Immunol 117(3):583–589

Nur Husna SM, Tan HT, Md Shukri N, Mohd Ashari NS, Wong KK (2022) Allergic rhinitis: a clinical and pathophysiological overview. Front Med 9:874114

Ogawa H, Fujimura M, Ohkura N, Satoh K, Makimura K (2014) Fungus-associated asthma: overcoming challenges in diagnosis and treatment. Expert Rev Clin Immunol 10(5):647–656

Oguma T, Asano K, Ishizaka A (2008) Role of prostaglandin D(2) and its receptors in the pathophysiology of asthma. Allergol Int 57(4):307–312

Okano M, Kariya S, Ohta N, Imoto Y, Fujieda S, Nishizaki K (2015) Association and management of eosinophilic inflammation in upper and lower airways. Allergol Int 64(2):131–138

Olaguibel JM, Alobid I, Alvarez Puebla M, Crespo-Lessmann A, Domínguez Ortega J, García-Rio F et al (2021) Functional examination of the upper and lower airways in asthma and respiratory allergic diseases: considerations in the post-SARS-CoV-2 era. J Investig Allergol Clin Immunol 31(1):17–35

Padem N, Saltoun C (2019) Classification of asthma. Allergy Asthma Proc 40(6):385–388

Paiva Ferreira LKD, Paiva Ferreira LAM, Monteiro TM, Bezerra GC, Bernardo LR, Piuvezam MR (2019) Combined allergic rhinitis and asthma syndrome (CARAS). Int Immunopharmacol 74:105718

Pakdaman MN, Corry DB, Luong A (2011) Fungi linking the pathophysiology of chronic rhinosinusitis with nasal polyps and allergic asthma. Immunol Investig 40(7–8):767–785

Pandya RJ, Solomon G, Kinner A, Balmes JR (2002) Diesel exhaust and asthma: hypotheses and molecular mechanisms of action. Environ Health Perspect 110(Suppl 1):103–112

Passalacqua G, Canonica GW (2001) Impact of rhinitis on airway inflammation: biological and therapeutic implications. Respir Res 2(6):320–323

Immunology of Allergic Airway Diseases 3

Mohit Agrawal, Anurag Mishra, Abhishek Dwivedi,
and Yogendra Singh

Abstract

Allergies have a similar immunological profile constituted up of recognizable cells, antibodies, and other biochemical and molecular structures. It is simple to identify eosinophils, neutrophils, and tissue mast cells using standard hematoxylin and eosin staining of histologic tissues. Symptoms of allergic respiratory disorders including hay fever (allergic rhinoconjunctivitis) and asthma include wet or itchy eyes, sneezing, nasal congestion, and coughing; these conditions are the result of a persistent inflammatory process. In theory, the best treatment for allergic illness would eliminate the triggers that cause it. Allergen-specific immunotherapy, which has been used for many years, is based in part on this principle. Omalizumab, a monoclonal antibody treatment that depletes circulating IgE to gradually lower the quantity of antibody attached to high-affinity IgE receptors on mast cells, is a promising example of the success that may be achieved by adopting a different approach. However, patient access to therapy in chronic illnesses affecting large populations is complicated by major payment hurdles

The original version of the chapter has been revised. A correction to this chapter can be found at https://doi.org/10.1007/978-981-97-1953-2_16

M. Agrawal
School of Medical & Allied Sciences, K.R. Mangalam University, Gurgaon, Haryana, India

A. Mishra
NIMS Institute of Pharmacy, NIMS University Rajasthan, Jaipur, India

A. Dwivedi
Department of Pharmaceutics, Maharishi Arvind College of Pharmacy,
Jaipur, Rajasthan, India

Y. Singh (✉)
Department of Pharmacology, Maharishi Arvind College of Pharmacy,
Jaipur, Rajasthan, India

P. Prasher et al. (eds.), *Understanding Allergic Airway Diseases*,
https://doi.org/10.1007/978-981-97-1953-2_3

associated with biologic medicines like omalizumab. The present study aims to examine the mechanisms by which a prominent allergen mediated allergic responses and explore potential interventions to mitigate the immune overactivation associated with allergic airway disorders caused by these allergens.

Keywords

Immunoglobulin (IgE) · Interleukin (IL) · Allergic asthma · Allergic airway disease

3.1 Introduction

Animals' immune systems likely developed through time so they could better defend themselves against the many germs they encountered. This immune response is not selective; it occurs in reaction to any foreign material, not only those produced by bacteria. To counteract these potentially harmful external factors (antigens), the immune system has evolved to become extremely specialized and intricate. Foreign chemicals entering the host may cause a broad variety of cellular harm; thus, the immune system creates highly specialized molecules (antibodies) and opt for cells like chemokines, adhesion molecules, cytokines, and the like. Therefore, the cells and cell products that are associated with each other serve as a protective system that collaborates with foreign agents to safeguard the host from harm (Bellanti et al. 1994).

- Immunity is the host's defense mechanism against invading pathogens or other non-self-substances. Most people have no negative repercussions from these occurrences; therefore, we classify them as normal.
- Immunology investigates how a host responds to non-self-organisms by attempting to restore internal stability.
- A modern definition of allergy includes type I immunoglobulin (Ig) E-mediated or cell-mediated responses, as well as any hypersensitivity that has deleterious effects on the host.

In allergy, human innate and adaptive defense mechanisms overreact to a normally harmless signal, leading to uncontrolled immune responses and persistent inflammation. Disease pathophysiology is driven by the secretion of soluble mediators of inflammation by many distinct cell types. Allergies have a similar immunological profile constituted up of recognizable cells, antibodies, and other biochemical and molecular structures. It is simple to identify eosinophils, neutrophils, and tissue mast cells using standard hematoxylin and eosin staining of histologic tissues. IgE- and G4-secreting B cells and T helper 2 (Th2) and Th17 cells, which secrete a wide range of inflammatory proteins involving interleukin (IL)-4/-5/-9/-10/-13 and -17A, communicate and stimulate various allergic resultant cells like eosinophils, mast cells, B cells, and innate immune cells. Allergic airway illnesses are often deadly and persistent, and although remissions may occur on their own, they are seldom cured. However, new information has significantly raised hopes for better treatment.

Omalizumab, a monoclonal antibody treatment that depletes circulating IgE to gradually lower the quantity of antibody attached to high-affinity IgE receptors on mast cells, is a promising example of the success that may be achieved by adopting a different approach. The introduction of omalizumab marked a dramatic shift from treating symptoms to making preventative, disease-modifying interventions. However, patient access to therapy in chronic illnesses affecting large populations is complicated by major payment hurdles associated with biologic medicines like omalizumab.

In theory, the best treatment for allergic illness would eliminate the triggers that cause it. Allergen-specific immunotherapy, which has been used for many years, is based in part on this principle. Until recently, the question of how allergens become allergenic—that is, by breaking immune tolerance and promoting IgE-directed immune responses—was surprisingly unfashionable in allergy, making it difficult to develop small-molecule interventions that target this underlying cause. Recent advances in allergen characterization, in tandem with a resurgence of interest in innate immunity, have provided fresh insight into the molecular basis of allergenicity. As a result, it may be possible to intervene at key nodes along these pathways to forestall the onset of allergic illness. Specifically, it is now conceivable to create a technique to target a significant asthma trigger at the root-cause level, which has the benefit of avoiding worries about the chronic implications of blocking checkpoints that are crucial to immunity. This novel approach has the benefit of blocking the innate immune responses that fuel the development of allergies, unlike allergen-specific immunotherapy. The present study aims to examine the mechanisms by which a prominent allergen mediated allergic responses and explore potential interventions to mitigate the immune overactivation associated with allergic airway disorders caused by these allergens.

Symptoms of allergic respiratory disorders including hay fever (allergic rhinoconjunctivitis) and asthma include wet or itchy eyes, sneezing, nasal congestion, and coughing; these conditions are the result of a persistent inflammatory process (Woloski et al. 2016). Additional signs and symptoms of allergic asthma include a constricted chest, a persistent cough, and a temporary decrease in airflow (Deng et al. 2020). Airway inflammation is the hallmark of allergic asthma (Murdoch and Lloyd 2010). Allergic airway inflammation in asthma is thought to result from the activation of the type 2 inflammatory cascade, which in turn recruit eosinophils, when vulnerable individuals are exposed to allergens (De Schryver et al. 2015; Kupryś-Lipińska et al. 2016). Support for this hypothesis comes from the correlation between elevated IgE levels and Th2 molecules such IL-4/IL-5/IL-13 in people with type 2 high asthma (Kupryś-Lipińska et al. 2016). These factors, as was discussed previously, are essential for the chronic upregulation of IgE synthesis and subsequent inflammation throughout disease.

The illness of asthma is believed to be heritable in approximately 30–50% of cases. Atopy has been shown to be the most important predictor for the onset of allergies in epidemiological investigations, particularly in connection to the formation of immunoglobulin E specific to allergens that exist (Backman et al. 2017; Del Giacco et al. 2017). Multiple genes have been linked to an elevated risk of illness, as shown by genome-wide association studies. Asthma, for instance, has been linked to a number of genes in the 5q region (Holloway et al. 2001; Bønnelykke et al. 2015). The genetic approaches outlined above are valuable tools for identifying the specific genes that play a role in the development of allergic asthma. Additionally, they provide valuable insights into how an individual's vulnerability to environmental variables that might trigger illness affects the development or protection against allergic asthma. In light of recent genetic discoveries, epithelial airway integrity and its immunological responses evoking a type 2 immune reaction have emerged as major indicators of the susceptibility to the occurrence of allergies including rhinoconjunctivitis and asthma.

3.2 Pathophysiology of Allergic Asthma and Airway Disease

Airway allergies are characterized by immunological responses that are often linked with the Th2 cytokine subset. High quantities of allergen-specific immunoglobulin E and IL-13/IL-4 and IL-5 cytokines are secreted in response to inflammation, which in turn attract and activate effector cells such as mast cells, basophils, eosinophils, and elevated levels of nitric oxide in the breath (Liu et al. 1991; Zhu et al. 1999; Anderson 2002; Woodruff et al. 2009; Cho et al. 2014; Agache and Akdis 2016; Chesné et al. 2016; Datta et al. 2016; Miskoff et al. 2019; Sze et al. 2020).

There is a significance of both the adaptive and innate immunological processes in the development of allergy-related asthma. Airway epithelial cells generate the thymic stromal lymphopoietin (TSLP), cytokines IL-25/IL-33, TRAIL, and MID1, which regulate the initiation of type 2 (T2) immune responses (Divekar and Kita 2015; Laulajainen-Hongisto et al. 2020). When it comes to allergic airway inflammation, innate lymphoid cells type 2 (ILC2s) are pivotal players in both its initiation and perpetuation by releasing the IL-13/IL-4 and IL-5 cytokines. In addition, ILC2s spread the inflammatory response and heal lung injury by interacting with various cell types and coordinating a network of cellular responses (Fig. 3.1). Different asthma phenotypes show that ILC2s can adapt to a wide variety of immunological niches thanks to their functional flexibility. ILC2 growth and plasticity are influenced by the transcription factor GATA-3, IL-4/IL-1/IL-12 (Ohne et al. 2016). To reduce allergic lung inflammation, researchers are focusing on controlling pulmonary ILC2s. Inducible T regulatory cells (iTregs) and inducible regulatory ILCs (ILCregs) fight against ILC2s (Fig. 3.1). Recent research has uncovered a novel anti-inflammatory and immunological regulation system wherein retinoic acid (RA) transforms ILC2s into ILCregs (Morita et al. 2019). The synthesis of enzymes for RA was seen to be elevated in epithelial cells of the airways in the presence of type 2 inflammation in vivo and when exposed to IL-13 in vitro. These findings suggest that the mechanism involving bronchial epithelial cells and ILC2s is crucial to serve the regulation of RA biosynthesis (Aron and Akbari 2017). ILC2 controls the production of IL-13/IL-5, two pro-inflammatory cytokines that are effectively inhibited by iTregs. In addition, IL-10 and transforming growth factor beta 1 greatly suppress cytokine release by ILC2s (Fig. 3.1). The level of protection provided by Toll-like receptor 4, or TLR4, against the development of allergic disease is based on CD14 allelic variation (Rodríguez et al. 2003; Haapakoski et al. 2013). The presence of cytokines (IL-13/IL-5/IL-4) and immune cells (CD4+ Th2 cells) that are specifically produced by allergens have characteristic of allergic asthma. The cytokines

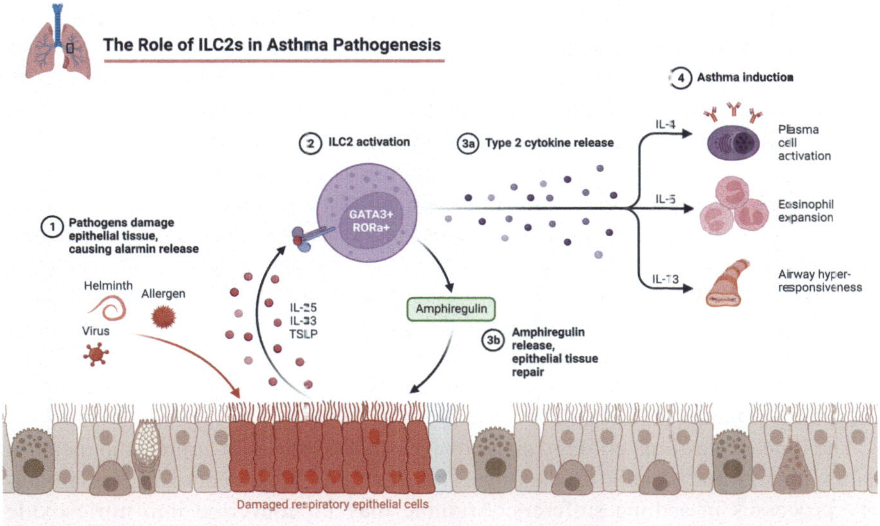

Fig. 3.1 Exhibits the immunological profile of ILCs and various other mediators

mentioned in the study were shown to have a role in both inflammation and airway hyperresponsiveness (AHR) at various stages. For instance, IL-4 was seen to aid the onset of inflammation, while IL-13/IL-5 acted as effector molecules toward the completion of the process. Eosinophils and Th2 cells communicate in both directions to control cytokine production in T cells. Eosinophil-derived IL-18 controls the generation of the cytokine IL-13. Eosinophils and ILC2 may also serve as antigen presentation cells, stimulating Th2 cells in an antigen-specific way to directly increase the inflammatory process associated with allergens. Tregs have been the subject of substantial research due to their potential to suppress allergic asthmatic inflammation. Tregs are thought to work through two primary channels: cell-cell contact, mediated by membrane-bound TGF-β or inhibitory chemicals like CTLA-4, and dissolved cytokine-dependent mechanisms including IL-10 and TGF-β. Foxp3, which belongs to a group of transcription regulators known as forkhead-winged helix, seems to play a pivotal role in the Tregs' ability to suppress immune responses. The clinical improvement in allergic reactions has been linked to the abundance of Foxp3-expressing IL-10, CD4+ CD25+ T cells, and TGF-beta-producing Tregs after immunotherapy (Kappen et al., 2017). Low-dose corticosteroids, vitamin D3, and other pharmacological and adjuvant interventions induces activation and proliferation of antigen-specific CD4+ CD25+ Tregs inside a living being presents a distinctive approach for the establishment of allergen-specific resistance in individuals with asthma (Cantorna et al. 2015). Recent research has shown that oral corticosteroids have a significant impact on the frequency and ability of individual allergic asthmatic people to release IL-10 (Fig. 3.1). This study also reported on the morphological and functional changes seen in Breg subgroups (Wiest et al. 2019).

3.3 Findings in Immunometabolism with Allergic Asthma and Airway Disease

Glucose is taken in by the cell and metabolized into pyruvate via glycolysis before being shuttled toward the cell's mitochondria for oxidative phosphorylation (OxPhos). Under aerobic circumstances, OxPhos is capable of effectively synthesizing 36–38 ATP units from a single glucose molecule for sustaining cell function. Although glycolysis can break down glucose to lactate and organic matter intermediates under anaerobic environments, it does so at a considerably lower efficiency than OxPhos (just 2 ATP molecules are synthesized from each glucose molecule). In addition to this, the participation of the adaptive immune mechanism through IL-1-induced glycolysis comes from studies in which house dust mite (HDM)-exposed animals missing mature B and T cells had entirely suppressed lactate concentrations (Qian et al. 2018).

The illness of asthma etiology also involves abnormalities in amino acid metabolism, with arginine breakdown playing a crucial role as an inducer of Th2 inflammatory processes in asthma sufferers. Arginine may be converted into nitric oxide (NO) and citrulline by a number of enzymes, including iNOS and ARG. Asthma patients' exhaled NO levels and illness severity have been shown to link with patients' iNOS and ARG1/2 concentrations (Xu et al. 2016). Exhaled nitric oxide (NO), arginine metabolism, and glycolysis have all been demonstrated to be interrelated processes, and people with asthma have been found to have elevated levels of all three compared to unaffected populations (Xu et al. 2017).

Synthesis of fatty acids and cholesterol requires acetyl-CoA. The sterol regulatory element-binding protein (SREBP) pathway mediates the metabolic processes of fatty acids and cholesterols, both of which are essential components of lipid metabolism. Phospholipids in cell membranes contain arachidonic acid, which may be converted into LTs and PGs through the 5-LOX and COX signaling mechanisms. Increased levels of lipid molecules, including pro-inflammatory PG-E2, were detected in the respiratory tract epithelium of asthmatics compared to controls (Johnson et al. 2022). Asthmatic patients' sputum has higher levels of lipid metabolite than those of healthy people, particularly eicosanoids, LTC4, LTD4, and LTE4 (Pavord et al. 1999). In response to FAs, macrophages become activated and differentiate, leading to an increase in the production a variety of pro-inflammatory molecules. These cytokines (TNF-α and IL-1β/IL-6) play a role in the development of allergic reactions and the gradual systemic inflammation seen in obese individuals (McNelis and Olefsky 2014).

3.4 Modern Immunometabolic Therapies for Asthma

Patients with poor sensitivity to conventional therapies, such as an adequate dosage of inhaled corticosteroids paired with long-acting β2 agonists or additional inhibitors, are more likely to have severe asthma (Schoettler and Strek 2020). Asthma is a chronic condition, and targeted biologics may have therapeutic benefits. Numerous biochemical mediators in various signaling networks are involved in the above-described immune cell metabolic activities in asthma. These signaling agents may serve as promising strategies to modulate aberrant reactions to allergens, suggesting new avenues for treating asthma.

3.5 Signaling Molecules in the TLR-NF-kb Cascade

TLRs and NF-kb are essential in allergic respiratory conditions because of their involvement in allergen recognition and the differentiation of CD4+ T cells into Th2 cells, respectively, which cause airway inflammation (Chen and Kolls 2013). Allergies was formerly assumed to originate from the differentiation and stimulation of CD4+ T cells, which was associated with aerobic glycolysis and metabolism of lipids (Wang et al. 2011). Accordingly, it is plausible to propose that the involvement of TLR and NF-kb in the differentiation of CD4+ T cells is also implicated in the transition from typical to aberrant metabolic profiles in allergies to asthma. This implies that modulation of the TLR-NF-kb signaling mechanism holds potential as a viable strategy for augmenting the immune-inflammatory response in allergic respiratory illness from a metabolic perspective. It is yet unclear, however, what pathways in particular metabolic illnesses they may be influencing.

In response to antigens, the TLR signaling pathway activates a cascade of events that results in the production of cytokines, chemokines, and co-stimulatory molecules (Takeda and Akira 2004). To mediate lung inflammation, the transcription factor NF-kb may cause the synthesis of COX-2 and iNOS (Tak and Firestein 2001; Athari 2019). In contrast, COX-2 and iNOS may help activate NF-kb, boosting inflammation in a cyclic fashion, indicating the enormous importance of managing NF-kb, COX-2, and iNOS in asthma treatment (Kim et al. 2007). The generation of iNOS and COX-2 may be suppressed by inhibiting the phosphorylation of the inhibitor kappa-B subunit alpha (IB), which in turn inhibits NF-kb activation. Adiponectin was shown to inhibit NF-kb and TNF-α-induced production of antiapoptotic Bcl-2 (Akita et al. 2005), hence encouraging death of inflammatory cells, indicating a possible function for adiponectin in inflammation management.

3.6 Substrates of the Interleukin-4/Interleukin-13/Janus Kinase/Stat-6 Network

More and more research over the last several years has connected irregularities in the Janus kinase (JAK)-STAT signaling axis to a metabolic defect, particularly alterations in the utilization of lipids. More specifically, the IL-4/IL-13/JAK/STAT-6 axis plays a pivotal role in regulating the etiology of asthma (Oh et al. 2010). Large amounts of allergic inflammatory markers are generated by asthmatics in response to stimuli, signaling through the JAK/STAT axis and contributing to the disease progression of asthma (O'Shea and Plenge 2012). Further research is needed to determine how the state of the route influences the metabolic processes and cellular dysfunction of asthma; however, it has been shown that several biochemical markers in the system connected to protein and lipid metabolism are strongly associated to asthmatic inflammatory conditions.

The heterodimeric IL-4R are recruited by the IL-4/IL-13-induced signaling mechanisms. The IL-4/IL-13 circuit may stimulate both forms of IL-4R, leading to increased generation of inflammatory mediators through phosphorylation of JAK 1, JAK 2, and tyrosine kinase 2 and activation of STAT-6 transcription factors (Ashino et al. 2014; Athari 2019; Iwaszko et al. 2021). In individuals with asthma, immune system response happens by the production of pro-inflammatory proteins, such as FA-binding protein 4 (FABP4) and IL-4/IL-13. These chemicals play a crucial role in promoting the adhesion and migration of eosinophils (Ge et al. 2018). However, STAT6 inhibition decreased IL-4/IL-13's ability to suppress FABP4 production, indicating that FABP4's function is governed by its binding to STAT6 and highlighting the significance of STAT6 in asthmatic inflammatory conditions (Shum et al. 2006). An effective strategy for blocking STAT-6 activation and acetylation is to prevent STAT-6-MAP kinase interacting with ERK1/2 and p38 (Ge et al. 2018). Transactivation of STAT-6 requires phosphorylation of serine residues, suggesting that serine kinases associated with protein and lipid metabolism might be viable targets for medicine for airway allergy illness. Pharmaceuticals focusing on the IL-4/L-13/JAK/STAT-6 axis may be able to prevent the p300/STAT-6 connection and reduce inflammatory chemokine release since serine phosphorylation of STAT-6 is inhibited by p38 and MEK1/2 inhibitors.

3.7 Conclusion

Despite having a far greater incidence, allergic airway illnesses provide a rare opportunity to investigate, treat, and control allergy phenotypes by focusing on HDM allergens because of their single, particular source. According to the foregoing, it seems that after a few sessions of allergen-specific immunotherapy (AIT), resistance to allergic reactions may be attained.

References

Agache I, Akdis CA (2016) Endotypes of allergic diseases and asthma: an important step in building blocks for the future of precision medicine. Allergol Int 65(3):243–252

Akita K, Kawata S, Shimotohno K (2005) p21WAF1 modulates NF-kappaB signaling and induces anti-apoptotic protein Bcl-2 in Tax-expressing rat fibroblast. Virology 332(1):249–257

Anderson GP (2002) The immunobiology of early asthma. Med J Aust 177:S47–S49

Aron J, Akbari O (2017) Regulatory T cells and type 2 innate lymphoid cell-dependent asthma. Allergy 72(8):1148–1155

Ashino S, Takeda K, Li H, Taylor V, Joetham A, Pine PR, Gelfand EW (2014) Janus kinase 1/3 signaling pathways are key initiators of TH2 differentiation and lung allergic responses. J Allergy Clin Immunol 133(4):1162–1174

Athari SS (2019) Targeting cell signaling in allergic asthma. Signal Transduct Target Ther 4:45–45

Backman H, Räisänen P, Hedman L, Stridsman C, Andersson M, Lindberg A, Lundbäck B, Rönmark E (2017) Increased prevalence of allergic asthma from 1996 to 2006 and further to 2016—results from three population surveys. Clin Exp Allergy 47(11):1426–1435
Bellanti JA, Kadlec JV, Escobar-Gutiérrez A (1994) Cytokines and the immune response. Pediatr Clin N Am 41(4):597–621
Bønnelykke K, Sparks R, Waage J, Milner JD (2015) Genetics of allergy and allergic sensitization: common variants, rare mutations. Curr Opin Immunol 36:115–126
Cantorna MT, Snyder L, Lin Y-D, Yang L (2015) Vitamin D and 1, 25 (OH) 2D regulation of T cells. Nutrients 7(4):3011–3021
Chen K, Kolls JK (2013) T cell-mediated host immune defenses in the lung. Annu Rev Immunol 31:605–633
Chesné J, Schmidt-Weber CB, von-Bieren JE (2016) The use of adjuvants for enhancing allergen immunotherapy efficacy. Immunol Allergy Clin 36(1):125–145
Cho K-S, Park M-K, Kang S, Park H-Y, Hong S-L, Park H-K, Yu H-S, Roh H-J (2014) Adipose-derived stem cells ameliorate allergic airway inflammation by inducing regulatory T cells in a mouse model of asthma. Mediat Inflamm 2014:436476
Datta A, Moitra S, Hazra I, Mondal S, Das PK, Singh MK, Chaudhuri S, Bhattacharya D, Tripathi SK, Chaudhuri S (2016) Specific allergen immunotherapy attenuates allergic airway inflammation in a rat model of Alstonia scholaris pollen induced airway allergy. Int Immunopharmacol 30:111–120
De Schryver E, Devuyst L, Derycke L, Dullaers M, Van Zele T, Bachert C, Gevaert P (2015) Local immunoglobulin e in the nasal mucosa: clinical implications. Allergy Asthma Immunol Res 7(4):321–331
Del Giacco SR, Bakirtas A, Bel E, Custovic A, Diamant Z, Hamelmann E, Heffler E, Kalayci Ö, Saglani S, Sergejeva S (2017) Allergy in severe asthma. Allergy 72(2):207–220
Deng S-Z, Jalaludin BB, Antó JM, Hess JJ, Huang C-R (2020) Climate change, air pollution, and allergic respiratory diseases: a call to action for health professionals. Chin Med J 133:13
Divekar R, Kita H (2015) Recent advances in epithelium-derived cytokines (IL-33, IL-25, and thymic stromal lymphopoietin) and allergic inflammation. Curr Opin Allergy Clin Immunol 15(1):98–103
Ge XN, Bastan I, Dileepan M, Greenberg Y, Ha SG, Steen KA, Bernlohr DA, Rao SP, Sriramarao P (2018) FABP4 regulates eosinophil recruitment and activation in allergic airway inflammation. Am J Physiol Lung Cell Mol Physiol 315(2):L227–L240
Haapakoski R, Karisola P, Fyhrquist N, Savinko T, Lehtimäki S, Wolff H, Lauerma A, Alenius H (2013) Toll-like receptor activation during cutaneous allergen sensitization blocks development of asthma through IFN-gamma-dependent mechanisms. J Invest Dermatol 133(4):964–972
Holloway J, Lonjou C, Beghé B, Peng Q, Gaunt T, Gomes I, Hall I, Dewar J, Wilkinson J, Thomas N (2001) Linkage analysis of the 5q31–33 candidate region for asthma in 240 UK families. Genes Immun 2(1):20–24
Iwaszko M, Biały S, Bogunia-Kubik K (2021) Significance of interleukin (IL)-4 and IL-13 in inflammatory arthritis. Cells 10(11):3000
Johnson RK, Manke J, Campbell M, Armstrong M, Boorgula MP, Pinheiro G, Santana CVN, Mathias RA, Barnes KC, Cruz A (2022) Lipid mediators are detectable in the nasal epithelium and differ by asthma status in female subjects. J Allergy Clin Immunol 150(4):965–971. e968
Kappen, J. H., S. R. Durham, H. I. t. Veen and M. H. Shamji (2017). "Applications and mechanisms of immunotherapy in allergic rhinitis and asthma." Therapeutic advances in respiratory disease 11(1):73–86.
Kim JB, Han AR, Park EY, Kim JY, Cho W, Lee J, Seo EK, Lee KT (2007) Inhibition of LPS-induced iNOS, COX-2 and cytokines expression by poncirin through the NF-kappaB inactivation in RAW 264.7 macrophage cells. Biol Pharm Bull 30(12):2345–2351
Kupryś-Lipińska I, Molińska K, Kuna P (2016) The effect of omalizumab on eosinophilic inflammation of the respiratory tract in patients with allergic asthma. Pneumonol Alergol Pol 84(4):232–243

Laulajainen-Hongisto A, Toppila-Salmi SK, Luukkainen A, Kern R (2020) Airway epithelial dynamics in allergy and related chronic inflammatory airway diseases. Front Cell Dev Biol 8:204

Liu MC, Hubbard WC, Proud D, Stealey BA, Galli SJ, Kagey-Sobotka A, Bleecker ER, Lichtenstein LM (1991) Immediate and late inflammatory responses to ragweed antigen challenge of the peripheral airways in allergic asthmatics. Cellular, mediator, and permeability changes. Am Rev Respir Dis 144(1):51–58

McNelis JC, Olefsky JM (2014) Macrophages, immunity, and metabolic disease. Immunity 41(1):36–48

Miskoff JA, Dewan A, Chaudhri M (2019) Fractional exhaled nitric oxide testing: diagnostic utility in asthma, chronic obstructive pulmonary disease, or asthma-chronic obstructive pulmonary disease overlap syndrome. Cureus 11(6):e4864

Morita H, Kubo T, Rückert B, Ravindran A, Soyka MB, Rinaldi AO, Sugita K, Wawrzyniak M, Wawrzyniak P, Motomura K (2019) Induction of human regulatory innate lymphoid cells from group 2 innate lymphoid cells by retinoic acid. J Allergy Clin Immunol 143(6):2190–2201. e2199

Murdoch JR, Lloyd CM (2010) Chronic inflammation and asthma. Mutat Res 690(1–2):24–39

Oh CK, Geba GP, Molfino N (2010) Investigational therapeutics targeting the IL-4/IL-13/STAT-6 pathway for the treatment of asthma. Eur Respir Rev 19(115):46–54

Ohne Y, Silver JS, Thompson-Snipes L, Collet MA, Blanck JP, Cantarel BL, Copenhaver AM, Humbles AA, Liu Y-J (2016) IL-1 is a critical regulator of group 2 innate lymphoid cell function and plasticity. Nat Immunol 17(6):646–655

O'Shea JJ, Plenge R (2012) JAK and STAT signaling molecules in immunoregulation and immune-mediated disease. Immunity 36(4):542–550

Pavord ID, Ward R, Woltmann G, Wardlaw AJ, Sheller JR, Dworski R (1999) Induced sputum eicosanoid concentrations in asthma. Am J Respir Crit Care Med 160(6):1905–1909

Qian X, Aboushousha R, van de Wetering C, Chia SB, Amiel E, Schneider RW, van der Velden JL, Lahue KG, Hoagland DA, Casey DT (2018) IL-1/inhibitory κB kinase ε–induced glycolysis augment epithelial effector function and promote allergic airways disease. J Allergy Clin Immunol 142(2):435–450. e410

Rodríguez D, Keller AC, Faquim-Mauro EL, de Macedo MS, Cunha FQ, Lefort J, Vargaftig BB, Russo M (2003) Bacterial lipopolysaccharide signaling through toll-like receptor 4 suppresses asthma-like responses via nitric oxide synthase 2 activity. J Immunol 171(2):1001–1008

Schoettler N, Strek ME (2020) Recent advances in severe asthma: from phenotypes to personalized medicine. Chest 157(3):516–528

Shum BO, Mackay CR, Gorgun CZ, Frost MJ, Kumar RK, Hotamisligil GS, Rolph MS (2006) The adipocyte fatty acid-binding protein aP2 is required in allergic airway inflammation. J Clin Invest 116(8):2183–2192

Sze E, Bhalla A, Nair P (2020) Mechanisms and therapeutic strategies for non-T2 asthma. Allergy 75(2):311–325

Tak PP, Firestein GS (2001) NF-kappaB: a key role in inflammatory diseases. J Clin Invest 107(1):7–11

Takeda K, Akira S (2004) TLR signaling pathways. Semin Immunol 16(1):3–9

Wang R, Dillon CP, Shi LZ, Milasta S, Carter R, Finkelstein D, McCormick LL, Fitzgerald P, Chi H, Munger J, Green DR (2011) The transcription factor Myc controls metabolic reprogramming upon T lymphocyte activation. Immunity 35(6):871–882

Wiest M, Upchurch K, Hasan MM, Cardenas J, Lanier B, Millard M, Turner J, Oh S, Joo H (2019) Phenotypic and functional alterations of regulatory B cell subsets in adult allergic asthma patients. Clin Exp Allergy 49(9):1214–1224

Woloski JR, Heston S, Escobedo Calderon SP (2016) Respiratory allergic disorders. Prim Care 43(3):401–415

Woodruff PG, Modrek B, Choy DF, Jia G, Abbas AR, Ellwanger A, Arron JR, Koth LL, Fahy JV (2009) T-helper type 2–driven inflammation defines major subphenotypes of asthma. Am J Respir Crit Care Med 180(5):388–395

Xu W, Comhair SAA, Janocha AJ, Lara A, Mavrakis LA, Bennett CD, Kalhan SC, Erzurum SC (2017) Arginine metabolic endotypes related to asthma severity. PLoS One 12(8):e0183066
Xu W, Ghosh S, Comhair SA, Asosingh K, Janocha AJ, Mavrakis DA, Bennett CD, Gruca LL, Graham BB, Queisser KA, Kao CC, Wedes SH, Petrich JM, Tuder RM, Kalhan SC, Erzurum SC (2016) Increased mitochondrial arginine metabolism supports bioenergetics in asthma. J Clin Invest 126(7):2465–2481
Zhu Z, Homer RJ, Wang Z, Chen Q, Geba GP, Wang J, Zhang Y, Elias JA (1999) Pulmonary expression of interleukin-13 causes inflammation, mucus hypersecretion, subepithelial fibrosis, physiologic abnormalities, and eotaxin production. J Clin Invest 103(6):779–788

Epidemiology of Allergic Airways Diseases

4

Tanisqa Mall, Mousmee Sharma, and Parteek Prasher

Abstract

The epidemiology of allergic airways diseases, encompassing asthma and allergic rhinitis, is a complex domain influenced by genetic predispositions, environmental factors, and lifestyle choices. This paper offers a comprehensive synthesis of current research, shedding light on the prevalence, risk factors, and global impact of these conditions. Epidemiological investigations demonstrate significant variability in disease occurrence across different regions, age brackets, and socio-economic backgrounds, underscoring the intricate interplay between genetic susceptibilities and environmental triggers. Furthermore, recent findings emphasize the profound effects of urbanization, air pollution, dietary habits, and early-life exposures on the initiation and progression of allergic airways diseases. An in-depth comprehension of the epidemiological patterns and determinants of these ailments is pivotal for guiding public health policies, resource allocation, and targeted interventions to alleviate their burden on individuals and healthcare systems globally.

Keywords

Allergic airways diseases · Asthma · Allergic rhinitis · Epidemiology · Prevalence · Risk factors · Global burden

T. Mall (✉) · P. Prasher
Department of Chemistry, University of Petroleum & Energy Studies, Dehradun, India

M. Sharma
Department of Chemistry, Uttaranchal University, Dehradun, India

P. Prasher et al. (eds.), *Understanding Allergic Airway Diseases*,
https://doi.org/10.1007/978-981-97-1953-2_4

4.1 Introduction

Epidemiology investigates the prevalence and distribution of diseases within populations. It encompasses considerations regarding disease definition, overall community morbidity and mortality, determinants contributing to disease onset, and the impact of interventions, including both therapeutic and preventive measures. As such, the emphasis lies on populations rather than individual patients (Rothman 2012). Epidemiological investigations conducted across various global geographical zones have consistently demonstrated a notable and robust correlation between elevated ambient pollutant concentrations and heightened manifestations of asthma and rhinitis symptoms (Salvi 2001). In its most basic form, epidemiology involves conducting surveys to assess disease incidence at a specific point in time within a particular population. These surveys may also unveil disease-associated factors that can be measured in terms of risk (Rothman et al. 2008). Some scientific observations propose that the occurrence of allergic disorders has potentially stabilized in affluent nations but continues to escalate in low- and middle-income regions (Asher et al. 2006). Typically, allergic ailments tend to initiate during childhood rather than adulthood, and the increased incidence of these conditions is largely driven by affected children rather than affected adults. Epidemiological data implies that not all instances of atopic dermatitis and asthma can be solely ascribed to atopic sensitization (Asher et al. 1995).

Allergic sensitization, marked by the existence of allergen-specific immunoglobulin E (IgE) in the bloodstream, can result from exposure to allergens via inhalation, ingestion, or skin contact. This exposure can lead to conditions like asthma, hay fever, eczema, and, in severe cases, systemic anaphylaxis. Over the last 150 years, there has been a swift and ongoing rise in the occurrence of allergic conditions. Presently, more than 30% of children experience allergies, approximately 10% suffer from asthma and allergic rhinitis, and 5–7% have developed food allergies. The exact reasons for the high prevalence of asthma remain unclear, but its swift emergence and expansion suggest a role for environmental or behavioral changes associated with Western lifestyles (Strachan 1989a). Allergic airways diseases encompass a group of respiratory conditions characterized by the immune system's hypersensitivity to airborne allergens. These conditions pose significant health burdens worldwide, affecting millions of individuals of all ages. The hallmark feature of allergic airways diseases is the inflammation and narrowing of the airways, leading to a range of symptoms and complications (Holgate and Polosa 2008). In individuals who have developed sensitivity to particular aeroallergens, exposure to these allergens commonly results in the expression of allergic rhinoconjunctivitis in the upper airways and allergic asthma in the lower airways (Dullaers et al. 2012).

The most prevalent and extensively studied allergic airways diseases are allergic rhinoconjunctivitis (commonly known as hay fever) and allergic asthma. Allergic rhinoconjunctivitis primarily affects the upper airways, including the nasal passages and conjunctiva of the eyes, leading to symptoms like sneezing, rhinorrhea, nasal congestion, pruritus, lacrimation, and conjunctival redness. In contrast, allergic asthma predominantly involves the lower airways, specifically the bronchial tubes, resulting in symptoms such as wheezing, dyspnea, coughing, and chest tightness. Asthma exacerbations can vary from mild to severe, with potential life-threatening implications, warranting significant public health concern. Allergic rhinoconjunctivitis and allergic asthma both have common underlying mechanisms marked by an excessive immune response to usually harmless environmental substances known as aeroallergens. Common aeroallergens encompass pollens from trees, grasses, and weeds, dust mites, pet dander, mold spores, and certain occupational allergens.

The incidence of allergic airway diseases has exhibited an upward trajectory, notably within urban environments, imposing substantial burdens on healthcare systems, productivity, and the overall well-being of affected individuals.

Moreover, these conditions frequently coexist with other allergic disorders, suggesting shared underlying pathways and potential systemic effects. Ongoing research into the epidemiology, pathophysiology, risk factors, and treatment modalities for allergic airways diseases contributes to a more profound comprehension of the intricate interplay between genetics, environmental exposures, and immune responses. This growing knowledge drives the development of innovative therapies and preventive approaches to alleviate the burden of allergic airways diseases and enhance the respiratory health of affected individuals. In this comprehensive review, we endeavor to explore various facets of allergic airways diseases, shedding light on recent research findings, advancements in clinical management, and implications for public health. By gaining enhanced insights into these conditions, we aspire to foster improved diagnostic methods, personalized therapeutic strategies, and augmented support for individuals living with allergic airways diseases.

4.2 Prevalence and Incidence

4.2.1 Global Prevalence

The incidence of allergic airway phenotypes has exhibited a substantial and noteworthy escalation across the majority of nations during recent decades. Projections indicate that by the year 2025, approximately 400 million individuals will experience asthma, while nearly 500 million will contend with allergic rhinitis (Masoli et al. 2004). Based on 2015 data from the World Health Organization (WHO), asthma represents a worldwide health crisis, impacting more than 1 billion individuals presently, with a projected escalation to affect a population of approximately 4 billion by the year 2050 (W. H. Organization 2015). Asthma and allergic rhinitis exhibit notably high prevalence rates among pediatric populations (Sears 2014; Pearce et al. 2007; Strachan et al. 1997) in developed nations, bearing a substantial burden (Ozdoganoglu and Songu 2012; Blaiss et al. 2018). Allergic conditions frequently manifest as comorbidities, with nearly two-thirds of individuals diagnosed with asthma concurrently exhibiting symptoms of allergic rhinitis (AR). Intriguingly, substantial geographic disparities are evident in the prevalence of allergic disorders, with urbanized and highly developed areas experiencing a higher burden of these conditions (Arshad et al. 2001; Huss et al. 2001).

The International Study of Asthma and Allergies in Childhood (ISAAC) systematically recorded escalating trends in the worldwide occurrence of allergic ailments, with a particular emphasis on the upsurge observed in highly developed countries (Kim and Hong 2012). Conversely, in low-income and middle-income countries (LMICs), the prevalence of these conditions is comparatively modest, albeit exhibiting a consistent upward trajectory in these regions (Bousquet et al. 2020). A comprehensive survey has consistently revealed heightened prevalence rates, which extend beyond the distinction between developed and developing nations. These variations also manifest within specific countries, highlighting disparities between urban and rural settings. Investigations into the epidemiology of allergic ailments have unveiled substantial heterogeneity in the prevalence of respiratory allergies. Prevalence rates of these conditions can change over time due to various factors, including changes in environmental exposures, genetics, and healthcare practices.

Numerous research investigations have revealed substantial disparities in asthma incidence across different nations, with a global affected population exceeding 300 million individuals (Masoli et al. 2004; Bousquet et al. 2008). The estimated prevalence of allergic rhinitis (AR) in the United States, United Kingdom, Germany, Switzerland, and Finland is within the range of 10% to 20% (Von Hertzen and Haahtela 2005; Bauchau and Durham 2004). Numerous published reports have highlighted substantial disparities in the occurrence of hay fever, atopy, and asthma in both pediatric and adult populations among Eastern and Western European regions. Notably, the incidence of asthma in Eastern European study areas was consistently lower when compared to their Western European counterparts. For example, among adolescents aged 13 to 14 years, the prevalence of wheezing varied from 7.6% to 8.59% in Estonia, Latvia, and Poland; 11.2% to 19.7% in Finland and Sweden; and 2.6% to 5.9% in Albania, Romania, Russia, Georgia, and Uzbekistan (excluding Samarkand) (Bjorksten et al. 1998; Nowak et al. 1996; Brozek et al. 2015; Bergmann et al. 2016; von Mutius et al. 1994).

The trend of increasing asthma symptom prevalence in developing areas has been substantiated by a Turkish study conducted in 1994 and 2004. This study, involving more than 5000 preschool children, reported the following rates of current asthma symptoms in rural and urban areas: 5.2% and 5.8% in 1994 and 8.6% and 12.1% in 2004, respectively. Moreover, numerous research studies have scrutinized the impact of diagnosing and treating asthma on the frequency of asthma symptoms. In Sweden, there was a notable increase in physician-diagnosed asthma from 5.7% to 7.4% between 1996 and 2006. This increase coincided with a rise in the usage of inhaled corticosteroids, reaching 55% in 1996 and 67% in 2006 (Bråbäck et al. 1994). This trend likely signifies improved diagnostic capabilities and therapeutic interventions, as evidenced by a concurrent reduction in the occurrence of severe asthma symptoms (Andersson et al. 2010; Ställberg et al. 2009). The practical objective of these investigations is to generate novel insights that can contribute to the prevention of asthma and allergic conditions.

4.2.2 Geographical Variations in the Prevalence of Asthma

4.2.2.1 Childhood Asthma

The International Study of Asthma and Allergies in Childhood (Weinmayr et al. 2014; Mallol et al. 2013) (ISAAC) initiated in 1991 was established to methodically investigate the globally recognized elevation in the occurrence of allergies and asthma. It aimed to address the lack of consistent and reliable data for assessing the magnitude of these conditions. ISAAC Phase One, spanning from 1992 to 1998, employed a standardized approach and a validated questionnaire. This initial phase encompassed a comprehensive study conducted across 56 countries, involving a substantial cohort of approximately 700,000 children between the age bracket of 6–7 years and 13–14 years, with the purpose of quantifying the global prevalence of rhinitis, hay fever, and asthma (Asher et al. 2012, 2020).

Asthma was characterized by an affirmative response to an inquiry concerning recent episodes of wheezing or chest sounds resembling whistling. Strikingly, a substantial 20-fold disparity in the global occurrence of asthma was seen (Weiland et al. 2004). Maximum rates were recorded in the New Zealand, United Kingdom, Ireland, and Australia, while the lowest rates were found in China, Eastern Europe, Greece, Indonesia, Taiwan, Ethiopia, and India (Hu et al. 2021; Uphoff et al. 2017). Interestingly, this disparity in asthma prevalence persisted even among populations that appeared to share genetic similarities. To investigate potential contributing factors to this variation, follow-up studies were conducted during ISAAC Phase Two. These studies explored various environmental factors such as diet, infections, indoor and outdoor environments, climate, and allergen exposure that could influence disease risk. This research primarily focuses on scrutinizing variations in asthma prevalence among children aged 10–12 years (Custovic 2017).

Comparative analyses were conducted among diverse populations across multiple research centers, employing quantifiable disease indicators and evaluations encompassing environmental parameters, lifestyle factors, and clinical interventions. Notwithstanding comprehensive inquiry, no singular overarching determinant surfaced as an explanation for the observed disparities in asthma prevalence.

In summary, global asthma prevalence exhibited substantial variation, with the highest rates prevalent in certain developed nations and the lowest rates prevalent in several Eastern European and developing countries. Investigations during ISAAC Phase Two examined numerous environmental variables, yet the identification of a singular, definitive causative factor for these divergences in asthma prevalence remained elusive (Mallol et al. 2013; Serebrisky and Wiznia 2019).

4.2.2.2 Adult Asthma

The European Community Respiratory Health Survey (ECRHS) is a multicenter research initiative meticulously structured to assess the geographic diversity in the occurrence, treatment approaches, and causative factors related with asthma and allergic conditions in a cohort of 140,000 adults ranging between the age of 22–44 years across 22 countries. This comprehensive investigation employs uniform instruments and standardized definitions for data collection and analysis

(Custovic 2017; Chinn et al. 1996). In this research, a rigorously validated survey was utilized to evaluate the occurrence of asthma and allergic ailments while concurrently gathering information on possible risk elements. The definition of "diagnosed current asthma" was predicated upon a positive response indicating either an asthma attack occurring within the preceding 12 months or the ongoing utilization of medication for asthma management (Fig. 4.1).

For a subset of participants chosen randomly from multiple research center, a more extensive assessment was conducted during Stage II, spanning from 1991 to 1993. This assessment encompassed detailed questionnaires, blood tests measuring total and specific IgE levels, methacholine challenge testing, spirometry, and skin prick test (Chinn et al. 1996).

Subsequently, ECRHS II was undertaken with a particular emphasis on examining the occurrence and the causal risk elements linked to the emergence of allergic disorders, atopy (a predisposition to allergic responses), and the accelerated decline in pulmonary function in middle-aged individuals. This stage also encompassed the acquisition of air pollution data and dust samples.

Following this, ECRHS III constitutes a subsequent survey study encompassing a cohort of more than 10,000 adults who were initially recruited between 1992 and 1994. The principal aims of this phase are centered on the recording of alterations in the occurrence of respiratory symptoms and the emergence of IgE sensitization in the adult cohort as they advance in age (Usmani et al. 2016).

4.2.2.3 Allergic Rhinitis

The clinical characterization of rhinitis poses challenges when applied in epidemiological contexts involving large populations. In such settings, individual visits or the acquisition of comprehensive laboratory evidence for each immune response is impractical. This situation suggests a potential overestimation of allergic rhinitis (AR) prevalence when relying solely on questionnaires. Moreover, it appears that slightly more than 50% of individuals diagnosed with AR based on questionnaires can attribute their condition to IgE-mediated allergy (Ozdoganoglu and Songu 2012).

4.3 Risk Factors

Significant insights into potential risk factors influencing the commencement of asthma in children have been acquired mostly through comprehensive birth cohort studies. Nonetheless, there has been comparatively sparse body of data concerning the initiation of asthma in adults. Multiple environmental exposures, encompassing indoor and outdoor allergen contact, air pollution, viral, exposure to tobacco smoke, and bacterial infections, as well as dietary factors, with a notable focus on obesity, exert significant influence on the origins and severity of asthma (Fig. 4.1) (Wang 2005).

Within developed regions exhibiting elevated prevalence rates of allergic diseases, notable patterns encompass the adoption of hygienic living environments, specific dietary practices, and the cultivation of smaller family units. These

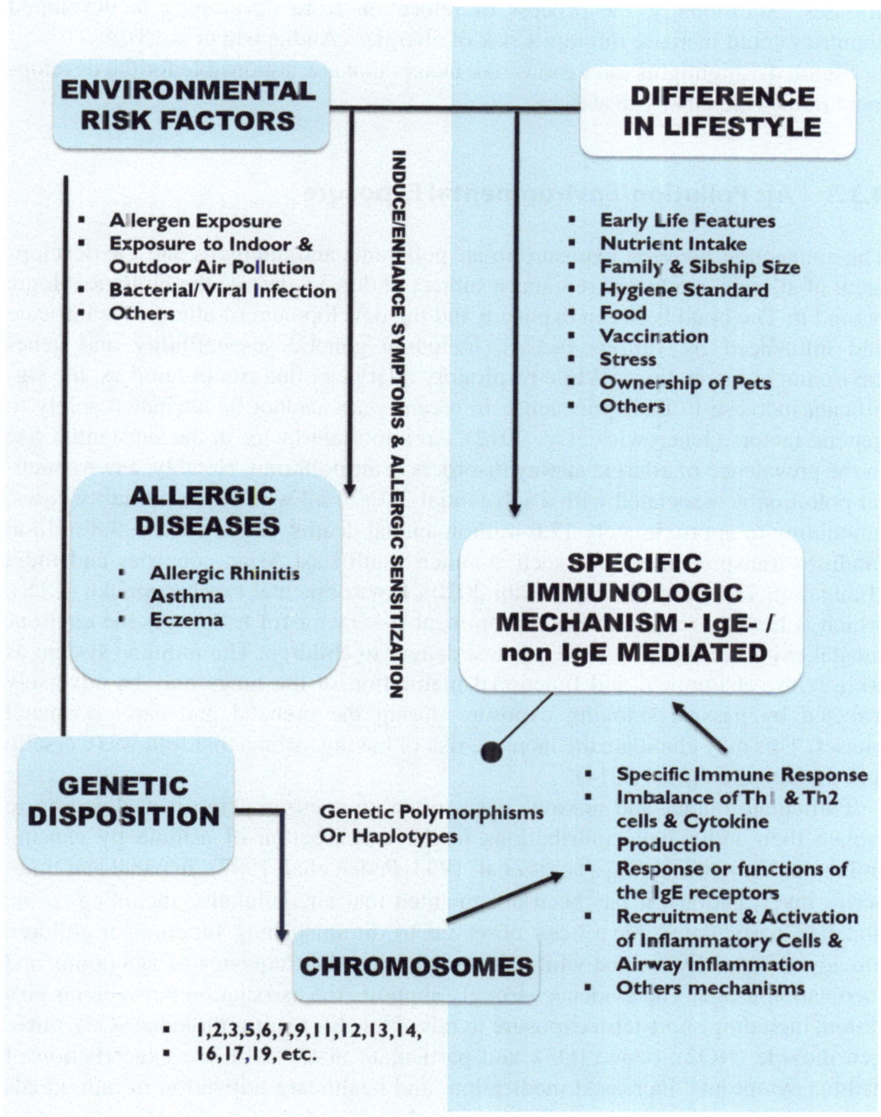

Fig. 4.1 Factors responsible for allergic airway diseases

environments are characterized by the consumption of highly processed and frequently sterilized foodstuffs, which are often fortified with additives and subject to alterations in nutrient composition (Fig. 4.1). To some extent, a segment of the inhabitants in developing countries has started to emulate aspects of the "Westernized" lifestyle, which may be linked to an increased incidence of allergic diseases. Additionally, the process of relocation from developing to developed countries could increase migrant's risk of allergies (Andiappan et al. 2014).

Figure 4.1 highlights the various risk factors that are responsible for the development of allergic airway diseases.

4.3.1 Air Pollution/Environmental Exposure

The connection between exposure to air pollutants and allergens and the development of allergic conditions remains a subject of debate, despite the biological logic behind it. The bond between exposure and the development of allergies is intricate and influenced by various factors, including genetic susceptibility and gene-environment interactions. While respiratory allergies often run in families, the significant increase in their prevalence in recent years cannot be attributed solely to genetic factors (Jenerowicz et al. 2012). An important factor in the substantial rise in the prevalence of allergic airway disorders is air pollution. Notably, environmental pollution is associated with a substantial 23% of all worldwide mortality cases, amounting to approximately 12.6 million annual deaths. Among these, 3.8 million fatalities transpire in regions such as other South East Asian countries and India (Bangalore Thimmadasiah and Joshi 2020). Environmental tobacco smoke (ETS), which is broadly recognized as a prominent risk factor for asthma, is the environmental exposure that poses the greatest danger to children. The immune system as well as the anatomical and functional maturation of the lungs may be adversely affected by passive smoking exposure during the prenatal and early postnatal phases. This may elucidate the increase risk of having asthma that follows (Corsello and Stefania La Grutta 2014).

Particulate matter and noxious gases often co-occur, making it challenging to isolate their individual contributions to the exacerbation of asthma by causing inflammation in the airways (Aris et al. 1993; Peden et al. 1997). In panel and time-series investigations, it has been documented that air pollutants, including ozone and fine particulates, have been observed to diminish lung function in children already afflicted by asthma while also amplifying the frequency of symptoms and medication usage. The evidence strongly supports the association between air pollution, including short-term exposure to raised levels of sulfur dioxide (SO2), nitrogen dioxide (NO2), ozone (O3), and particulate matter with the exacerbation of asthma symptoms, increased medication, and healthcare utilization in individuals already diagnosed with the condition and a decline in lung function (D'amato et al. 2015). For a significant duration, there has existed a hypothesis suggesting that ozone and various pollutants could elevate the susceptibility of individuals with allergies to the antigens they are sensitized to. There is supporting evidence from animal studies to substantiate this hypothesis. Moreover, there have been documented accounts suggesting a link between ozone exposure and an elevated propensity for asthma development in children engaged in outdoor athletic activities in California. Therefore, it is hypothesized that the combination of air pollution and outdoor physical exertion may play a pivotal role in the initiation of asthma in children by amplifying airway inflammation and reactivity (Yanai et al. 1990; McConnell et al. 2002). Essentially, the interaction of air pollution, specifically ozone exposure during outdoor physical activity, has the potential to play an important role in the initiation of asthma in children. This is achieved by accentuating airway inflammation and reactivity in individuals who are already sensitized to allergens (Holm and Balmes 2022). The existing data suggests that the adverse consequences of exposure to traffic-related factors may exert a substantial influence on the initiation of asthma in contrast to allergic sensitization. Additionally, indoor pollutants, notably exposure to environmental tobacco smoke indoors, contribute to morbidity associated with asthma (Zu et al. 2018).

Traffic-related air pollution encompasses a complex blend of initial gaseous emissions, particularly nitrogen oxides, which instigate the formation of secondary pollutants such as ozone, organic aerosols, and nitrates and particulate matter. Evidence suggests that residing in proximity to heavily trafficked roads is linked to compromised respiratory well-being. Road traffic, which emits both gaseous substances and particulate matter, presently stands as the primary source of air pollution in the majority of urban regions and is anticipated to maintain this status for an extended period (Wjst et al. 1993). The likelihood of asthma prevalence rises in direct proportion to heightened exposure to vehicular emissions, implying that continuous exposure to these pollutants may elevate the susceptibility to asthma development (Fuertes and Heinrich 2015). Air pollution stemming from motor vehicle emissions has been identified as a potential contributor to the escalated occurrence of allergic diseases observed in recent years (Gasana et al. 2012). Exposure to various combinations of pollutants can exhibit variations contingent on factors such as geographical location, proximity to roadways, seasonal patterns, and additional influencing factors. The levels of pollution associated with traffic proximity decrease significantly as one moves farther away from roadways. To elicit adverse effects, closeness within the range of 300–500 m from roadways seems to have the most pronounced impact. A significant 30–45% of the population in North American urban center lives within this crucial radius, highlighting the significant impact of traffic-related air pollution. Cities in Europe and the developing world may have even closer access to major roads. Notably, low-income households are frequently located close to busy roads, and additional risk factors like stress and smoking may worsen the impacts already shown. Living in close proximity to major roads during pregnancy is associated with an increased likelihood of asthma and atopic eczema, hence suggesting a correlation between residing near major roads and the development of allergic disorders in infants. Accumulating epidemiological data suggests that infants exposed to elevated levels of air pollution during prenatal development may face an increased susceptibility to the development of respiratory ailments.

4.3.2 Hygiene Hypothesis

Over the past decade, there has been a growing inquiry into the factors behind the increasing prevalence of atopic diseases. This rise in prevalence appears to be inversely related to the incidence of communicable diseases (Wang 2005; Fujieda et al. 1998). It is postulated that this phenomenon is associated with advancements in healthcare policies and elevated hygiene standards, particularly in developed nations. The concept of the "hygiene hypothesis" in relation to allergic diseases emerged following a pivotal study conducted by David Strachan. This study revealed an inverse relationship between size of the family and the occurrence of hay fever, suggesting that early-life infections could potentially serve as a protective factor against allergies (Strachan 1989b). Strachan initially postulated that infections and exposure to less hygienic conditions might offer a protective effect against the onset of allergies. The hygiene hypothesis suggests that the immune system's direction toward a nonallergic Th1 response is influenced by the early-life environment, whereas an environment considered "sterile" tends to encourage the development of allergies.

To elucidate the hygiene hypothesis, it is crucial to consider the maturation of the immune system in early development. In the prenatal environment, certain cytokines resembling those of T helper cells (Th2), such as IL-5, IL-13, IL-5, IL-4, are produced and can trigger Th2-like responses in a newborn's nascent immune system. This initial immune predisposition could increase the probability of exposure to allergens during both postnatal and possibly prenatal periods, resulting in the production of allergen-specific IgE. In accordance with the hygiene hypothesis, infections caused by potentially other intracellular organisms and viruses also play a role in molding the developing immune system. In response to these infections, the immune system generates Th1-like cytokines like IFN-γ and IL-12, which have a downregulating effect on Th2 responses. Therefore, the fundamental idea behind the hygiene hypothesis is that microbial-triggered Th1 immune responses can counteract allergen-triggered Th2 immune responses. Additionally, the "endotoxin switch" hypothesis posits that the quantity of exposure to bacterial products in the environment is dose-dependent and can influence the outcome of the immune response. Such exposure might also influence the equilibrium of Th1/Th2 cytokine levels. The hypothesis posits that recurrent viral infections, excluding those affecting the lower respiratory tract, during initial life may stimulate the developing immune system to favor a Th1 phenotype. This, in turn, may reduce the risk of developing asthma during childhood. However, the precise mechanisms underlying these protective immunological effects remain to be fully elucidated. The diminished exposure to infections in childhood resulting from improved hygiene and reduced social interaction with peers may signify the absence of a protective element, thereby potentially increasing vulnerability to the onset of allergic conditions (Garn and Renz 2007). Strachan initially postulated that infections and exposure to less hygienic conditions might offer a protective effect against the onset of allergies. The hygiene hypothesis posits that the early-life environment influences the immune system toward a Th1 (nonallergic) response, whereas an environment characterized as "sterile" tends to endorse the development of allergic reactions. Subsequent to this initial discovery, several investigations have documented a negative correlation between atopic conditions and factors such as crowded living conditions, interaction with peers (often assessed through attendance at day-care facilities), and indicators of socioeconomic well-being. However, the precise mechanisms underlying these associations remain unclear. Table 4.1 highlights the various factors the influence the airway allergies and evidence to support the claim.

While it is undeniable that mechanisms related to the balance between Th1 and Th2 immune responses are crucial in allergy development, they may not fully account for the effects associated with the hygiene hypothesis. To comprehensively understand the impacts of microbial exposures, we must consider mechanisms that operate at various stages of allergy development. Serum levels of immunoglobulin E (IgE), along with other antibody isotypes, are influenced not only by class-switch recombination to IgE but also by factors that regulate the final maturation of previously switched B cells and the rate at which IgE is secreted. Notably, the pro-inflammatory cytokine interleukin-6 (IL-6) is a significant factor in governing these

Table 4.1 Factors responsible for manifestation of airway allergies and the supporting evidence

Factors influencing allergies	Findings
Overcrowding	Numerous investigations indicate a negative association between atopic diseases and overcrowded living conditions, suggesting that lower household density might elevate the susceptibility to allergies. Nevertheless, the precise underlying mechanisms for this phenomenon remain enigmatic
Pet ownership	Early exposure to household pets, particularly dogs, seems to confer a protective effect against the onset of allergic sensitization
Gut microbiota	Experimental research in animals has proposed that the gut microbiota may play an important role in the onset of allergies. It appears that proper colonization of the gastrointestinal tract by bacteria is vital for maintaining normal immune responses. Discrepancies in gut flora have been noticed between children with allergies and those without allergies, although conclusive evidence about its role in human allergies is still lacking. Intervention studies utilizing prebiotics and probiotics have generated mixed outcomes, with some suggesting advantages in preventing eczema. However, uncertainties remain concerning their effects on allergic sensitization, asthma, and rhinitis
Contact with other children	Engagement in day-care programs during the first year of life has been correlated with a decline in the incidence of atopic asthma and allergic sensitization in most studies. Nevertheless, it is important to highlight that in specific cases, such attendance was linked with an elevated risk of early childhood wheezing
Microbial exposure	Overcrowded household environments, early enrollment in day-care facilities, and dog ownership could be linked to increased exposure to environmental microbial elements, such as endotoxins. These factors may be indicative of greater microbial diversity in the environment, which could potentially contribute to their protective effects. However, the specific mechanisms driving these associations require further clarification and investigation

processes. Furthermore, dendritic cells, working in conjunction with T regulatory cells, assume a crucial role in overseeing immune responses, and their importance in the context of allergy development is well established. Additionally, mechanisms inherent to the innate immune system, particularly those linked to endotoxin tolerance, may also contribute to mediating the effects triggered by microbial exposure.

4.3.3 Exposure to Allergens

Allergens are antigens that elicit an allergic response by binding to specific immunoglobulin E (IgE) antibodies. They have diverse origins, encompassing various animals, mites, insects, plants, and fungi. Indoor allergens predominantly arise from sources like house dust mites, particularly allergenic proteins found in their excrement, as well as allergens from furry pets, with a primary focus on dog and cats. Additionally, molds, cockroaches, certain plants, and rodents contribute to indoor allergen exposure. On the other hand, outdoor allergens primarily derive from plants, yeast, molds, and fungi (Bousquet et al. 2008; Hulin et al. 2012).

Sensitization to an allergen necessitates exposure to that specific allergen. Early-life exposure may be linked to an elevated risk of sensitization compared to exposure occurring later in life. However, there is currently no conclusive evidence to support the notion that reducing allergen levels in the homes of young children mitigates sensitization or the onset of allergic diseases. Once sensitization has taken place, repeated exposure to the specific allergen is likely to provoke symptomatic responses. Instances of asthma outbreaks have been observed in situations where elevated levels of allergens were present in the surrounding ambient air (Jarvis and Burney 1998a). Studies exploring how air pollution may enhance the sensitization and heightened reactivity to allergens in individuals with allergies have observed that pollen grains collected from industrial areas or in close proximity to major roadways exhibit elevated pro-inflammatory properties (Fujieda et al. 1998; Nel et al. 1998). Many research studies on primary prevention have aimed to ascertain whether reducing exposure to allergens during the initial life can decrease the likelihood of sensitization and commencement of asthma (Renz and Skevaki 2021). Nevertheless, the clinical outcomes from these studies have been inconsistent and at times perplexing. Certainly, the UK Isle of Wight study observed a substantial reduction in asthma prevalence and house dust mite sensitization from initial phase of childhood to age 18 years. In contrast, findings from the Manchester study suggested that intensive measures to reduce mite allergen exposure may have contributed to an elevated incidence of mite-specific sensitization. The efficacy of these interventions often exhibited variability depending on the age at which assessments were conducted, and specific intervention trials conducted in Australia and Netherlands reported no visible influence of allergen avoidance on the studied outcomes (Jarvis and Burney 1998b; Custovic et al. 2001).

4.3.3.1 Indoor Allergen Exposure

The indoor environment constitutes a significant reservoir of health-related risk factors, notably due to the fact that a substantial majority of individuals allocate more than 90% of their time within indoor spaces. The quality of the indoor environment is contingent upon the inflow of external air and the existence of indoor sources of air pollution (Wahn et al. 1997; Maciag and Phipatanakul 2022). Exposure to indoor allergens within the household environment is a prevalent trigger for perennial allergic rhinitis. This is particularly pertinent for young children who predominantly live in indoor spaces. Elevated concentrations of one or more allergens within the home environment are linked to an augmented risk of allergic sensitization during early life. In infants at a heightened risk for allergies, prenatal exposure to mite allergens originating from domestic dust, including the living room and household mattress, has been correlated with elevated total serum IgE levels at birth (Schonberger et al. 2005). The predominant indoor allergens associated with individuals suffering from atopic diseases encompass allergens stemming from house dust mites (HDM), domestic animals such as dogs, cockroaches, various fungi, rodents like mice, and cats (Beck et al. 2015; Matsui et al. 2010; Illi et al. 2006).

4.3.3.1.1 House Dust Mites

The house dust mite (HDM) is ubiquitously found in human habitation and serves as one of the most important contributors to the development of allergic rhinitis and asthma. These characteristics collectively establish it as one of the foremost origins of indoor allergenic substances (Richardson et al. 2005; Sporik et al. 1990). House dust mites (HDM) are frequently encountered in conjunction with bedding materials, carpets, drapes, and textile items. Sensitivity to HDM is notably linked to the presence of persistent allergic rhinitis (PAR). An initial longitudinal investigation conducted by Sporik et al. monitored 65 children from the day they were born to 11 years of age and revealed an elevated likelihood of allergic sensitization and asthma in children exposed to substantial levels of house dust mite (HDM) during their first year of life (Sporik et al. 1990). Subsequent investigations have provided further substantiating evidence, confirming a causal link between sensitization to house dust mites (HDM) during early life and the subsequent onset of continuous wheezing, allergic rhinitis, and asthma (Leger et al. 2017; Celedón et al. 2007). Tovey and colleagues (Tovey et al. 2008) propose a nonlinear relationship between exposure to house dust mite (HDM) allergens and the development of allergic conditions. Their results indicate that both extremely low and exceedingly high levels of exposure to house dust mites (HDM) are correlated with a diminished likelihood of sensitization and asthma, whereas exposure to moderate levels of allergens is associated with an increased risk. It remains unclear whether certain host or environmental factors can modify this relationship, particularly at the extremely low and high levels of allergen exposure. Furthermore, exposure to moderate levels of house dust mite (HDM) allergens has been linked to the severity of atopic diseases (Baïz et al. 2017; Huang et al. 2001).

International studies have shown that *Dermatophagoides pteronyssinus* (Der p) and *Dermatophagoides farinae* (Der f) are one of the most prevalent species of mites found worldwide (Bousquet et al. 2001). The German Multicentre Allergy Study, which monitored a cohort of newborn children during their first 3 years of life, observed a progressively higher incidence of allergic conditions as exposure to the predominant house dust mite allergens, Der p 1 and Der f 1, increased over time (Wahn et al. 1997). An investigation into house dust mite (HDM) sensitization in the initial 3 years of life revealed a low rate during infancy (0.5%), followed by an elevation during the second (1.4%) and third (1.9%) years of life. The study proposed that interventions aimed at primary prevention of sensitization should be initiated as early as possible, preferably during infancy (Sporik et al. 1990). Although it is a widely accepted notion that exposure to house dust mite (HDM) allergens during infancy elevates the risk of asthma development, a recent investigation failed to find an association between the level of house dust mite (HDM) allergen exposure in children's bedrooms during early childhood and the development of bronchial hyperresponsiveness or physician-diagnosed asthma by the age of 6–7 years (Carter et al. 2003). The relationships between household characteristics and the concentrations of allergens, specifically Der f 1 and Der p 1, in both air and dust are complex. The levels of Der f 1 allergen in household dust tend to rise with factors such as increased occupancy and higher relative humidity. However, they decrease in homes using forced air heating systems. On the other hand, concentrations of Der p 1 allergen are lower in newer residences and during periods of forced air heating, but higher in homes with elevated relative humidity and those with dogs. A study comparing indoor allergen (Der p 1 and Der f 1) and endotoxin levels in urban and rural settings highlighted the significance of indoor environmental factors, notably dampness, in influencing these levels. However, certain lifestyle choices such as the absence of mattress covers appeared to have less impact on these factors (Wickens et al. 2004; Peterson et al. 2003).

4.3.3.1.2 Ownership of Pets

Ownership of pets was associated with a notable rise in the likelihood of developing sensitization to pets (Al-Mousawi et al. 2004). Furthermore, the occurrence of conditions such as rhinitis, skin allergies, and asthma were comparatively higher in households with pets compared to those without them (Bener et al. 2004). Research investigating the association between owning pets and the likelihood of developing atopic diseases has primarily concentrated on the most common household pets, specifically cats and dogs, or has considered pet ownership in a broader context. The widespread presence of allergens from cats and dogs has posed challenges in epidemiological studies aiming to assess the risk associated with "exposure." (Arbes et al. 2004; McNicholas et al. 2005; Díaz-Perales et al. 2013).

4.3.4 Genetic Character

A substantial genetic predisposition, as evidenced by a family history of atopic disorders, has consistently been identified as the predominant factor for the onset of allergic manifestations. The influence of genetics in this context has received substantial backing from twin studies, commencing with the work of Edfors-Lubs in 1971 (Hopp et al. 1984), which underscored the hereditary aspect of allergic rhinitis in twins. Subsequent research by Hopp and colleagues (Hopp et al. 1984) revealed a higher degree of concordance for allergic manifestations among monozygotic (identical) twins compared to dizygotic (fraternal) twins. This correspondence was apparent in aspects such as serum total IgE levels, skin test response, and bronchial responsiveness to inhaled methacholine. Atopy plays an important role in the pathogenesis of allergic disorders, predominantly through the IgE-mediated pathway. This genetic condition typically appears during childhood when individuals develop sensitivity and produce IgE antibodies in response to common allergens. It involves the genetic control of bronchial hyperreactivity, specific IgE production, and total IgE concentration and displays varying levels of genetic impact.

In the realm of human genetic research, numerous candidate genes have been identified through methods such as position cloning and linkage analysis (Peden 2002; Toda and Ono 2002; Moffatt and Cookson 1999). Genome-wide investigations have revealed associations between specific allergic disease phenotypes (such as asthma and rhinitis) and genetic markers located on more than 14 pairs of chromosomes, including chromosomes 1, 2, 3, 5, 6, 7, 9, 11, 12, 13, 14, 16, 17, and 19, among others. Chromosome 5 appears to play a role in controlling total IgE levels, chromosome 11q is associated with the atopic phenotype characterized by elevated total IgE or specific IgE to common allergens, and chromosome 14 is linked to eczema. Some of these genes play roles in regulating the specific immune response (e.g., STAT6, TCR, HLA-D, CD14, toll-like receptor), total IgE response, IgE receptor functions (e.g., FcεRIβ, FcepsilonRI, IL-4, IL-4R), Th1/Th2 cell differentiation, and genes involved in the inflammatory process (e.g., IL-3, IFN-γ) (Jarvis and Burney 1998b; Weidinger 2004; Shek et al. 2001; Nishiyama et al. 2004; Park et al. 2004; Deichmann et al. 1999; Noguchi et al. 1997; Barnes et al. 1996; William et al. 1989; Marsh et al. 1994; Baldini et al. 1999). While genetic susceptibility plays a significant role in allergic diseases, it is improbable that genetic factors alone can account for the substantial disparities in disease prevalence observed among populations with similar genetic backgrounds, or for the notable rise in allergic diseases witnessed over recent decades (Simpson et al. 2006).

4.3.5 Rural-Urban Differences and Modification of Lifestyle

Numerous investigations have documented significant disparities in the occurrence of allergic conditions and allergic sensitization when comparing rural and urban regions within the same country, as evidenced in studies conducted in various nations such as China, Mongolia, Germany, Ghana, Poland, Canada, and Greece, among others. Some research even suggests that individuals who migrate from rural to urban environments may partially retain a level of protection against these conditions. In rural European settings, studies have revealed notably lower rates of atopic sensitization and hay fever among individuals residing on farms compared to those living in the same rural areas but not engaged in farming activities (Ernst and

Cormier 2000; Omland et al. 1999). In Europe, research emphasis has shifted toward investigations that delve into distinctions between rural farming and rural nonfarming populations, given that the lowest prevalence of asthma and allergic conditions has been observed within rural farming communities. Children born to farming families exhibit a lower prevalence of allergic rhinitis (AR) compared to their counterparts, suggesting that a rural way of life may potentially offer protective benefits against the onset of allergic conditions in children. The observed "protective" farm effect was associated with agricultural activities involving livestock and, consequently, exposure to microbial agents. Within rural settings, a significant dose-dependent inverse correlation was established between children's exposure to endotoxin present in mattress dust and the incidence of atopic diseases (Leynaert et al. 2001; Kilpelainen et al. 2002; Braun-Fahrländer 2003; Wickens et al. 2002).

Research conducted in Europe (von Mutius et al. 1994), South Africa (Masekela et al. 2018), and Central America (Celedón et al. 2001) has demonstrated an elevated prevalence of allergic rhinitis (AR) in urban locales when compared to rural settings. This discrepancy is particularly pronounced in cases of pollinosis despite the fact that pollen counts generally tend to be greater in urban areas than in rural ones (Nicolaou et al. 2005). Initial investigations conducted in African countries, including South Africa, Kenya, Ghana, and Ethiopia, highlighted that individual residing in rural environments, devoid of urban or Western lifestyle influences, exhibited a notably reduced prevalence of allergic diseases (Ng'ang'a et al. 1998; Addo Yobo et al. 1997; Yemaneberhan et al. 1997). Researchers investigated the potential protective factors of a conventional rural lifestyle and the influence of parasitic infections. Multiple studies conducted in Asian regions, including China, Korea, India, Saudi Arabia, and Japan, substantiated the urban-rural disparity and explored the impact of various factors such as allergen exposure, air quality, socioeconomic status, and dietary patterns on the development of allergies (Addo Yobo et al. 1997; Miyamoto 1997; Wong et al. 2004).

4.4 Public Health Implication

4.4.1 Burden on Healthcare System

In recent decades, there has been a significant increase in the prevalence of asthma and other atopic conditions like atopic dermatitis, food allergies, and allergic rhinitis. Collectively, these conditions now affect roughly 20% of the globe's population (Zheng 2014). While atopic dermatitis and allergic rhinitis are more common, the majority of epidemiological research in the field of allergic diseases has primarily focused on asthma. For example, a comprehensive examination of systematic reviews in the field of allergic epidemiology demonstrated that atopic eczema and allergic rhinitis were discussed in only 24.5% and 15.7% of the studies, respectively, while 72.9% of these studies included information on asthma (Genuneit et al. 2017). The economic repercussions linked to asthma are estimated to be among the most substantial when compared to other chronic illnesses. This is primarily attributable to the considerable healthcare resources and services required for managing this ailment. Extensive research endeavors have been undertaken to assess the financial impact of asthma at both societal and individual levels (Morgan and Khan 2003; Gergen 2001).

Asthma represents a chronic condition characterized by variable degrees of severity over a patient's lifetime, and it exerts a notable societal and economic burden. This ailment can lead to restrictions in both the physical and social dimensions of daily life, particularly when symptoms are inadequately managed. On a global scale, the economic costs associated with asthma are considerable and exhibit substantial disparities among nations. These variations are influenced by multiple factors, including the nature of the healthcare system, public health funding, and methodologies employed for data collection (Nunes et al. 2017).

Studies investigating the economic ramifications of asthma can be broadly categorized into two groups: those utilizing population-based sampling frameworks or administrative databases to provide cost assessments at geographical area or countries and those employing clinical-based sampling frameworks. Population-based studies provide broader applicability, whereas clinical-based studies offer a higher level of diagnostic certainty and often include data on disease severity, which is especially pertinent for understanding the economic consequences associated with asthma (Cisternas et al. 2003). In summary, the worldwide economic expenditures linked to asthma are substantial and exhibit considerable divergence among nations. These disparities are contingent upon a multitude of determinants, encompassing the healthcare system's structure, the allocation of financial resources in public health, and the methodologies employed for data gathering (Stróżek et al. 2019).

Direct costs encompass expenses associated with asthma management, comprising hospital admissions, medications (including prescription, over the counter, alternative medicines), outpatient visits (inclusive of healthcare professionals like doctors, physician, home care, drugs, nursing), emergency service visits, complementary diagnostic investigations or treatments (e.g., blood tests, lung function assessments, MRI scan), and other miscellaneous costs (e.g., occupational or domestic preventive measures, transportation to medical appointments).

Additionally, it encompasses early mortality, which refers to premature death attributed to asthma-related factors.

Over the past two decades, a substantial body of research has been dedicated to assessing the influence of asthma on labor productivity, absenteeism, temporary disability characterized by partial or complete workdays lost, early disability, and permanent disability, and the associated economic burdens. Additionally, it encompasses early mortality, which refers to premature death attributed to asthma-related factors. These investigations have examined multifaceted costs incurred by various stakeholders, encompassing employers, patients, caregivers, healthcare providers, and parents of children affected by asthma.

Intangible costs represent unquantifiable losses, encompassing factors such as diminished quality of life, increased suffering or pain, alterations in employment status, and limitations in physical activities. Regrettably, the literature on asthma costs has not yet systematically addressed these intangible costs (Tables 4.2 and 4.3).

Table 4.2 Direct costs of allergic rhinitis, asthma, and atopic dermatitis in various countries

Country	Study detail	Prevalence (%)	Direct cost per patient-year (currency)	Main cost component
USA (2003)	*Law* et al. (Law et al. 2003) calculated the total medical cost of allergic rhinitis at $3.4 billion, with 46.6% attributed to prescription medication.	Not specified	$3.4 billion	Prescription medication
France (2010)	*Belhassen* et al. (Belhassen et al. 2017) analyzed the healthcare data for people ranging between 18 and 65 years. The lower direct costs in this study can be attributed to the inclusion of patients based on general practitioner (GP) diagnoses and prescription data. Additionally, in France, many nonsteroidal anti-inflammatory drugs (NSAID) are available over the counter, potentially leading to underestimation of medical resource usage and associated costs.	6.8%	€159 (rhinitis only), €266–€375 (rhinitis with asthma)	GP diagnoses and OTC medication
India	*Ghoshal* et al. (Ghoshal et al. 2016) in his study included patients over 18 years who sought routine healthcare, encompassing respiratory conditions such as allergic rhinitis, asthma, rhinosinusitis, and COPD. Moreover, the study observed a significant utilization of alternative medicine, which could be indicative of factors such as limited access to evidence-based medicine, constrained resources, or cultural beliefs.	80%	$215	Medication and alternative medicine
Sweden	*Cardell* et al. (Cardell et al. 2016) studied for the population residing between 18 and 65 years and he found the direct costs based on the ARIA guidelines.	24%	€210	ARIA Guidelines

(continued)

Table 4.2 (continued)

Country	Study detail	Prevalence (%)	Direct cost per patient-year (currency)	Main cost component
South Korea	*Kim* et al. (Kim et al. 2010) The study collected insurance claims data from the National Health Insurance Corporation in Korea, which included a cohort of 4,068,517 individuals diagnosed with allergic rhinitis.	8.4%	$224 million	Korea National Health Insurance Data
Turkey	*Celik* et al. (Celik et al. 2004) The study provides details regarding the financial burden on patients experiencing allergic rhinitis without additional health conditions during the pollen season, which amounted to $79 USD. In contrast, individuals with allergic rhinitis who also had coexisting asthma and/or conjunctivitis experienced direct costs amounting to $139 USD.	Not specified	$79 (rhinitis), $139 (rhinitis with comorbidity)	Seasonal variation and comorbidity
USA (2018)	*Nurmagambetov* et al. examined 10,237 asthma patients and determined the following per-person medical cost breakdown resulting in an estimated total cost of $3266 per person and an overall expenditure of $57.9 billion in 2013.	4.9%	$3266	Medication, office visits, emergency room visits, hospitalizations, and outpatient visits

(continued)

Table 4.2 (continued)

Country	Study detail	Prevalence (%)	Direct cost per patient-year (currency)	Main cost component
Europe (2017–2018)	In a recent study by *Zink* et al. (Zink et al. 2019) encompassing 9 European countries and involving a cohort of 1189 individuals with atopic dermatitis, participants incurred an additional healthcare expenditure of €927 per annum. This primarily consisted of €332 for emollients and moisturizers, €213 for medication, and €104 for doctors and hospitals.	Not specified	€927	Emollients and moisturizers, medication, doctors, and hospitals
Korea (2015)	*Kim* et al. (Kim et al. 2015) performed a cost analysis in Korea, utilizing home accounting records of individuals suffering from atopic dermatitis. Their evaluation encompassed direct expenses, encompassing medical expenditures for consultations, laboratory examinations, medical procedures, hospital admissions, pharmaceuticals, dispensing fees, and expenses related to medication storage.		KRW 510,777 (€358)	Hospital stays, medical procedures, drugs, dispensing fees, consultation fee, and laboratory examinations
USA (2006)	In 2006, the American Academy of Dermatology and the Society for Investigative Dermatology conducted an extensive economic analysis of atopic dermatitis in the United States. The study calculated that the overall direct expenditures associated with atopic dermatitis in the United States in the year 2004 amount to $1.009 billion (Bickers et al. 2006)	~5%	$1009 billion	Not specified (excludes OTC products)

Table 4.3 Indirect costs of atopic dermatitis in various countries

Country	Study detail	Prevalence (%)	Indirect cost per patient-year (currency)	Main cost component
Japan	*Arima* et al. (Loh and Tang 2018) The study involved a comparison of 634 individuals who self-diagnosed with atopic dermatitis with 1268 matched controls. The findings indicated that employed individuals with atopic dermatitis exhibited significantly higher levels of productivity loss in comparison to the matched controls.	28.6% vs 22.4%	–	Higher productivity loss
South Korea	*Kim* et al. (Kim et al. 2015) The study in Korea conducted an estimation of indirect costs among working-age adults with atopic dermatitis, revealing 3 days of premature departure from work and average of 4.5 days of work absenteeism.	–	KRW1,507,068 (€1055)	Work absence and early leave

4.5 Prevention and Control Strategies

4.5.1 Primary, Secondary, and Tertiary Prevention of Airways Allergic Diseases

Due to the significant prevalence of allergic disorders, substantial attention has been dedicated to prevention strategies. These strategies can be categorized into three levels:

1. Primary Prevention: This involves measures aimed at averting the initial onset of allergic diseases.
2. Secondary Prevention: This pertains to interventions designed to halt the progression of allergic symptoms or manifestations in individuals who have already exhibited some allergic signs.
3. Tertiary Prevention: This encompasses efforts to prevent the deterioration of the condition and alleviate symptoms in allergic patients undergoing regular treatment (G. I. for Asthma 2013).

4.5.2 Allergy Vaccination

Allergy vaccination, also known as immunotherapy, has been a long-standing therapeutic approach for managing allergic airway diseases. The advancement of pharmacological treatments has led to a reduction in the utilization of allergy vaccination. Nevertheless, notable enhancements have been made in standardizing allergen extracts used in immunotherapy and characterizing the allergenic components within these extracts (Backman et al. 1991; Carlsen 2004). Immunotherapy has found extensive application in the treatment of allergic rhinitis, and it is recommended by the ARIA Guidelines when conventional pharmacotherapy proves insufficient. However, the effectiveness of allergy vaccination in treating asthma has yielded mixed results. Notably, there is a risk of systemic reactions when administering immunotherapy to individuals with uncontrolled asthma. A meta-analysis of allergy vaccination in asthma has suggested that its application should be reserved for carefully selected patients (Bousquet et al. 1990).

4.5.3 Treatment Aimed at Mitigating Inflammation at an Early Stage

Inhaled corticosteroids have shown significant effectiveness in the management of childhood asthma, resulting in a decrease in nonspecific bronchial reactivity (Bousquet et al. 1990) and exercise-induced asthma, even when administered at small dosages (Jónasson et al. 2000; Henriksen and Dahl 1983). Children who initiated inhaled corticosteroid therapy within 2 years of their asthma diagnosis exhibited significantly greater annual improvements in lung function compared to those who commenced treatment 5 years or more after diagnosis. Although the study lacked a control group, these findings were subsequently corroborated by Ko¨nig (Konig and Shaffer 1996).

The Childhood Asthma Management Program Research Group (CAMP) conducted a study involving 1041 children having moderate to mild asthma, treated for 4–6 years, comparing the effects of inhaled budesonide and nedocromil with a placebo. The budesonide-treated group initially demonstrated significantly higher baseline measures than the other two groups although this difference dissipated after bronchodilator administration. However, it is noteworthy that, at the commencement of the study, patients had been grappling with asthma for approximately 5 years, a duration similar to that observed in Agertoft and Pedersen's cohort (Szefler et al. 2000).

4.6 Conclusion

Over the past few decades, there has been a significant rise in the occurrence of these conditions, affecting individuals of all age groups and varying disease severity. Recent data indicates that this increase may have started to decline for certain conditions like asthma in specific age groups within developed countries. Nevertheless, in less developed regions, the prevalence of these conditions continues to rise, potentially leading to a reduction in global disparities. Allergic rhinitis and asthma represent significant early-onset noncommunicable diseases that impose a considerable socioeconomic health burden. The rising prevalence of these united airway diseases can be primarily attributed to environmental factors such as outdoor and indoor air pollution, global warming, and climate change. Epidemiological research has consistently demonstrated a substantial and coherent connection between environmental concentrations of diverse pollutants, including particulate matter, ozone, and nitrogen dioxide primarily originating from automobile emissions, and the rising incidence of allergic conditions. Several environmental factors have undergone thorough examination; however, a definitive explanation for the escalating trends has remained elusive. Upcoming endeavors involve elucidating the intricate relationship between environmental factors and genetic predispositions.

References

Addo Yobo EO, Custovic A, Taggart SC, Asafo-Agyei AP, Woodcock A (1997) Exercise induced bronchospasm in Ghana: differences in prevalence between urban and rural schoolchildren. Thorax 52(2):161–165. https://doi.org/10.1136/thx.52.2.161

Al-Mousawi MSH, Lovel H, Behbehani N, Arifhodzic N, Woodcock A, Custovic A (2004) Asthma and sensitization in a community with low indoor allergen levels and low pet-keeping frequency. J Allergy Clin Immunol 114(6):1389–1394. https://doi.org/10.1016/j.jaci.2004.09.005

Andersson M, Bjerg A, Forsberg B, Lundbäck B, Rönmark E (2010) The clinical expression of asthma in schoolchildren has changed between 1996 and 2006. Pediatr Allergy Immunol 21(5):859–866

Andiappan AK et al (2014) Allergic airway diseases in a tropical urban environment are driven by dominant mono-specific sensitization against house dust mites. Allergy 69(4):501–509. https://doi.org/10.1111/all.12364

Arbes SJ, Cohn RD, Yin M, Muilenberg ML, Friedman W, Zeldin DC (2004) Dog allergen (Can f 1) and cat allergen (Fel d 1) in US homes: results from the National Survey of Lead and Allergens in Housing. J Allergy Clin Immunol 114(1):111–117. https://doi.org/10.1016/j.jaci.2004.04.036

Aris RM, Christian D, Hearne PQ, Kerr K, Finkbeiner WE, Balmes JR (1993) Ozone-induced airway inflammation in human subjects as determined by airway lavage and biopsy. Am Rev Respir Dis 148:1363

Arshad SH, Tariq SM, Matthews S, Hakim E (2001) Sensitization to common allergens and its association with allergic disorders at age 4 years: a whole population birth cohort study. Pediatrics 108(2):e33–e33

Asher MI, García-Marcos L, Pearce NE, Strachan DP (2020) Trends in worldwide asthma prevalence. Eur Respir J 56(6):2002094

Asher MI et al (1995) International study of asthma and allergies in childhood (ISAAC): rationale and methods. Eur Respir J 8(3):483–491

Asher MI et al (2006) Worldwide time trends in the prevalence of symptoms of asthma, allergic rhinoconjunctivitis, and eczema in childhood: ISAAC Phases One and Three repeat multicountry cross-sectional surveys. Lancet 368(9537):733–743

Asher MI et al (2012) Changes over time in the relationship between symptoms of asthma, rhinoconjunctivitis and eczema: a global perspective from the International Study of Asthma and Allergies in Childhood (ISAAC). Allergol Immunopathol (Madr) 40(5):267–274

Backman A, Belin L, Dreborg S, Halvorsen R, Malling H-J, Weeke B (1991) Standardization of allergenic preparations. Allergy 46(2):81–84. https://doi.org/10.1111/j.1398-9995.1991.tb00548.x

Baïz N et al (2017) Early oral exposure to house dust mite allergen through breast milk: a potential risk factor for allergic sensitization and respiratory allergies in children. J Allergy Clin Immunol 139(1):369–372.e10. https://doi.org/10.1016/j.jaci.2016.07.021

Baldini M, Carla Lohman I, Halonen M, Erickson RP, Holt PG, Martinez FD (1999) A polymorphism* in the 5 ′ flanking region of the CD14 gene is associated with circulating soluble CD14 levels and with total serum immunoglobulin E. Am J Respir Cell Mol Biol 20(5):976–983. https://doi.org/10.1165/ajrcmb.20.5.3494

Bangalore Thimmadasiah N, Joshi TK (2020) India: country report on children's environmental health. Rev Environ Health 35(1):27–39

Barnes KC et al (1996) Linkage of asthma and total serum IgE concentration to markers on chromosome 12q: evidence from Afro-Caribbean and Caucasian populations. Genomics 37(1):41–50. https://doi.org/10.1006/geno.1996.0518

Bauchau V, Durham SR (2004) Prevalence and rate of diagnosis of allergic rhinitis in Europe. Eur Respir J 24(5):758–764

Beck AF et al (2015) Allergen sensitization profiles in a population-based cohort of children hospitalized for asthma. Ann Am Thorac Soc 12(3):376–384. https://doi.org/10.1513/AnnalsATS.201408-376OC

Belhassen M et al (2017) Costs of perennial allergic rhinitis and allergic asthma increase with severity and poor disease control. Allergy 72(6):948–958. https://doi.org/10.1111/all.13098

Bener A, Mobayed H, Sattar HA, Al-Mohammed AA, Ibrahimi AS, Sabbah A (2004) Pet ownership: its effect on allergy and respiratory symptoms. Eur Ann Allergy Clin Immunol 36(8):306–310

Bergmann K-C, Heinrich J, Niemann H (2016) Current status of allergy prevalence in Germany: position paper of the Environmental Medicine Commission of the Robert Koch Institute. Allergo J Int 25:6–10

Bickers DR et al (2006) The burden of skin diseases: 2004. J Am Acad Dermatol 55(3):490–500. https://doi.org/10.1016/j.jaad.2006.05.048

Bjorksten B et al (1998) Prevalence of childhood asthma, rhinitis and eczema in Scandinavia and Eastern Europe. Eur Respir J 12(2):432–437

Blaiss MS, Hammerby E, Robinson S, Kennedy-Martin T, Buchs S (2018) The burden of allergic rhinitis and allergic rhinoconjunctivitis on adolescents: a literature review. Ann Allergy Asthma Immunol 121(1):43–52

Bousquet J, Hejjaoui A, Michel F (1990) Specific immunotherapy in asthma. J Allergy Clin Immunol 86(3):292–305. https://doi.org/10.1016/S0091-6749(05)80091-0

Bousquet J, van Cauwenberge P, Khaltaev N (2001) Allergic rhinitis and its impact on asthma. J Allergy Clin Immunol 108(5):S147–S334. https://doi.org/10.1067/mai.2001.118891

Bousquet J et al (2008) Allergic rhinitis and its impact on asthma (ARIA) 2008. Allergy 63:8–160

Bousquet J et al (2020) Allergic rhinitis. Nat Rev Dis Primers 6(1):95. https://doi.org/10.1038/s41572-020-00227-0

Bråbäck L, Breborowicz A, Dreborg S, Knutsson A, Pieklik H, Björksten B (1994) Atopic sensitization and respiratory symptoms among Polish and Swedish school children. Clin Exp Allergy 24(9):826–835

Braun-Fahrländer C (2003) Environmental exposure to endotoxin and other microbial products and the decreased risk of childhood atopy: evaluating developments since April 2002. Curr Opin Allergy Clin Immunol 3(5):325–329. https://doi.org/10.1097/01.all.0000092600.10871.e0

Brozek G, Lawson J, Szumilas D, Zejda J (2015) Increasing prevalence of asthma, respiratory symptoms, and allergic diseases: four repeated surveys from 1993-2014. Respir Med 109(8):982–990

Cardell L-O et al (2016) TOTALL: high cost of allergic rhinitis—a national Swedish population-based questionnaire study. NPJ Prim Care Respir Med 26(1):15082. https://doi.org/10.1038/npjpcrm.2015.82

Carlsen K-H (2004) Therapeutic strategies for allergic airways diseases. Paediatr Respir Rev 5(1):45–51. https://doi.org/10.1016/j.prrv.2003.09.010

Carter PM, Peterson EL, Ownby DR, Zoratti EM, Johnson CC (2003) Relationship of house-dust mite allergen exposure in children's bedrooms in infancy to bronchial hyperresponsiveness and asthma diagnosis by age 6 to 7. Ann Allergy Asthma Immunol 90(1):41–44
Celedón JC, Soto-Quiros ME, Silverman EK, Hanson LA, Weiss ST (2001) Risk factors for childhood asthma in Costa Rica. Chest 120(3):785–790. https://doi.org/10.1378/chest.120.3.785
Celedón JC et al (2007) Exposure to dust mite allergen and endotoxin in early life and asthma and atopy in childhood. J Allergy Clin Immunol 120(1):144–149. https://doi.org/10.1016/j.jaci.2007.03.037
Celik GE, Bavbek S, Paşaoğlu G, Mungan D, Abadoğlu O, Harmanci E, Misirligil Z (2004) Direct medical cost of asthma in Ankara, Turkey. Respiration. 71:587–593
Chinn S et al (1996) Variations in the prevalence of respiratory symptoms, selfreported asthma attacks, and use of asthma medication in the European Community Respiratory Health Survey (ECRHS). Eur Respir J 9:687–695
Cisternas MG et al (2003) A comprehensive study of the direct and indirect costs of adult asthma. J Allergy Clin Immunol 111(6):1212–1218. https://doi.org/10.1067/mai.2003.1449
Corsello G, Stefania La Grutta MD (2014) Smoke exposure as a risk factor for asthma in childhood: a review of current evidence. Allergy Asthma Proc 35:454–461
Custovic A (2017) Epidemiology of allergic diseases. In: *Middleton's allergy essentials*. Elsevier, pp 51–72
Custovic A, Simpson BM, Simpson A, Kissen P, Woodcock A (2001) Effect of environmental manipulation in pregnancy and early life on respiratory symptoms and atopy during first year of life: a randomised trial. Lancet 358(9277):188–193
D'amato G et al (2015) Effects on asthma and respiratory allergy of climate change and air pollution. Multidiscip Respir Med 10(1):1–8
Deichmann KA et al (1999) Linkage and association studies of atopy and the chromosome 11q13 region. J Med Genet 36(5):379–382. https://doi.org/10.1136/jmg.36.5.379
Díaz-Perales A, González-de-Olano D, Pérez-Gordo M, Pastor-Vargas C (2013) Allergy to uncommon pets: new allergies but the same allergens. Front Immunol 4:492
Dullaers M, De Bruyne R, Ramadani F, Gould HJ, Gevaert P, Lambrecht BN (2012) The who, where, and when of IgE in allergic airway disease. J Allergy Clin Immunol 129(3):635–645
Ernst P, Cormier Y (2000) Relative scarcity of asthma and atopy among rural adolescents raised on a farm. Am J Respir Crit Care Med 161(5):1563–1566. https://doi.org/10.1164/ajrccm.161.5.9908119
Fuertes E, Heinrich J (2015) The influence of childhood traffic-related air pollution exposure on asthma, allergy and sensitization. Allergy. 70:1350–1351
Fujieda S, Diaz-Sanchez D, Saxon A (1998) Combined nasal challenge with diesel exhaust particles and allergen induces in vivo IgE isotype switching. *Am J Respir Cell Mol Biol* 19(3):507–512
G. I. for Asthma (2013) GINA workshop report, global strategy for asthma management and prevention. Revised 2013
Garn H, Renz H (2007) Epidemiological and immunological evidence for the hygiene hypothesis. Immunobiology 212(6):441–452. https://doi.org/10.1016/j.imbio.2007.03.006
Gasana J, Dillikar D, Mendy A, Forno E, Vieira ER (2012) Motor vehicle air pollution and asthma in children: a meta-analysis. Environ Res 117:36–45
Genuneit J et al (2017) Overview of systematic reviews in allergy epidemiology. Allergy 72(6):849–856. https://doi.org/10.1111/all.13123
Gergen PJ (2001) Understanding the economic burden of asthma. J Allergy Clin Immunol 107(5):S445–S448. https://doi.org/10.1067/mai.2001.114992
Ghoshal A et al (2016) The burden of segregated respiratory diseases in India and the quality of care in these patients: results from the Asia-Pacific burden of respiratory diseases study. Lung India 33(6):611. https://doi.org/10.4103/0970-2113.192878
Henriksen JM, Dahl R (1983) Effects of inhaled budesonide alone and in combination with low-dose terbutaline in children with exercise-induced asthma. Am Rev Respir Dis 128(5):993–997
Holgate ST, Polosa R (2008) Treatment strategies for allergy and asthma. Nat Rev Immunol 8(3):218–230

Holm SM, Balmes JR (2022) Systematic review of ozone effects on human lung function, 2013 through 2020. Chest 161(1):190–201. https://doi.org/10.1016/j.chest.2021.07.2170

Hopp RJ, Bewtra AK, Watt GD, Nair NM, Townley RG (1984) Genetic analysis of allergic disease in twins. J Allergy Clin Immunol 73(2):265–270. https://doi.org/10.1016/S0091-6749(84)80018-4

Hu Y-B et al (2021) Increasing prevalence and influencing factors of childhood asthma: a cross-sectional study in Shanghai, China. World J Pediatr 17(4):419–428

Huang J-L, Chen C-C, Kuo M-L, Hsieh K-H (2001) Exposure to a high concentration of mite allergen in early infancy is a risk factor for developing atopic dermatitis: a 3-year follow-up study. Pediatr Allergy Immunol 12(1):11–16. https://doi.org/10.1034/j.1399-3038.2001.012001011.x

Hulin M, Simoni M, Viegi G, Annesi-Maesano I (2012) Respiratory health and indoor air pollutants based on quantitative exposure assessments. Eur Respir J 40(4):1033–1045. https://doi.org/10.1183/09031936.00159011

Huss K, Adkinson NF Jr, Eggleston PA, Dawson C, Van Natta ML, Hamilton RG (2001) House dust mite and cockroach exposure are strong risk factors for positive allergy skin test responses in the childhood asthma management program. J Allergy Clin Immunol 107(1):48–54

Illi S, von Mutius E, Lau S, Niggemann B, Grüber C, Wahn U (2006) Perennial allergen sensitisation early in life and chronic asthma in children: a birth cohort study. Lancet 368(9537):763–770. https://doi.org/10.1016/S0140-6736(06)69286-6

Jarvis D, Burney P (1998a) ABC of allergies: the epidemiology of allergic disease. BMJ: Br Med J 316(7131):607

Jarvis D, Burney P (1998b) ABC of allergies: the epidemiology of allergic disease. BMJ 316(7131):607–610. https://doi.org/10.1136/bmj.316.7131.607

Jenerowicz D, Silny W, Danczak-Pazdrowska A, Polanska A, Osmola-Mankowska A, Olek-Hrab K (2012) Environmental factors and allergic diseases. Ann Agric Environ Med 19(3):475

Jónasson G, Carlsen K, Hultquist C (2000) Low-dose budesonide improves exercise-induced bronchospasm in schoolchildren. Pediatr Allergy Immunol 11(2):120–125. https://doi.org/10.1034/j.1399-3038.2000.00067.x

Kilpelainen M, Terho EO, Helenius H, Koskenvuo M (2002) Childhood farm environment and asthma and sensitization in young adulthood. Allergy 57(12):1130–1135. https://doi.org/10.1034/j.1398-9995.2002.23341.x

Kim B-J, Hong S-J (2012) Ambient air pollution and allergic diseases in children. Korean J Pediatr 55(6):185

Kim C et al (2015) Economic impact of atopic dermatitis in Korean patients. Ann Dermatol 27(3):298. https://doi.org/10.5021/ad.2015.27.3.298

Kim SY, Yoon S-J, Jo M-W, Kim E-J, Kim H-J, Oh I-H (2010) Economic burden of allergic rhinitis in Korea. Am J Rhinol Allergy 24(5):e110–e113. https://doi.org/10.2500/ajra.2010.24.3513

Konig P, Shaffer J (1996) The effect of drug therapy on long-term outcome of childhood asthma: a possible preview of the international guidelines. J Allergy Clin Immunol 98(6):1103–1111. https://doi.org/10.1016/S0091-6749(96)80198-9

Law AW, Reed SD, Sundy JS, Schulman KA (2003) Direct costs of allergic rhinitis in the United States: estimates from the 1996 medical expenditure panel survey. J Allergy Clin Immunol 111(2):296–300. https://doi.org/10.1067/mai.2003.68

Leger D, Bonnefoy B, Pigearias B, de La Giclais B, Chartier A (2017) Poor sleep is highly associated with house dust mite allergic rhinitis in adults and children. Allergy, Asthma Clin Immunol 13:1–9

Leynaert B, Neukirch C, Jarvis D, Chinn S, Burney P, Neukirch F (2001) Does living on a farm during childhood protect against asthma, allergic rhinitis, and atopy in adulthood? Am J Respir Crit Care Med 164(10):1829–1834. https://doi.org/10.1164/ajrccm.164.10.2103137

Loh W, Tang M (2018) The epidemiology of food allergy in the global context. Int J Environ Res Public Health 15(9):2043. https://doi.org/10.3390/ijerph15092043

Maciag MC, Phipatanakul W (2022) Update on indoor allergens and their impact on pediatric asthma. Ann Allergy Asthma Immunol 128(6):652–658

Mallol J et al (2013) The International Study of Asthma and Allergies in Childhood (ISAAC) phase three: a global synthesis. Allergol Immunopathol (Madr) 41(2):73–85

Marsh DG et al (1994) Linkage analysis of *IL4* and other chromosome 5q31.1 markers and total serum immunoglobulin E concentrations. Science (1979) 264(5162):1152–1156. https://doi.org/10.1126/science.8178175

Masekela R et al (2018) The increasing burden of asthma in South African children: a call to action. S Afr Med J 108(7):537. https://doi.org/10.7196/SAMJ.2018.v108i7.13162

Masoli M, Fabian D, Holt S, Beasley R, G. I. for A. (GINA) Program (2004) The global burden of asthma: executive summary of the GINA dissemination committee report. Allergy 59(5):469–478

Matsui EC et al (2010) Allergen-specific IgE as a biomarker of exposure plus sensitization in inner-city adolescents with asthma. Allergy 65(11):1414–1422. https://doi.org/10.1111/j.1398-9995.2010.02412.x

McConnell R et al (2002) Asthma in exercising children exposed to ozone: a cohort study. Lancet 359(9304):386–391. https://doi.org/10.1016/S0140-6736(02)07597-9

McNicholas J, Gilbey A, Rennie A, Ahmedzai S, Dono J-A, Ormerod E (2005) Pet ownership and human health: a brief review of evidence and issues. BMJ 331(7527):1252–1254

Miyamoto T (1997) Epidemiology of pollution-induced airway disease in Japan. Allergy 52:30–34. https://doi.org/10.1111/j.1398-9995.1997.tb04867.x

Moffatt MF, Cookson WO (1999) Genetics of asthma and inflammation: the status. Curr Opin Immunol 11(6):606–609. https://doi.org/10.1016/S0952-7915(99)00024-2

Morgan M, Khan DA (2003) Asthma: epidemiology, burden, and quality of life. Adv Psychosom Med 24:1–15

von Mutius E, Martinez FD, Fritzsch C, Nicolai T, Roell G, Thiemann H-H (1994) Prevalence of asthma and atopy in two areas of West and East Germany. Am J Respir Crit Care Med 149(2):358–364

Nel AE, Diaz-Sanchez D, Ng D, Hiura T, Saxon A (1998) Enhancement of allergic inflammation by the interaction between diesel exhaust particles and the immune system. J Allergy Clin Immunol 102(4):539–554

Ng'ang'a LW et al (1998) Prevalence of exercise induced bronchospasm in Kenyan school children: an urban-rural comparison. Thorax 53(11):919–926. https://doi.org/10.1136/thx.53.11.919

Nicolaou N, Siddique N, Custovic A (2005) Allergic disease in urban and rural populations: increasing prevalence with increasing urbanization. Allergy 60(11):1357–1360. https://doi.org/10.1111/j.1398-9995.2005.00961.x

Nishiyama C et al (2004) Polymorphisms in the *FcεRI* β promoter region affecting transcription activity: a possible promoter-dependent mechanism for association between *FcεRI* β and atopy. J Immunol 173(10):6458–6464. https://doi.org/10.4049/jimmunol.173.10.6458

Noguchi E et al (1997) Evidence for linkage between asthma/atopy in childhood and chromosome 5q31-q33 in a Japanese population. Am J Respir Crit Care Med 156(5):1390–1393. https://doi.org/10.1164/ajrccm.156.5.9702084

Nowak D et al (1996) Prevalence of respiratory symptoms, bronchial hyperresponsiveness and atopy among adults: west and East Germany. Eur Respir J 9(12):2541–2552

Nunes C, Pereira AM, Morais-Almeida M (2017) Asthma costs and social impact. Asthma Res Pract 3(1):1–11

Omland Ø, Sigsgaard T, Hjort C, Pedersen OF, Miller MR (1999) Lung status in young Danish rurals: the effect of farming exposure on asthma-like symptoms and lung function. Eur Respir J 13(1):31–37. https://doi.org/10.1183/09031936.99.13103199

Ozdoganoglu T, Songu M (2012) The burden of allergic rhinitis and asthma. Ther Adv Respir Dis 6(1):11–23

Park BL et al (2004) Interleukin 3 (IL3) polymorphisms associated with decreased risk of asthma and atopy. J Hum Genet 49(10):517–527

Pearce N et al (2007) Worldwide trends in the prevalence of asthma symptoms: phase III of the international study of asthma and allergies in childhood (ISAAC). Thorax 62(9):758–766

Peden DB (2002) Influences on the development of allergy and asthma. Toxicology 181–182:323–328. https://doi.org/10.1016/S0300-483X(02)00301-3

Peden DB, Boehlecke B, Horstman D, Devlin R (1997) Prolonged acute exposure to 0.16 ppm ozone induces eosinophilic airway inflammation in asthmatic subjects with allergies. J Allergy Clin Immunol 100(6):802–808

Peterson EL, Ownby DR, Johnson CC (2003) The relationship of housing and household characteristics to the indoor concentrations of Der f 1, Der p 1, and Fel d 1 measured in dust and air samples. Ann Allergy Asthma Immunol 90(5):564–571. https://doi.org/10.1016/S1081-1206(10)61852-2

Renz H, Skevaki C (2021) Early life microbial exposures and allergy risks: opportunities for prevention. *Nat Rev Immunol* 21(3):177–191. Accessed: Sep. 18, 2023. [Online]. Available: https://www.nature.com/articles/s41577-020-00420-y

Richardson G, Eick S, Jones R (2005) How is the indoor environment related to asthma?: literature review. J Adv Nurs 52(3):328–339. https://doi.org/10.1111/j.1365-2648.2005.03591.x

Rothman KJ (2012) Epidemiology: an introduction. Oxford University Press

Rothman KJ, Greenland S, Lash TL (2008) *Modern epidemiology*, vol 3. Wolters Kluwer Health/Lippincott Williams & Wilkins, Philadelphia

Salvi S (2001) Pollution and allergic airways disease. Curr Opin Allergy Clin Immunol 1(1):35–41

Schonberger HJ, Dompeling E, Knottnerus JA, Kuiper S, Weel C, Schayck CP (2005) Prenatal exposure to mite and pet allergens and total serum IgE at birth in high-risk children. Pediatr Allergy Immunol 16(1):27–31. https://doi.org/10.1111/j.1399-3038.2005.00243.x

Sears MR (2014) Trends in the prevalence of asthma. Chest 145(2):219–225

Serebrisky D, Wiznia A (2019) Pediatric asthma: a global epidemic. Ann Glob Health 85:1

Shek LP-C, Tay AH, Chew FT, Goh DLM, Lee BW (2001) Genetic susceptibility to asthma and atopy among Chinese in Singapore - linkage to markers on chromosome 5q31-33. Allergy 56(8):749–753. https://doi.org/10.1034/j.1398-9995.2001.056008749.x

Simpson A et al (2006) Endotoxin exposure, CD14, and allergic disease. Am J Respir Crit Care Med 174(4):386–392. https://doi.org/10.1164/rccm.200509-1380OC

Sporik R, Holgate ST, Platts-Mills TAE, Cogswell JJ (1990) Exposure to house-dust mite allergen (*Der p* I) and the development of asthma in childhood. N Engl J Med 323(8):502–507. https://doi.org/10.1056/NEJM199008233230802

Ställberg B, Lisspers K, Hasselgren M, Janson C, Johansson G, Svärdsudd K (2009) Asthma control in primary care in Sweden: a comparison between 2001 and 2005. Prim Care Respir J 18(4):279–286

Strachan D et al (1997) Worldwide variations in prevalence of symptoms of allergic rhinoconjunctivitis in children: the International Study of Asthma and Allergies in Childhood (ISAAC). Pediatr Allergy Immunol 8(4):161–168

Strachan DP (1989a) Hay fever, hygiene, and household size. BMJ: Br Med J 299(6710):1259

Strachan DP (1989b) Hay fever, hygiene, and household size. BMJ 299(6710):1259–1260. https://doi.org/10.1136/bmj.299.6710.1259

Stróżek J et al (2019) The indirect costs of allergic diseases. Int J Occup Med Environ Health 32(3):281–290. https://doi.org/10.13075/ijomeh.1896.01275

Szefler S et al (2000) Long-term effects of budesonide or nedocromil in children with asthma. N Engl J Med 343(15):1054–1063. https://doi.org/10.1056/NEJM200010123431501

Toda M, Ono SJ (2002) Genomics and proteomics of allergic disease. Immunology 106(1):1–10. https://doi.org/10.1046/j.1365-2567.2002.01407.x

Tovey ER, Almqvist C, Li Q, Crisafulli D, Marks GB (2008) Nonlinear relationship of mite allergen exposure to mite sensitization and asthma in a birth cohort. J Allergy Clin Immunol 122(1):114–118 .e5. https://doi.org/10.1016/j.jaci.2008.05.010

Uphoff EP et al (2017) Variations in the prevalence of childhood asthma and wheeze in MeDALL cohorts in Europe. ERJ Open Res 3(3):00150

Usmani OS, Singh D, Spinola M, Bizzi A, Barnes PJ (2016) The prevalence of small airways disease in adult asthma: a systematic literature review. Respir Med 116:19–27. https://doi.org/10.1016/j.rmed.2016.05.006

Von Hertzen L, Haahtela T (2005) Signs of reversing trends in prevalence of asthma. Allergy 60(3):283–292
W. H. Organization (2015) Reducing global health risks through mitigation of short-lived climate pollutants. Scoping report for policy-makers
Wahn U et al (1997) Indoor allergen exposure is a risk factor for sensitization during the first three years of life. J Allergy Clin Immunol 99(6):763–769
Wang D-Y (2005) Risk factors of allergic rhinitis: genetic or environmental? Ther Clin Risk Manag 1(2):115–123
Weidinger S (2004) Association of a STAT 6 haplotype with elevated serum IgE levels in a population based cohort of white adults. J Med Genet 41(9):658–663. https://doi.org/10.1136/jmg.2004.020263
Weiland SK, Björkstén B, Brunekreef B, Cookson WOC, Von Mutius E, Strachan DP (2004) Phase II of the International Study of Asthma and Allergies in Childhood (ISAAC II): rationale and methods. Eur Respir J 24(3):406–412
Weinmayr G et al (2014) Overweight/obesity and respiratory and allergic disease in children: international study of asthma and allergies in childhood (ISAAC) phase two. PLoS One 9(12):e113996
Wickens K et al (2002) Farm residence and exposures and the risk of allergic diseases in New Zealand children. Allergy 57(12):1171–1179. https://doi.org/10.1034/j.1398-9995.2002.t01-1-23644.x
Wickens K et al (2004) The determinants of dust mite allergen and its relationship to the prevalence of symptoms of asthma in the Asia-Pacific region. Pediatr Allergy Immunol 15(1):55–61. https://doi.org/10.1046/j.0905-6157.2003.00100.x
William OC, Cookson M, Faux JA, Sharp PA, Hopkin JM (1989) Linkage between immunoglobulin e responses underlying asthma and rhinitis and chromosome 11q. Lancet 333(8650):1292–1295. https://doi.org/10.1016/S0140-6736(89)92687-1
Wjst M et al (1993) Road traffic and adverse effects on respiratory health in children. BMJ 307(6904):596–600. https://doi.org/10.1136/bmj.307.6904.596
Wong GWK et al (2004) Factors associated with difference in prevalence of asthma in children from three cities in China: multicentre epidemiological survey. BMJ 329(7464):486. https://doi.org/10.1136/bmj.329.7464.486
Yanai M et al (1990) Ozone increases susceptibility to antigen inhalation in allergic dogs. J Appl Physiol 68(6):2267–2273. https://doi.org/10.1152/jappl.1990.68.6.2267
Yemaneberhan H, Bekele Z, Venn A, Lewis S, Parry E, Britton J (1997) Prevalence of wheeze and asthma and relation to atopy in urban and rural Ethiopia. Lancet 350(9071):85–90. https://doi.org/10.1016/S0140-6736(97)01151-3
Zheng T (2014) The atopic march: progression from atopic dermatitis to allergic rhinitis and asthma. J Clin Cell Immunol 05:2. https://doi.org/10.4172/2155-9899.1000202
Zink A et al (2019) Out-of-pocket costs for individuals with atopic eczema: a cross-sectional study in nine European countries. Acta Derm Venereol 99(3):263–267. https://doi.org/10.2340/00015555-3102
Zu K, Shi L, Prueitt RL, Liu X, Goodman JE (2018) Critical review of long-term ozone exposure and asthma development. Inhal Toxicol 30(3):99–113. https://doi.org/10.1080/08958378.2018.1455772

Nanotechnology in the Diagnosis of Allergic Airway Diseases

5

Nisha Gulati, Lokesh Nagar, Kamal Dua, and Harish Dureja

Abstract

The prevalence and burden of allergic airway diseases like asthma, chronic obstructive pulmonary disease (COPD), chronic tonsillitis, allergic rhinitis, and chronic rhinosinusitis are continuously rising worldwide and have the biggest public health burden. Further, these diseases are also directly or indirectly linked to noncommunicable diseases like cardiovascular diseases, obesity, and autoimmune diseases that lead to increased morbidity and expenditures. Nanotechnology, a rapidly growing sector with continually developing revolutionary technologies, offers new hope to clinicians for diagnosis of allergic airway diseases. It provides advancements in imaging, early detection, and diagnosis at the cellular level by providing new as well as improving existing clinically relevant technologies. It also helps in reducing the problems of toxicity, time-consuming, less accurate, and precise and off-target effects associated with the conventional diagnostic method. For example, with the help of nanotechnology and cell tracking agents, specific detection of allergic airway diseases with CT scan and MRI is now possible. Nanotechnology has improved breathalyzers' ability to detect nitric oxide (NO), a biomarker used to diagnose asthma. Spirometry in conjunction with functionalized nanoparticles may produce a synergy of state-of-the-art clinical imaging techniques and a quantifiable test for the early to late identification of allergic

The original version of the chapter has been revised. A correction to this chapter can be found at https://doi.org/10.1007/978-981-97-1953-2_16

N. Gulati · L. Nagar · H. Dureja (✉)
Department of Pharmaceutical Sciences, Maharshi Dayanand University, Rohtak, India
e-mail: harishdureja@mdurohtak.ac.in

K. Dua
Discipline of Pharmacy, University of Technology, Sydney, NSW, Australia

P. Prasher et al. (eds.), *Understanding Allergic Airway Diseases*,
https://doi.org/10.1007/978-981-97-1953-2_5

disease. Despite all the ongoing studies and progress in developing numerous new targeted agents, more research is still needed to determine the early diagnosis of allergic airway diseases and lead to more futuristic studies. In the present chapter, we will highlight the burden of allergic airway diseases on public health and disadvantages of conventional diagnostic method with great attention on the role of the nanotechnology, i.e., nanoparticles, quantum dots, gold nanoparticles, carbon tubes, nanocrystals, and hydrogel in the diagnosis of allergic airway diseases. We also discuss how the nano technology revolutionizes the field of respiratory medicine by helping in modifying pointing, improve biosensors resolution and sensitivity of existing imaging modalities.

Keywords

Allergic airway diseases · Diagnosis · Point-of-care testing · Nanobiosensor

5.1 Introduction

Allergic airway diseases are anticipated to rank among the world's leading causes of death by 2030, accounting for one in every six fatalities worldwide (World Health Organization 2023c). Allergic airway diseases are a group of respiratory disorders that are characterized by inflammation and narrowing of the airways, typically triggered by allergic reactions to specific environmental substances or allergens (Chung et al. 2019). These conditions often lead to symptoms such as coughing, wheezing, shortness of breath, and chest tightness (World Health Organization 2023c). The most prevalent allergic airway diseases are asthma, chronic obstructive pulmonary disease (COPD), chronic tonsillitis, allergic rhinitis, and chronic rhinosinusitis (Feshchenko et al. 2017). In most industrialized nations, the prevalence of allergic airway disorders has drastically increased in the past few decades (Gutowska-Ślesik et al. 2023). Both allergy sensitization and allergic airway disorders are on the rise though the rise in children's prevalence of allergic airway diseases has moderated (Devereux et al. 2014). Rural areas of low-income countries seem to have a lower prevalence of allergic airway disease despite the fact that allergic sensitization can be frequent there (Douwes and Pearce 2014). Despite being connected to urbanization, there has not been much success in explaining why allergic airway disorders are becoming more common (Robinson et al. 2011). The drastic fluctuations in prevalence and distribution found strongly imply that the environment plays a significant influence in disease development (Devereux et al. 2014). Environmental factors that affect allergies include food; exposure to a healthy, diversified microbiome; infections; exposure to air pollution; and occupational exposures. Asthma is linked to both death rates and lung function decline although allergies are not (Murrison et al. 2019).

5.1.1 Examples of Some Allergic Airway Diseases

Asthma is a diverse disease entity characterized by the presence of persistent lung inflammation that causes symptoms like coughing, wheezing, chest tightness, and blockage of airway and hyperactivity. According to a WHO research, asthma killed 4,55,000 people in 2019 and impacted an estimated 262 million people. Asthma is still a poorly understood disease despite significant progress in our understanding of its etiology (World Health Organization 2023a; Bobolea 2016).

The airway eosinophilia syndrome, which is one of the other allergic airway disease syndromes, is characterized by increase in the presence of eosinophils in the airways. Eosinophils are a type of white blood cell that plays a role in the immune response, particularly in allergic reactions and inflammation (George and Brightling 2015). The most common pulmonary eosinophilia syndromes include helminth infections; allergic drug reactions; negative reactions to inhaled toxins; lymphoid and myeloid hypereosinophilic syndromes (HES), which are typically premalignant or downright malignant disorders; and eosinophilic granulomatosis with polyangiitis (previously known as Churg-Strauss syndrome) (Salahuddin et al. 2020).

A heterogeneous illness known as **chronic rhinosinusitis (CRS)** is characterized by inflammation of the nasal passages and the sinuses that lasts for an extended period of time, typically at least 12 weeks (Stevens et al. 2016). It can cause a range of symptoms including nasal congestion, facial pain or pressure, headache, loss of smell, postnasal drip, and fatigue. Chronic rhinosinusitis can significantly impact a person's quality of life and daily functioning (World Health Organization 2023b). Clinically, CRS impacts more than 35 million Americans and has a social cost of more than $12.8 billion yearly, which includes direct healthcare expenses and indirect costs associated to lost productivity (Agboola et al. 2018). Patients with CRS consequently commonly seek medical care (Agboola et al. 2018).

In 1952, three individuals with asthma-like allergic respiratory tract disease with concurrent fungal airway infections were given the diagnosis of **allergic bronchopulmonary aspergillosis (ABPA)**. Since the middle of the 1800s, there have been infrequent reports of allergic bronchopulmonary mycosis, which is brought on by lung fungal infections such aspergillus and others (Asano et al. 2018; Snen et al. 2021).

The bacterial infectious disease known as **cystic fibrosis**, an autosomal recessive inflammatory disease of the lungs, is brought on by airway infection with mucoid strains of *Pseudomonas aeruginosa* and similar organisms. Also closely related to airway mycosis is cystic fibrosis (Malhotra et al. 2019). The cystic fibrosis transmembrane conductance regulator (CFTR) gene, which is expressed on airway ionocyte cells and regulates chloride ion transport as its principal function, has decreased or non-existent activity, resulting in abnormally thick and sticky mucus released into the cystic fibrosis airway (Hanssens et al. 2021). A major element that encourages the growing of bacteria behind the blockages is probably the widespread obstruction of airways caused by thick mucus (Fahy and Dickey 2010). However, when the best sputum culture techniques are used, sputum cultures from cystic fibrosis patients typically yield fungus, of which thousands of species have been discovered in this condition, and the incidence of ABPA and fungal bronchitis that complicate cystic fibrosis can also be quite high (Mortensen et al. 2011). Cystic fibrosis is similar to all other allergic airway illnesses in this sense, including their high incidence of airway mycosis. Because immunodeficiency underlies the majority of these conditions, cystic fibrosis should also be seen as an immunodeficiency in which the defective airway encourages the expansion of many bacteria (Redondo et al. 2015).

Patients with symptoms for at least 1–3 months are considered to have **chronic pulmonary aspergillosis (CPA)**, a set of primary fungal-associated airway illnesses that are characterized by lung inflammation and damage that is thought to be caused by progressing cavitary aspergillus infections (Denning et al. 2016). In comparison to noninvasive airway mycosis, CPA represents a more serious disease process, involving a semi-invasive necrotic infection of the airways and lung parenchyma but without the overt fungal angioinvasive illness found in very lethal cases of invasive aspergillosis (Prasad 2016).

5.2 Diagnosis of Allergic Airway Diseases

The diagnosis of allergic airway diseases involves a combination of clinical assessment, medical history, physical examination, and various tests. As with any infectious condition, it is crucial to have precise, highly sensitive diagnostic tools that are easily accessible to clinicians (Ansotegui et al. 2020). However, despite recent improvements, such techniques have never been available for identifying allergic airway illness (Douwes and Pearce 2014). The general acceptance of allergic airway illness as a legitimate diagnostic category and an entity requiring therapy has been significantly hampered by the lack of such methods, or worse, by the availability of only insensitive diagnostic methods. Technology gives medical professionals fresh hope for the detection of allergic airway disorders (Gerber et al. 2009). It advances imaging, early detection, and diagnosis at the cellular level by introducing new technologies and enhancing already-existing ones that are clinically pertinent. Additionally, it aids in minimizing the drawbacks of current diagnostic methods, such as their toxicity, time-consuming nature, lack of precision, and side effects that are unintended (Pulumati et al. 2023). For instance, targeted identification of allergic airway illnesses with CT scan and MRI is now achievable because of nanotechnology and cell tracking agents (Peserico et al. 2022). The various advantages and limitations of conventional diagnosis methods for allergic airway diseases is shown in Table 5.1.

5.2.1 Emerging Techniques in Diagnosis of Allergic Airway Diseases

Over the years, the traditional screening methods have undergone various changes in an effort to create a more accurate and effective method for identifying allergic airway disease. Nanotechnology-based approaches are one of the emerging methodologies. These strategies are not only widely employed, but they also have a great deal of potential to replace the current ones (Homayoonnia et al. 2021).

Nanotechnology is at the leading edge of the rapidly developing field of new therapeutic and diagnostic concepts in all areas of medicine (Laroui et al. 2013). Nanotechnology has the potential to revolutionize the diagnosis of allergic airway diseases by offering several advantages that can significantly improve the accuracy

Table 5.1 Advantages and limitations of conventional diagnosis methods for allergic airway diseases

Sr. no.	Techniques	Advantages	Limitations	Reference
1.	CT scan	A painless and well-known method	Radiation, an allergy to the contrast medium, and misunderstanding	Baerlocher et al. (2010)
2.	Sputum cytology	Highly sensitive and minimally invasive for centrally situated lung cancer	Produces false positives and makes patients uncomfortable when they induce sputum	Arimanagi et al. (2012)
3.	Pleural fluid cytology	Avoiding misdiagnosis is preferable to using radiology.	Quite intrusive	Pairman et al. (2022)
4.	Biopsy	Data that is exact and in-depth is acquired.	Highly invasive, uncomfortable, and time-consuming procedure	Musani and Dutau (2017)
5.	Bronchoscopy	Detection of a specific tumor tissue quickly	Invasive and inexperienced practices may result in airway and laryngeal scrapes.	Paradis et al. (2016)
6.	Biosensors and biomarkers	Highly perceptive and quick to react	Preparation of the sample and superior analytical skill	de Oliveira et al. (2020)
7.	Radiography	Nonintrusive and accurate	Time-consuming	Mamchur et al. (2022)

and efficiency of detection and monitoring (Malik et al. 2023). The various advantages of nanotechnology in diagnosis of allergic airway diseases include:

1. **Early detection with high specificity:** Nanoscale materials and devices can detect even trace amounts of allergens or disease markers in the airway, providing greater sensitivity than traditional diagnostic methods (Testera-Montes et al. 2021). The high sensitivity of nanoscale diagnostic devices can lead to the early detection of allergic airway diseases, allowing for proactive interventions and potentially preventing disease progression or exacerbations. Nanotechnology-based sensors can be engineered to have high specificity for particular allergens or disease biomarkers, reducing the risk of false-positive or false-negative results (Swierczewska et al. 2012).
2. **Miniaturization:** Nanoscale diagnostic devices can be miniaturized, allowing for portable and point-of-care testing. Patients can perform tests at home or in a clinic, reducing the need for laboratory visits and providing real-time results (Nayak et al. 2017).
3. **Multiplexing:** Nanosensors can simultaneously detect multiple allergens or biomarkers, enabling a comprehensive assessment of the patient's condition in a single test. This can be particularly useful for identifying coexisting allergies or monitoring disease progression (Jarockyte et al. 2020).
4. **Noninvasive sampling and rapid results:** Some nanotechnology-based diagnostic methods allow for noninvasive or minimally invasive sample collection, such as breath analysis. This makes testing more comfortable for patients, espe-

cially children and those with respiratory issues. Nanoscale diagnostic devices can provide rapid results, often within minutes or even seconds, compared to traditional laboratory tests that may take hours or days. This speed can facilitate timely clinical decision-making (Moabelo et al. 2021).

5. **Lower costs:** Nanotechnology can enable the production of cost-effective diagnostic devices, reducing healthcare expenses for patients and healthcare systems (Haleem et al. 2023).
6. **Remote monitoring:** Nanosensors can be integrated with wireless technology, enabling remote monitoring of patients' allergic airway conditions. Healthcare providers can track patients' health status in real time and provide timely interventions when necessary (Vegesna et al. 2017).
7. **Personalized medicine:** Nanotechnology can enable the development of personalized diagnostic approaches by tailoring tests and treatments to an individual's specific allergens and disease characteristics. Nanotechnology-based diagnostic methods often require smaller sample volumes and less invasive procedures, making the testing experience more comfortable for patients (Alghamdi et al. 2022).
8. **Research and drug development:** Nanotechnology can also aid in the development of new therapeutic agents and drug delivery systems for allergic airway diseases, contributing to more effective treatments (Luo et al. 2021).

The various advantages of nanotechnology in diagnosis of allergic airway diseases are shown in Fig. 5.1.

5.3 Approaches of Nanotechnology in the Diagnosis of Allergic Airway Diseases

Nanotechnology has the potential to revolutionize the diagnosis of allergic airway diseases by offering improved sensitivity, specificity, and noninvasive methods for detection (Anjum et al. 2021). Nanotechnology offers a wide range of approaches for the diagnosis of allergic airway diseases, from improved imaging and detection methods to targeted drug delivery and personalized medicine strategies. These advancements have the potential to enhance the accuracy of diagnosis for individuals suffering from allergic airway diseases (Sim and Wong 2021). The various approaches of nanotechnology in the diagnosis of allergic airway diseases are shown in Fig. 5.2.

5.3.1 Nanoparticle-Based Imaging

Nanoparticles can be engineered to carry contrast agents, fluorescent dyes, or other imaging markers. These nanoparticles can be used in various imaging techniques, such as magnetic resonance imaging (MRI), computed tomography (CT), and positron emission tomography (PET), to visualize the inflammation and structural changes in the airways associated with allergic diseases (Han

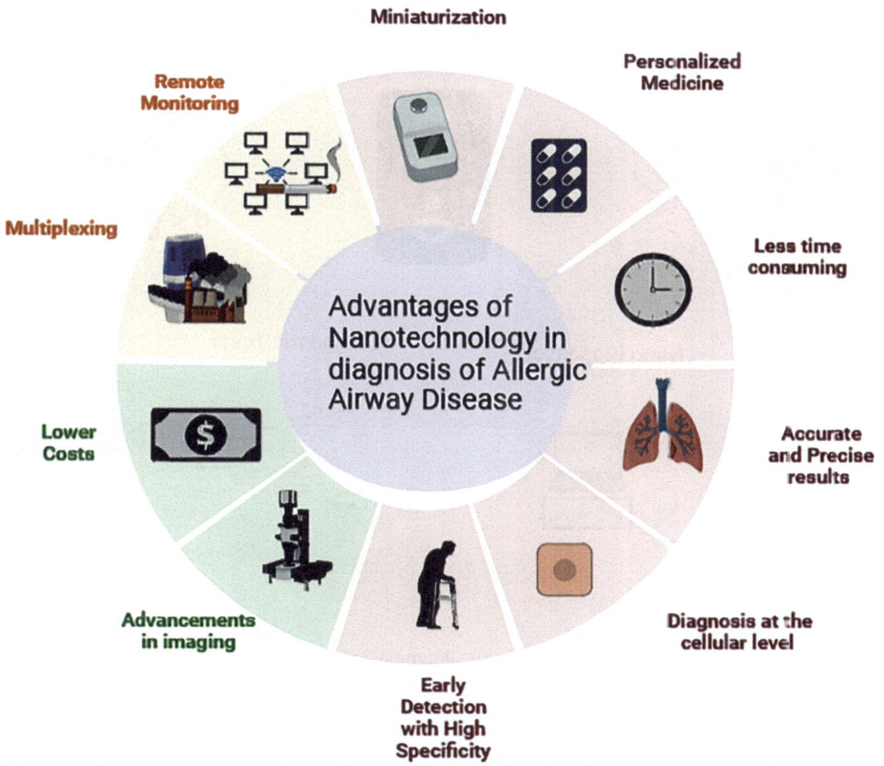

Fig. 5.1 Advantages of nanotechnology in the diagnosis of allergic airway diseases

et al. 2019). This enhanced imaging can provide more detailed information for accurate diagnosis and help researchers and clinicians identify airway remodeling and changes in airway dimensions that occur in allergic airway diseases. Nanoparticles can be designed to target and accumulate in specific cells or tissues associated with allergic inflammation. This could enable noninvasive imaging techniques to visualize the presence and extent of inflammation in the airways (Lv et al. 2021).

Magnetic nanoparticles have gained significant attention in various fields, including medicine, due to their unique properties and potential applications. Magnetic nanoparticles can be utilized for different purposes such as imaging, diagnostics, and targeted therapy and enabling earlier and more accurate interventions that can improve patient outcomes (Farinha et al. 2021).

Al Faraj et al. showed that combining iron oxide nanoparticles with an antibody that is specific to a certain subpopulation of macrophages could provide a promising method for the improved and earlier identification of several inflammatory illnesses using noninvasive magnetic resonance imaging. The particles concentrate on a particular subpopulation of macrophages in the lung of a COPD animal model,

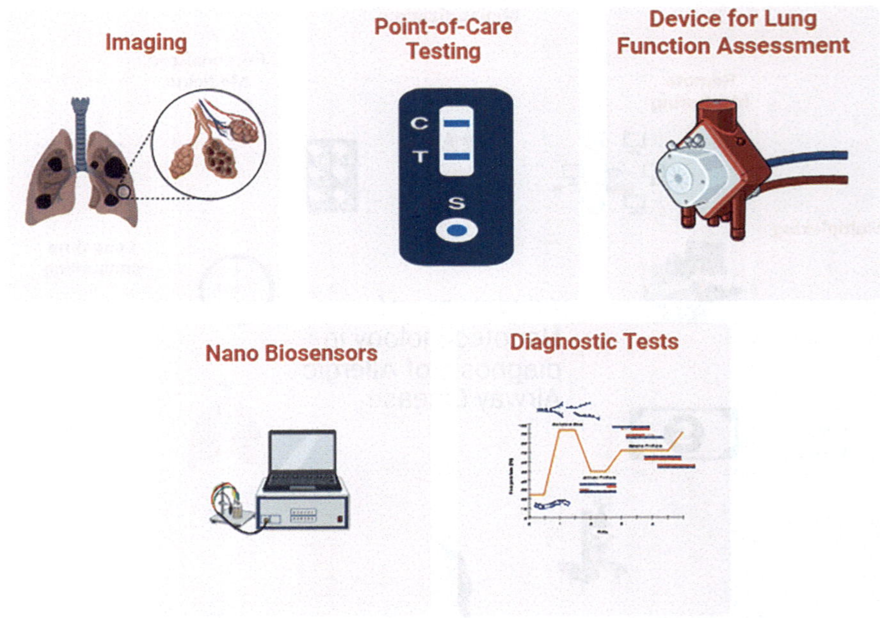

Fig. 5.2 Approaches of nanotechnology in the diagnosis of allergic airway diseases

allowing noninvasive surveillance of those cells using a free-breathing magnetic resonance imaging procedure (Al Faraj et al. 2014).

The use of super-paramagnetic iron oxide nanoparticles in imaging raises the possibility of unintended damage and inflammation. Super-paramagnetic iron oxide nanoparticles coated with the neutral natural dietary amino acid glycine have been shown by Chakraborty et al. to support noninvasive lung imaging, determine particle biodistribution, and comprehend the effects of these nanoparticles' interactions with lung immune cells. Three-dimensional (3D) ultrashort echo-time magnetic resonance imaging has been shown to be a useful tool for determining the biodistribution of nanoparticles in the lung (Chakraborty et al. 2020).

5.3.2 Nanotechnology-Enabled Point-of-Care Testing

Nanotechnology-based diagnostic tools can be designed for use at the point of care, allowing for rapid and convenient testing without the need for extensive laboratory facilities. This is particularly important for diagnosing and monitoring allergic airway diseases in remote or resource-limited settings (Thwala et al. 2023).

Quantum dots (QD) are tiny semiconducting crystals that are composed of elements such as tellurium (Te), selenium (Se), and cadmium sulfide (CdS). Because of the challenges posed by toxicity in QD, efforts are underway to construct quantum dots using cadmium (Cd), silicon, and carbon as alternatives to overcome this limitation (Gidwani et al. 2021). Along with having good electrochemical and optical capabilities, Cd-based QD has also demonstrated excellent photoluminescent properties. Different wavelengths of light are absorbed and emitted depending on the size of QD (Sakthivel et al. 2023). The greatest size of the QD will have a red color, while the smallest size will have a green color. Each size of the QD has a separate color assigned to it. The QD is utilized in a variety of applications, including biosensing, diagnostics, imaging, and as biomarkers, and its size depends on the ingredients used in its creation (2–50 nm) (Singh et al. 2020).

The quantum dots-based lateral flow immunoassay with image analysis was prepared by Zhao et al. IgE antibodies in mites can be identified by examining the produced dots. IgE might be measured semiquantitatively in a linear range of 0.2–10 U/mL using the suggested detection technique. For in-home testing and rural locations, the suggested detection technology is easy to understand, trustworthy, and simple to use (Zhao et al. 2017).

Similar to another study conducted by Liang et al., in order to identify serum-specific IgE against the two common mite allergens in China, Dermatophagoides pteronyssinus (Der-p) and Dermatophagoides farina (Der-f), in the serum of patients with allergic rhinitis, a portable fluorescence immunoassay chip detector (PFICD) and quantum dot-based lateral flow immunoassay (QD-LFIA) were used. When compared to previous approaches, the QD-LFIA strategy has a number of advantages, including high detection sensitivity and specificity; ease of use in hospitals, pharmacies, and at home; and lower costs and quicker processing times (Liang et al. 2020).

5.3.3 Nanomaterial-Based Device for Lung Function Assessment

Nanoscale materials can be integrated into devices for assessing lung function. For instance, nanocomposites could be incorporated into devices that measure airway resistance, lung capacity, and airflow dynamics. These devices could provide real-time information about the functional status of the airways in individuals with allergic respiratory diseases (Brashier and Salvi 2015).

Electronic nose (e-nose) contains nanosensor, a cutting-edge device that allows for the automated detection and identification of certain volatile organic compounds by scent. An electronic nose (e-nose) is a device designed to mimic the human olfactory system's ability to detect and identify odors and volatile compounds (Borowik et al. 2020). It consists of an array of chemical nanosensors that respond to different odor molecules and generate a pattern of signals, which can then be analyzed to identify the specific odors or compounds present in a sample. The sensors detect and respond to the VOCs in the breath sample, generating a unique pattern of sensor responses (Ramgir 2013). This pattern is then compared to a database of known patterns associated with different lung diseases. Based on the analysis, the electronic nose can provide information about the presence, type, and potentially even the severity of lung disease. It could aid in early diagnosis, monitoring disease progression, and assessing treatment effectiveness (van der Sar et al. 2021).

Numerous VOCs, including pentane, ethane, 8-isoprostane, NO, ammonia, and H_2O_2, can be found as a VOC marker in asthma. As a biomarker for the identification of chronic obstructive pulmonary disease, NO, ethane, and isoprene can be found (Das et al. 2016).

Nanocrystals are valuable tools for improving diagnostic techniques due to its size-dependent optical and electronic properties. Nanocrystals have shown great potential in various fields, including breath analysis for medical diagnostics (Kaloumenou et al. 2022). Breath analysis, also known as breathomics or volatile organic compounds (VOCs) analysis, involves detecting and analyzing the unique combination of gases and compounds present in a person's breath to gain insights into their health status. Using nanocrystals for breath analysis in asthma patients is a promising application that can help in the diagnosis, monitoring, and management of asthma (Keogh and Riches 2022).

Hanh and the co-worker demonstrated that the VOC gas sensor is based on hollow cubic assembled nanocrystal Zn_2SnO_4 for breath analysis in asthmatic patients. The hydrothermal one-step approach was used to create the hollow cubic assembled nanocrystal Zn_2SnO_4. The hollow cubic sensor displayed great acetone sensitivity and selectivity, outstanding stability, and a 175 ppb detection limit (Hanh et al. 2021).

5.3.4 Nanorobots, Nanobiosensors, and Nanoscale Detectors

Nanotechnology enables the development of highly sensitive biosensors and detectors capable of identifying specific biomarkers associated with allergic airway diseases. These nanoscale devices can detect allergen-specific antibodies, inflammatory cytokines, or other molecules in bodily fluids like blood or sputum. Such biosensors could provide rapid and precise diagnostic information, allowing for early intervention and personalized treatment (Swierczewska et al. 2012).

Nanotechnology can be combined with the inhalable devices that patients can use to monitor their airway health in real time. These devices may provide instant feedback on allergen exposure or inflammation levels (Kumar et al. 2022).

Nanorobots are extremely small, and only fictitious models have been created so far. They benefit from targeted intervention, and they may be set up as you choose using computer algorithms (Hu et al. 2020). According to a study, nanorobots embedded with fluorescent or other biosensing material can identify allergic airway illness in its early stages. Acoustic signals and sound waves can be utilized by nanorobots in therapeutic applications to treat specific tumors (Aggarwal and Kumar 2022). Utilizing several bots directed toward a cancer tissue is the idea and mechanism. In its most basic form, a nanorobot has tentacles for moving through the body's blood and fluid compartments and a microprocessor chip as its primary functional component (Kong et al. 2023). Nanorobots are useful for specialized and focused action since they can automatically recognize malignant cells and release the agents to eliminate them. At the moment, the nanorobot theory is simply conceptualized and not fully implemented in the work (Saadeh and Vyas 2014).

Gold and silver nanoparticles are frequently employed for diagnostic uses due to their optical properties. They can be functionalized with certain ligands, antibodies, or aptamers to bind to the target biomolecules. This binding interaction may alter the nanoparticles' localized surface plasmon resonance properties, and this alteration can be recognized and quantified. This strategy is used in assays to detect infections and biomarkers (Rabiee et al. 2022).

Liu et al. demonstrated that the nanostructured biosensor can identify a patient's level of allergy by measuring the presence of IgE in patient samples. Nanosensors can provide real-time information about allergen exposure and inflammatory responses. This enables continuous monitoring of a patient's condition, allowing for immediate adjustments to treatment plans if necessary. The advantages of nanobiosensors include sample requirements, a short sample preparation time, and speedy processing. Silver nanoparticles (SNPs) were placed on the GNP layer to improve the biosensor's charge transfer efficiency. An affinity-purified goat antihuman IgE antibody was further immobilized to the IgE to ensure the specificity of detection. Electrochemical impedance spectroscopy (EIS) analysis was used to analyze the concentration of the target. For confirmation of the results, blood serum samples with known allergy levels were identified by the commercially available ImmunoCAP (Liu et al. 2012).

Monitoring oxidative stress and pulmonary illnesses may be done by measuring the nitric oxide gas in exhaled human breath. According to Gouma et al., a tungsten oxide-based nanosensing probe can specifically find low levels of nitric oxide gas. The nitric oxide nanoprobe is anticipated to provide noninvasive, single-breath sampling that is portable and inexpensive, as well as NO diagnostics (Gouma and Kalyanasundaram 2008).

5.3.5 Nanotechnology-Enhanced Diagnostic Tests

Nanoparticles can be incorporated into diagnostic assays, such as enzyme-linked immunosorbent assays (ELISAs) or polymerase chain reactions (PCRs), to increase their sensitivity. Polymerase chain reaction (PCR) can be improved using nanoparticles to enhance the amplification of specific genes associated with allergic airway diseases (Zhu et al. 2022). This allows for highly sensitive and specific molecular diagnosis. Functionalized nanoparticles can bind to target allergen biomarkers and amplify the signal, leading to more accurate diagnosis. Furthermore, nanoparticles can be used in air quality monitoring devices to capture and detect allergenic particles in the environment. This information can help individuals with allergies avoid allergen exposure (Gómez-Arribas et al. 2018).

5.4 Challenges of Nanotechnology in Diagnosis of Allergic Airway Diseases

There are several challenges that need to be addressed to fully realize the potential of nanotechnology in diagnosis of allergic airway diseases:

1. **Biocompatibility:** Nanoparticles used in diagnostic applications must be biocompatible to ensure they do not elicit an immune response or other adverse reactions when introduced into the body. Achieving precise targeting of nanoparticles to specific cells or tissues in the respiratory system can be challenging (Kyriakides et al. 2021). Ensuring that nanoparticles reach the intended target, such as inflamed airway cells, is crucial for accurate diagnosis. Nanoscale diagnostic tools need to be highly sensitive and specific to detect biomarkers associated with allergic airway diseases. False positives or negatives can lead to misdiagnosis or delayed treatment (Hu et al. 2011).
2. **Sample Collection:** Collecting samples from the respiratory system can be challenging, especially in pediatric patients or those with severe symptoms. Developing noninvasive or minimally invasive techniques for sample collection is essential (Carroll 2002).
3. **Safety Concerns:** There is a need to thoroughly assess the long-term safety of nanomaterials used in diagnostics. Understanding how these materials interact with the body over time is crucial. Furthermore, patients may be hesitant or skeptical about using new nanotechnology-based diagnostic tools. Educating patients and healthcare providers about the benefits and safety of these technologies is important (Ray et al. 2009).
4. **Regulatory Approval and Ethical and Privacy Concerns:** New nanotechnology-based diagnostic tools may face regulatory hurdles. Demonstrating safety and efficacy to regulatory agencies can be time-consuming and costly. As with any advanced medical technology, there are ethical and privacy concerns related to the collection and use of patient data in nanotechnology-based diagnostics. Proper safeguards and regulations are needed to address these issues (Farjadian et al. 2019).
5. **Integration with Healthcare Systems:** Integrating nanotechnology-based diagnostics into existing healthcare systems and workflows can be challenging. Ensuring that healthcare providers can effectively use these tools is essential for their widespread adoption (Junaid et al. 2022).

5.5 Conclusion

Nanotechnology holds promising potential for advancing the diagnosis of allergic airway diseases by providing more sensitive, specific, and efficient tools for detecting allergens, measuring inflammation, and monitoring disease progression. Nanotechnology offers several significant advantages for improving the diagnosis of allergic airway diseases like early detection, improved imaging, and reduced sample volume. Early detection, personalized treatment, and efficient monitoring can lead to better disease management, fewer hospitalizations, and overall reduced healthcare costs. Research is ongoing to develop more accurate and personalized diagnostic methods that can address some of these challenges and provide better insights into the underlying mechanisms of these conditions. Future research should focus on the rigorous research, testing, and validation to ensure the safety, efficacy, and reliability of nanotechnology-based tools for proper diagnostic of allergic airway diseases.

References

Agboola AA, Esan OT, Afolabi OT et al (2018) Economic burden of the therapeutic management of mental illnesses and its effect on household purchasing power. PLoS One 13:e0202396. https://doi.org/10.1371/journal.pone.0202396

Aggarwal M, Kumar S (2022) The use of nanorobotics in the treatment therapy of cancer and its future aspects: a review. Cureus 14. https://doi.org/10.7759/cureus.29366

Al Faraj A, Sultana Shaik A, Afzal S et al (2014) MR imaging and targeting of a specific alveolar macrophage subpopulation in LPS-induced COPD animal model using antibody-conjugated magnetic nanoparticles. Int J Nanomedicine 9:1491. https://doi.org/10.2147/ijn.s59394

Alghamdi MA, Fallica AN, Virzì N et al (2022) The promise of nanotechnology in personalized medicine. J Pers Med 12:673. https://doi.org/10.3390/jpm12050673

Ammanagi A, Miskin A, Sangolli S et al (2012) Sputum cytology in suspected cases of carcinoma of lung (sputum cytology a poor man's bronchoscopy!). Lung India 29:19. https://doi.org/10.4103/0970-2113.92356

Anjum S, Ishaque S, Fatima H et al (2021) Emerging applications of nanotechnology in healthcare systems: grand challenges and perspectives. Pharmaceuticals 14:707. https://doi.org/10.3390/ph14080707

Ansotegui IJ, Melioli G, Canonica GW et al (2020) IgE allergy diagnostics and other relevant tests in allergy, a World Allergy Organization position paper. The World Allergy Organization Journal 13:100080. https://doi.org/10.1016/j.waojou.2019.100080

Asano K, Kamei K, Hebisawa A (2018) Allergic bronchopulmonary mycosis—pathophysiology, histology, diagnosis, and treatment. Asia Pac Allergy 8:e24. https://doi.org/10.5415/apallergy.2018.8.e24

Baerlocher MO, Asch M, Myers A (2010) Allergic-type reactions to radiographic contrast media. Can Med Assoc J 182:1328–1328. https://doi.org/10.1503/cmaj.090371

Bobolea I (2016) Asthma-Copd Overlap Syndrome (Acos) Current Understanding and Future Perspectives. IntechOpen, [s l.]

Borowik P, Adamowicz L, Tarakowski R et al (2020) Odor detection using an E-nose with a reduced sensor array. Sensors 20:3542. https://doi.org/10.3390/s20123542

Brashier B, Salvi S (2015) Measuring lung function using sound waves: role of the forced oscillation technique and impulse oscillometry system. Breathe 11:57–65. https://doi.org/10.1183/20734735.020514

Carroll KC (2002) Laboratory diagnosis of lower respiratory tract infections: controversy and conundrums. J Clin Microbiol 40:3115–3120. https://doi.org/10.1128/jcm.40.9.3115-3120.2002

Chakraborty A, Royce S, Selomulya C, Plebanski M (2020) A novel approach for non-invasive lung imaging and targeting lung immune cells. Int J Mol Sci 21:1613. https://doi.org/10.3390/ijms21051613

Chung KF, Israel E, Gibson PG (2019) Severe asthma. European Respiratory Society

Das S, Pal S, Mitra M (2016) Significance of exhaled breath test in clinical diagnosis: a special focus on the detection of diabetes mellitus. J Med Biol Eng 36:605–624. https://doi.org/10.1007/s40846-016-0154-6

Denning DW, Cadranel J, Beigelman-Aubry C et al (2016) Chronic pulmonary aspergillosis: rationale and clinical guidelines for diagnosis and management. Eur Respir J 47:45–68. https://doi.org/10.1183/13993003.00583-2015

Devereux G, Matsui EC, Burney P (2014) Epidemiology of asthma and allergic airway diseases. Elsevier eBooks, pp 754–789. https://doi.org/10.1016/b978-0-323-08593-9.00049-8
Douwes J, Pearce N (2014) Epidemiology of respiratory allergies and asthma. Handbook of Epidemiology, pp 2263–2319. https://doi.org/10.1007/978-0-387-09834-0_50
Fahy JV, Dickey BF (2010) Airway mucus function and dysfunction. N Engl J Med 363:2233–2247. https://doi.org/10.1056/nejmra0910061
Farinha P, Coelho JMP, Reis CP, Gaspar MM (2021) A comprehensive updated review on magnetic nanoparticles in diagnostics. Nano 11:3432. https://doi.org/10.3390/nano11123432
Farjadian F, Ghasemi A, Gohari O et al (2019) Nanopharmaceuticals and nanomedicines currently on the market: challenges and opportunities. Nanomedicine 14:93–126. https://doi.org/10.2217/nnm-2018-0120
Feshchenko YI, Iashyna L, Nugmanova D et al (2017) Chronic obstructive pulmonary disease, bronchial asthma and allergic rhinitis in the adult population within the commonwealth of independent states: rationale and design of the CORE study. BMC Pulm Med 17. https://doi.org/10.1186/s12890-017-0471-x
George L, Brightling CE (2015) Eosinophilic airway inflammation: role in asthma and chronic obstructive pulmonary disease. Therapeutic Advances in Chronic Disease 7:34–51. https://doi.org/10.1177/2040622315609251
Gerber MA, Baltimore RS, Eaton CB et al (2009) Prevention of rheumatic fever and diagnosis and treatment of acute streptococcal pharyngitis. Circulation 119:1541–1551. https://doi.org/10.1161/circulationaha.109.191959
Gidwani B, Sahu V, Shukla SS et al (2021) Quantum dots: Prospectives, toxicity, advances and applications. J Drug Deliv Sci Technol 61:102308. https://doi.org/10.1016/j.jddst.2020.102308
Gómez-Arribas L, Benito-Peña E, Hurtado-Sánchez M, Moreno-Bondi M (2018) Biosensing based on nanoparticles for food allergens detection. Sensors 18:1087. https://doi.org/10.3390/s18041087
Gouma P-I, Kalyanasundaram K (2008) A selective nanosensing probe for nitric oxide. Appl Phys Lett 93. https://doi.org/10.1063/1.3050524
Gutowska-Ślesik J, Samoliński B, Krzych-Fałta E (2023) The increase in allergic conditions based on a review of literature. Adv Dermatol Allergol/Postępy Dermatologii i Alergologii 40:1. https://doi.org/10.5114/ada.2022.119009
Haleem A, Javaid M, Singh RP et al (2023) Applications of nanotechnology in medical field. Glob Health J 7. https://doi.org/10.1016/j.glohj.2023.02.008
Han X, Xu K, Taratula O, Farsad K (2019) Applications of nanoparticles in biomedical imaging. Nanoscale 11:799–819. https://doi.org/10.1039/c8nr07769j
Hanh NT, Ngoc TM, Van Duy L et al (2021) A comparative study on the VOCs gas sensing properties of Zn2SnO4 nanoparticles, hollow cubes, and hollow octahedra towards exhaled breath analysis. Sensors Actuators B Chem 343:130147–130147. https://doi.org/10.1016/j.snb.2021.130147
Hanssens LS, Duchateau J, Casimir GJ (2021) CFTR protein: not just a Chloride Channel? Cells 10:2844. https://doi.org/10.3390/cells10112844
Homayoonnia S, Lee Y, Andalib D et al (2021) Micro/nanotechnology-inspired rapid diagnosis of respiratory infectious diseases. Biomed Eng Lett 11:335–365. https://doi.org/10.1007/s13534-021-00206-8
Hu M, Ge X, Chen X et al (2020) Micro/Nanorobot: a promising targeted drug delivery system. Pharmaceutics 12:665. https://doi.org/10.3390/pharmaceutics12070665
Hu Y, Fine DH, Tasciotti E et al (2011) Nanodevices in diagnostics. Wiley Interdiscip Rev Nanomed Nanobiotechnol 3:11–32. https://doi.org/10.1002/wnan.82
Jarockyte G, Karabanovas V, Rotomskis R, Mobasheri A (2020) Multiplexed Nanobiosensors: current trends in early diagnostics. Sensors 20:6890. https://doi.org/10.3390/s20236890
Junaid SB, Imam AA, Balogun AO et al (2022) Recent advancements in emerging Technologies for Healthcare Management Systems: a survey. Healthcare 10:1940. https://doi.org/10.3390/healthcare10101940

Kaloumenou M, Skotadis E, Lagopati N et al (2022) Breath analysis: a promising tool for disease diagnosis—the role of sensors. Sensors 22:1238. https://doi.org/10.3390/s22031238

Keogh RJ, Riches JC (2022) The use of breath analysis in the Management of Lung Cancer: is it ready for primetime? Curr Oncol 29:7355–7378. https://doi.org/10.3390/curroncol29100578

Kong X, Gao P, Wang J et al (2023) Advances of medical Nanorobots for future cancer treatments. J Hematol Oncol 16. https://doi.org/10.1186/s13045-023-01463-z

Kumar R, Mehta P, Shankar KR et al (2022) Nanotechnology-assisted metered-dose inhalers (MDIs) for high-performance pulmonary drug delivery applications. Pharm Res 39. https://doi.org/10.1007/s11095-022-03286-y

Kyriakides TR, Raj A, Tseng TH et al (2021) Biocompatibility of nanomaterials and their immunological properties. Biomed Mater (Bristol, England) 16. https://doi.org/10.1088/1748-605X/abe5fa

Laroui H, Rakhya P, Xiao B et al (2013) Nanotechnology in diagnostics and therapeutics for gastrointestinal disorders. Dig Liver Dis 45:995–1002. https://doi.org/10.1016/j.dld.2013.03.019

Liang Z-Y, Deng Y-Q, Tao Z-Z (2020) A quantum dot-based lateral flow immunoassay for the rapid, quantitative, and sensitive detection of specific IgE for mite allergens in sera from patients with allergic rhinitis. Anal Bioanal Chem 412:1785–1794. https://doi.org/10.1007/s00216-020-02422-0

Liu Y-F, Tsai J-J, Chin Y-T et al (2012) Detection of allergies using a silver nanoparticle modified nanostructured biosensor. Sensors Actuators B Chem 171-172:1095–1100. https://doi.org/10.1016/j.snb.2012.06.039

Luo M-X, Hua S, Shang Q-Y (2021) Application of nanotechnology in drug delivery systems for respiratory diseases (review). Mol Med Rep 23. https://doi.org/10.3892/mmr.2021.11964

Lv C, Li H, Cui H et al (2021) Solid lipid nanoparticle delivery of rhynchophylline enhanced the efficiency of allergic asthma treatment via the upregulation of suppressor of cytokine signaling 1 by repressing the p38 signaling pathway. Bioengineered 12:8635–8649. https://doi.org/10.1080/21655979.2021.1988364

Malhotra S, Hayes D, Wozniak DJ (2019) Cystic fibrosis and Pseudomonas aeruginosa: the host-microbe Interface. Clin Microbiol Rev 32. https://doi.org/10.1128/CMR.00138-18

Malik S, Muhammad K, Waheed Y (2023) Nanotechnology: a revolution in modern industry. Molecules 28:661. https://doi.org/10.3390/molecules28020661

Mamchur D, Peksa J, Le Clainche S, Vinuesa R (2022) Application and advances in radiographic and novel technologies used for non-intrusive object inspection. Sensors (Basel, Switzerland) 22:2121. https://doi.org/10.3390/s22062121

Moabelo KL, Martin DR, Fadaka AO et al (2021) Nanotechnology-based strategies for effective and rapid detection of SARS-CoV-2. Materials 14:7851. https://doi.org/10.3390/ma14247851

Mortensen KL, Jensen RH, Johansen HK et al (2011) Aspergillus species and other molds in respiratory samples from patients with cystic fibrosis: a laboratory-based study with focus on aspergillus fumigatus azole resistance. J Clin Microbiol 49:2243–2251. https://doi.org/10.1128/jcm.00213-11

Murrison LB, Brandt EB, Myers JB, Hershey GKK (2019) Environmental exposures and mechanisms in allergy and asthma development. J Clin Invest 129:1504–1515. https://doi.org/10.1172/jci124612

Musani AI, Dutau H (2017) Advances in interventional pulmonology. Bentham Science Publishers

Nayak S, Blumenfeld NR, Laksanasopin T, Sia SK (2017) Point-of-care diagnostics: recent developments in a connected age. Anal Chem 89:102–123. https://doi.org/10.1021/acs.analchem.6b04630

de Oliveira WF, dos Santos Silva PM, Coelho LCBB, dos Santos Correia MT (2020) Biomarkers, biosensors and biomedicine. Curr Med Chem 27:3519–3533. https://doi.org/10.2174/0929867326666190124103125

Pairman L, Beckert LEL, Dagger M, Maze MJ (2022) Evaluation of pleural fluid cytology for the diagnosis of malignant pleural effusion: a retrospective cohort study. Intern Med J 52:1154–1159. https://doi.org/10.1111/imj.15725

Paradis TJ, Dixon J, Tieu BH (2016) The role of bronchoscopy in the diagnosis of airway disease. J Thorac Dis 8:3826–3837. https://doi.org/10.21037/jtd.2016.12.68
Peserico A, Di Berardino C, Russo V et al (2022) Nanotechnology-assisted cell tracking. Nanomaterials 12:1414. https://doi.org/10.3390/nano12091414
Prasad A (2016) Pulmonary Aspergillosis: What CT can Offer Before it is too Late! J Clin Diagn Res 10. https://doi.org/10.7860/jcdr/2016/17141.7684
Pulumati A, Pulumati A, Dwarakanath BS et al (2023) Technological advancements in cancer diagnostics: improvements and limitations. Cancer Rep 6. https://doi.org/10.1002/cnr2.1764
Rabiee N, Ahmadi S, Iravani S, Varma RS (2022) Functionalized silver and gold nanomaterials with diagnostic and therapeutic applications. Pharmaceutics 14:2182. https://doi.org/10.3390/pharmaceutics14102182
Ramgir NS (2013) Electronic nose based on nanomaterials: issues, challenges, and prospects. ISRN Nanomaterials 2013:1–21. https://doi.org/10.1155/2013/941581
Ray PC, Yu H, Fu PP (2009) Toxicity and environmental risks of nanomaterials: challenges and future needs. J Environ Sci Health C 27:1–35. https://doi.org/10.1080/10590500802708267
Redondo M, Keyt H, Dhar R, Chalmers JD (2016) Global impact of bronchiectasis and cystic fibrosis. Breathe 12:222–235. https://doi.org/10.1183/20734735.007516
Robinson CL, Baumann LM, Romero K et al (2011) Effect of urbanisation on asthma, allergy and airways inflammation in a developing country setting. Thorax 66:1051–1057. https://doi.org/10.1136/thx.2011.158956
Saadeh Y, Vyas D (2014) Nanorobotic applications in medicine: current proposals and designs. Am J Robot Surg 1:4–11. https://doi.org/10.1166/ajrs.2014.1010
Sakthivel P, Kumar P, Dhavamurthy M et al (2023) Structural, morphological, optical, Photoluminescent and electrochemical performance of ZnS quantum dots: influence of Mn2+ and La3+ ions. J Mol Struct 1288:135723–135723. https://doi.org/10.1016/j.molstruc.2023.135723
Salahuddin M, Anjum F, Cherian SV (2020) Pulmonary eosinophilia. In: PubMed. https://www.ncbi.nlm.nih.gov/books/NBK470600/
van der Sar IG, Wijbenga N, Nakshbandi G et al (2021) The smell of lung disease: a review of the current status of electronic nose technology. Respir Res 22. https://doi.org/10.1186/s12931-021-01835-4
Sim S, Wong N (2021) Nanotechnology and its use in imaging and drug delivery (review). Biomedical Reports 14. https://doi.org/10.3892/br.2021.1418
Singh S, Dhawan A, Karhana S et al (2020) Quantum dots: an emerging tool for point-of-care testing. Micromachines 11:1058. https://doi.org/10.3390/mi11121058
Snen H, Kallel A, Blibech H et al (2021) Case report: allergic bronchopulmonary aspergillosis revealing asthma. Front Immunol 12:695954. https://doi.org/10.3389/fimmu.2021.695954
Stevens WW, Schleimer RP, Kern RC (2016) Chronic rhinosinusitis with nasal polyps. J Allergy Clin Immunol Pract 4:565–572. https://doi.org/10.1016/j.jaip.2016.04.012
Swierczewska M, Liu G, Lee S, Chen X (2012) High-sensitivity nanosensors for biomarker detection. Chem Soc Rev 41:2641–2655. https://doi.org/10.1039/c1cs15238f
Testera-Montes A, Jurado R, Salas M et al (2021) Diagnostic tools in allergic rhinitis. Front Allergy 2. https://doi.org/10.3389/falgy.2021.721851
Thwala LN, Ndlovu SC, Mpofu KT et al (2023) Nanotechnology-based diagnostics for diseases prevalent in developing countries: current advances in point-of-care tests. Nano 13:1247–1247. https://doi.org/10.3390/nano13071247
Vegesna A, Tran M, Angelaccio M, Arcona S (2017) Remote patient monitoring via non-invasive digital technologies: a systematic review. Telemed e-Health 23:3–17. https://doi.org/10.1089/tmj.2016.0051
World Health Organization (2023a) Asthma. In: www.who.int. https://www.who.int/news-room/fact-sheets/detail/asthma#:~:text=Asthma%20is%20a%20chronic%20lung

World Health Organization (2023b) Influenza (seasonal). In: www.who.int. https://www.who.int/health-topics/influenza-seasonal?gclid=CjwKCAjwr_CnBhA0EiwAci5siomFgNPzyb_cP6hgzz1RVzZ6fHuExzTc7-M4cFal-23jTcGT8Uu6txoC0j8QAvD_BwE#tab=tab_1. Accessed 10 Sep 2023
World Health Organization (2023c) Chronic obstructive pulmonary disease (COPD). World Health Organization: WHO. https://www.who.int/news-room/fact-sheets/detail/chronic-obstructive-pulmonary-disease-(copd)
Zhao Y, Zhang Q, Meng Q et al (2017) Quantum dots-based lateral flow immunoassay combined with image analysis for semiquantitative detection of IgE antibody to mite. Int J Nanomedicine 12:4805–4812. https://doi.org/10.2147/ijn.s134539
Zhu Y, Shen B-G, Yue L et al (2022) Application of nanomaterials to enhance polymerase chain reaction. Molecules 27:8854–8854. https://doi.org/10.3390/molecules27248854

Metal and Metal Oxide Nanoparticles for the Management of Allergic Airway Diseases

6

Francisco Castillo and Flavia C. Zacconi

Abstract

In the intricate realm of respiratory health, allergic airways diseases stand as a significant challenge, affecting millions globally with conditions such as asthma and allergic rhinitis, and pose substantial challenges to individuals and healthcare systems worldwide. In the intricate tapestry of human health, the respiratory system stands as a sentinel of life, facilitating the exchange of vital gases and ensuring the sustenance of every cell. Yet, this intricate machinery is not immune to disruptions, and among the myriad challenges it faces, allergic airways diseases emerge as significant contenders. Asthma and allergic rhinitis, two prominent entities within this spectrum, orchestrate a symphony of immunological responses that, when dysregulated, induce chronic inflammation, and compromise respiratory function. Amid the quest for innovative therapeutic strategies, the application of metal and metal oxide nanoparticles emerges as a promising frontier of exploration. This chapter endeavors to unravel the dynamic interplay between allergic airways diseases and the therapeutic potential of metal and metal oxide nanoparticles, delving into the underlying mechanisms, recent advancements, and the promising prospects for future clinical applications in this evolving field.

F. Castillo
School of Chemistry, Faculty of Chemistry and of Pharmacy, Pontificia Universidad Católica de Chile, Santiago, Chile

F. C. Zacconi (✉)
School of Chemistry, Faculty of Chemistry and of Pharmacy, Pontificia Universidad Católica de Chile, Santiago, Chile

Institute for Biological and Medical Engineering, Schools of Engineering, Medicine and Biological Sciences, Pontificia Universidad Católica de Chile, Santiago, Chile

P. Prasher et al. (eds.), *Understanding Allergic Airway Diseases*, https://doi.org/10.1007/978-981-97-1953-2_6

Keywords

Metal nanoparticle · Metal oxide nanoparticles · Allergic airway diseases

6.1 Introduction

Allergic airways diseases, characterized by chronic inflammation of the respiratory mucosa triggered by immune responses to environmental allergens, present a diverse clinical spectrum (Chen et al. 2022). Asthma, with its bronchoconstriction and airway hyperresponsiveness, and allergic rhinitis, marked by nasal congestion and rhinorrhea, collectively impact the quality of life for millions (Brożek et al. 2017; Global Initiative for Asthma 2022). Despite significant advances in understanding their immunopathogenesis, effective and targeted treatments remain a pressing need, motivating exploration into novel therapeutic approaches.

Metal and metal oxide nanoparticles serve as inorganic nanocarriers capable of transporting both small drug molecules and larger entities such as DNA or proteins throughout the body. To enhance stability in bodily fluids, the nanoparticle surface can be coated with a protective layer and adorned with various ligands on the exterior, facilitating preferential attachment to cancer cell receptors (Han et al. 2023; Zuo et al. 2022).

Metal nanoparticles, owing to their unique physicochemical properties, have garnered attention as potential therapeutic agents in various medical fields, including respiratory diseases. Their small size, large surface area, and tunable reactivity make them promising candidates for targeted drug delivery, imaging, and modulation of immune responses. In the context of allergic airways diseases, metal nanoparticles offer a novel avenue for precision medicine, aiming to address the underlying inflammatory processes with enhanced efficacy and reduced side effects (Fig. 6.1).

Gold (Au) plays a crucial role as a nanomaterial in both the electronics and medical sectors. Utilized extensively in these fields, gold nanoparticles (Au-NPs) stand out as excellent carriers for biological molecules, facilitating their transport into cells. This characteristic positions gold nanoparticles as an appealing platform for medication and drug delivery (Hussain et al. 2013). The entry of nanoparticles into the body occurs through various routes, such as inhalation, ingestion, and skin diffusion. Gold has been traditionally recognized as a noble and inert metal with non-cytotoxic properties, exhibiting therapeutic and medical benefits. Its ability to form conjugates with peptides and proteins allows gold nanoparticles to be directed toward specific interaction partners. In the context of a chronic obstructive pulmonary disease (COPD) animal model, gold NPs demonstrated successful transport to alveolar epithelial cells, showcasing their potential in targeted drug delivery. As another example, gold nanoparticles may undergo PEGylation (Batool et al. 2021; Urso et al. 2023) or be functionalized with fluorescent labels, proteins, or other targeting agents, enabling them to traverse cell membranes while delivering DNA or drugs.

By the other hand, silver nanoparticles have the potential to mitigate antigen-induced airway inflammation and hyperresponsiveness, with their antioxidant

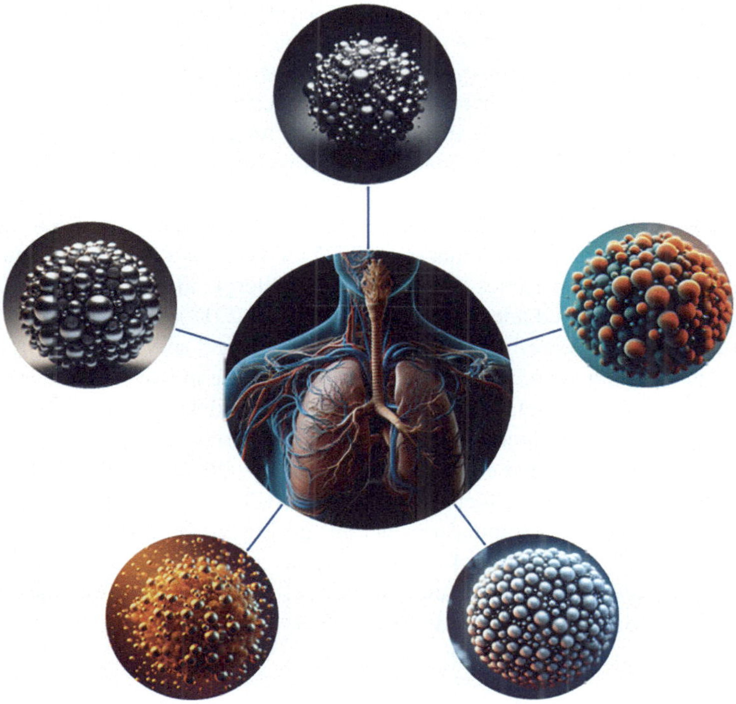

Fig. 6.1 Metal and metal oxides nanoparticles and respiratory system

effects being a potential molecular basis in a murine model of asthma (Park et al. 2010; Yaqoob et al. 2020). These discoveries offer insights into a potential molecular mechanism by which silver nanoparticles could be involved in the prevention or treatment of asthma.

Furthermore, titanium dioxide nanoparticles (TiO_2-NP) intensified the disruption of the AJC (adherens junction complex) induced by RSV (respiratory syncytial virus). Exposure to nanoparticles heightened viral infection in epithelial cells. TiO_2-NP triggered the generation of reactive oxygen species (ROS) (Sengul and Asmatulu 2020), and pretreatment with the antioxidant N-acetylcysteine reversed the resulting barrier dysfunction. In vivo, mice given TiO_2-NP displayed increased RSV-induced injury and AJC disruption in the lungs. This was accompanied by exacerbated airway inflammation as indicated by heightened white blood cell infiltration into the BAL (bronchoalveolar lavage) and an exaggeration of peribronchial inflammation and AJC disruption. These findings highlight that TiO_2-NP exposure worsens RSV-induced AJC dysfunction and inflammation through mechanisms involving ROS generation.

Moreover, respiratory exposure to either pristine or modified CuO materials does not significantly alter classical indicators of allergic airway inflammation. However, these materials contribute to the exacerbation of the condition by introducing an additional inflammatory characteristic—pulmonary neutrophilia—to the existing allergic state. Transcriptomic analysis of lung tissue revealed that CuO PEG exhibits significantly lower potential in inducing changes in healthy and asthmatic lungs compared to CuO and its carboxylated and methylaminated derivatives.

Additionally, the unique physicochemical properties of metal oxide nanoparticles, including their size, surface reactivity, and tunable characteristics, position them as promising candidates for innovative therapeutic interventions (Siddiqi et al. 2018). With a focus on metal oxides such as titanium dioxide (TiO_2), zinc oxide (ZnO) (Faisal et al. 2021; Jiang et al. 2018), copper oxide (CuO), and cerium oxide (CeO_2) (Casals et al. 2020; Nemmar et al. 2017), this chapter aims to scrutinize the potential of these nanoparticles in the context of allergic airways diseases (Raliya and Tarafdar 2013; Abdelkader et al. 2022; Tsai et al. 2018). Their ability to modulate immune responses, deliver therapeutic payloads, and serve as imaging agents makes them intriguing candidates for precision medicine in respiratory care.

Notably, metal and metal oxide nanoparticles exhibit modified physical and chemical properties compared to their native bulk materials, particularly in terms of surface, optical, thermal, and electrical characteristics (Roach et al. 2019).

The synthesis of metal and metal oxide nanoparticles involves the introduction of reducing or oxidizing/precipitating factors during their respective processes (Negrescu et al. 2022). Various factors contribute to nanoparticle reactivity with biomolecules, including nanoparticle size, shape, surface properties, stability, and the fabrication process (Taghavizadeh Yazdi et al. 2023).

6.2 Mechanisms of Action

To harness the therapeutic potential of metal and metal oxide nanoparticles, a comprehensive understanding of their mechanisms of action and possible toxicity is imperative (Bi et al. 2023; Elbakary et al. 2018; Geiser et al. 2013; Seabra and Durán 2015; Smallcombe et al. 2020). These mechanisms may involve direct interactions with immune cells, modulation of inflammatory signaling pathways (see Chaps. 2 and 3), and the ability to encapsulate or deliver therapeutic agents to specific cellular targets within the respiratory system (see Chaps. 8–15) (Agarwal et al. 2019; Zhong et al. 2021). Exploring these interactions provides valuable insights into the development of targeted and efficient treatment strategies (Ilves et al. 2019; Sirelkhatim et al. 2015).

6.3 Metal Nanoparticles

The most common metal used to obtain metallic nanoparticles are nickel, gold, silver, copper, iron, aluminum, zinc, and zirconium. This order was obtained by a simple search using SciFinder and the term "metal nanoparticle," where metal is replaced by the corresponding element, and subsequently, the results were filtered using the terms "preparation" and "synthetic preparation." The results can be seen in Table 6.1, along with the results from two other databases: Scopus and PubMed.

Other metals used for nanoparticles are platinum, palladium, rhenium, ruthenium, cobalt, and cadmium.

These results show that the most referenced metals are nickel, silver, and gold. However, the utilization of them in airways diseases treatments only consider mainly gold and silver metals.

Metal nanoparticles have characteristic properties due to their oscillating free electrons present in the interface between the metal and a dielectric material or, in other words, their localized surface plasmon resonance features. Normally, these nanoparticles have a broad absorption band in the visible region. They have different sizes and shapes, commonly controlled by synthesis, and their uses depend on them. Other common characteristics are color, interparticle interaction, atomic arrangement, high reactivity, and applications like optoelectrical, catalytic, antimicrobial, anticancer, and antiviral (Saleh 2022).

Nickel nanoparticles are reactive nanoparticles due to their redox activity, mostly used in fuel investigations as a catalyst. Modern publication suggests that these nanoparticles can be used in the purification on recombinant proteins (Minkner et al. 2020), and in the photocatalytic reduction of carbon dioxide (Li et al. 2024).

Additionally, nickel nanoparticles have been used to hydrophobized anti-inflammatory steroids obtaining nanoparticles for the treatment of bronchial asthma and allergic rhinitis (Yamashita et al. 2006), and nickel-cobalt nanocrystals have been used to inhibit the activation of inflammasomes on divers inflammatory responses (Lin et al. 2023). A patent shows a nickel-based magnetic nanoparticle for the delivery of irinotecan to the lungs, for the treatment of lung cancer and other lung diseases (Rajadura et al. 2018).

By the other hand, gold nanoparticles are widely used due to their optical and electrical properties, which are shape and size dependent (Khare et al. 2019), and molecular-recognition capabilities, with a wide range of uses like electron microscopy, electronics, nanotechnology, materials sciences, and biomedicine (Granmayeh Rad et al. 2011).

New publications suggest that gold nanoparticles can be used to augment catalytic degradation of organic dyes used as antifungal and antibacterial compounds (Parmar and Sanyal 2024).

Recent studies regarding allergic airways diseases show that gold nanoparticles can be used as drug delivery systems in the treatment of asthma (Mehrabi Nasab et al. 2023) and, conjugated with cortistatin, can be beneficial as anti-inflammatory in the treatment of allergic asthma (Nasab et al. 2023).

Two interesting studies link the gold nanoparticles with biomarkers of COPD in a population of smokers (Swinton et al. 2019), and the effect of the intranasal application of these nanoparticles on pulmonary tissue (Porto et al. 2022), that show good anti-inflammatory effect.

Furthermore, silver nanoparticles properties can be manipulated synthetically the same way as the other nanoparticles, but they have specific characteristic, for example, small size, high surface area to volume ratio, and the ability to scatter light in the visible and near infrared range (Shenashen et al. 2014). Silver nanoparticles have been used for multiple applications; for example, recent discoveries show that

Table 6.1 Articles found in diverse databases (SciFinder, Scopus, and PubMed) for metal nanoparticles and its preparation and use as treatment for allergic airway diseases, obtained November 2023

	Coincidences by using "metal[a] nanoparticle" term			Coincidences when filtered by "preparation" or "synthetic preparation" or "synthesis"			Coincidences when "keyword[b]" and "treatment" are added to the search		
Metal	SciFinder	Scopus	PubMed	SciFinder	Scopus	PubMed	SciFinder	Scopus	PubMed
Nickel (Ni)	798,852	35,413	90	61,354	5210	55	334	441	0
Gold (Au)	77,412	475,279	7791	13,584	50,512	3119	3646	9532	11
Silver (Ag)	60,921	348,790	3360	13,891	40,387	1674	2946	7731	4
Copper (Cu)	9538	72,613	250	2559	9291	130	258	1468	0
Iron (Fe)	6612	48,806	169	1214	6908	83	196	753	0
Aluminum (Al)	1686	11,317	54	115	674	13	21	155	0
Zinc (Zn)	1161	5273	18	121	674	10	61	216	0
Zirconium (Zr)	144	531	0	15	78	0	3	11	0

[a]The word "metal" is replaced by the corresponding element in each search
[b]Keywords: Asthma, allergic rhinitis, chronic obstructive pulmonary disease, and respiratory function

these can interact with other anthropogenic contaminants, allowing the separation of water pollutants (Obare and Bonner 2022).

In the case of silver metal, it has been commonly used in the treatment of allergic airways diseases; for example, silver nanoparticles functionalized by quercetin are biocompatible and can be used in the treatment of chronic bronchitis and asthma (Aiswarriya et al. 2023). In another study, biocompatible silver nanoparticles synthesized by green methods are considered to be a significant anti-asthma agent (Mubaraki et al. 2023). Finally, bio fabricated silver nanoparticles can be used in the treatment of allergic rhinitis in nursing care, improving the allergic inflammatory response (Xiaoshang et al. 2021).

Considering the utilization importance of abundant and nontoxic metals, it is well known that copper nanoparticles have an extraordinary performance as antibacterial and antimicrobial agents (Majzlik et al. 2011). In fact, the Santiago METRO (subway) in Chile uses copper in its handrails as a bactericide to avoid possible cross-infection in a place that has millions of daily visits (CODELCO 2011). Recent studies have shown that copper nanoparticles have multiple application in medicine and health beside antibacterial properties, for example, as nanocarriers for anticancer compounds (Akash et al. 2024) and as a catalyst, in the reduction of carbon dioxide (Zhao et al. 2024).

Regarding advances in allergic airways diseases using copper nanoparticles, only one approach has been made through the functionalization of copper nanoparticles with *Symplocos Racemosa* bark, but its effectiveness in the treatment of asthma has not been proven, where the plant is known for its therapeutic action (Vaishnavi et al. 2020). Copper nanoparticles have been used as contrast agent in photodynamic therapy for treating lung tumors (Cremillieux et al. 2012).

One of the most common and important characteristics of iron nanoparticles is that they possess a high magnetic nature. This characteristic has been widely explored due to the number of applications in biology and medical sciences (Umair et al. 2016). Other characteristics for iron nanoparticles are its stability and abundance and price of prime materials. One important characteristic for small iron nanoparticles (between 10 and 20 nm), it is a property called superparamagnetism, a phenomenon in which their magnetization can flip direction under the influence of temperature, and it is used in magnet-driven drug delivery and release (Kievit et al. 2012).

New studies show thar iron nanoparticles can be used to remove heavy metals from aqueous solutions, having also antimicrobial properties (Mohamed et al. 2023).

The same patents described for nickel also apply to iron. First, an iron-based magnetic nanoparticle shows that it can also deliver irinotecan to the lungs to treat lung cancer and other lung diseases (Rajadura et al. 2018), and second, iron-based nanoparticles containing anti-inflammatory steroids can treat bronchial asthma (Yamashita et al. 2006).

Aluminum nanoparticles show high reactivity, making them a perfect candidate to be used in high-energy compositions, hydrogen production, and synthesis of alumina (Lerner et al. 2016). The plasmonic characteristics of these nanoparticles allow them to be used in application such as surface-enhanced Raman spectroscopy

(Ekinci et al. 2008) and plasmon-enhanced solar desalination (Zhou et al. 2016). New studies show that they can be also used to enhance array-based solar cells (Rautela and Kumar 2023). Interestingly, these kinds of metal nanoparticles are not used in treating respiratory diseases because of their toxic properties. However, they could play a crucial role as effective component to the delivery systems for vaccine adjuvants, defending against pathogens that invade hosts through the respiratory system (Wang et al. 2020).

Zinc nanoparticles are well-known semiconductor nanoparticles that can also be used to remove pollutants from wastewater (Paulkumar et al. 2021). Recent studies show that biosynthesized lactoferrin-coated zinc nanoparticles are promising candidates against COVID-19 (El-Fakharany et al. 2023). Moreover, due to the significant impact of this metal on human health, particularly its role in supplementation and therapeutic responses to conditions such as pneumonia, acute lower respiratory infection, and the common cold, as well as its function in helping to regulate homeostasis, among others (Prasad 2009). Considering these properties, the opportunity to create innovative nanoparticle systems using zinc remains open for further clinical applications in the prevention and treatment of respiratory diseases (Luan et al. 2023).

Zirconium nanoparticles have been obtained using common chemical methods like redox reactions (Eshed et al. 2011) and more novel approaches like fungus biosynthesis (Golnaraghi Ghomi et al. 2019). These nanoparticles have shown antimicrobial and antioxidant properties (Chau et al. 2022), and new studies show that they have near-IR absorptions, allowing its use in phototheranostic applications (Chau et al. 2022). This kind of metal nanoparticles are currently under study due to their potential utilization to reduce inflammation caused by the H5N1 influenza virus (Huo et al. 2020).

6.4 Metal Oxide Nanoparticles

The most common metal oxide nanoparticles contain oxides of silver, iron, aluminum, zinc, and zirconium and can also be made of manganese, titanium, cerium, and magnesium.

Silver oxide nanoparticles have been used in diverse fields, including textile, food, electronics, construction (Iqbal et al. 2019), pharmacy, cosmetics, and antimicrobial (Gudkov et al. 2022). These nanoparticles have great potential toward biomedical applications (Rahimi et al. 2023) due to gene expression and apoptosis induction properties and toward photocatalytic applications (Shume et al. 2020) due to their performance under visible light.

Along with copper and copper oxide nanoparticles, silver oxide nanoparticles have been used in the fight against human viruses, like SARS-CoV-2 (Singh 2023; Tortella et al. 2022).

Iron oxide nanoparticles have many bioapplications due to the excess reactive oxygen species (ROS) formation when cells are exposed to these nanoparticles (Valdiglesias et al. 2016). These have been used in cancer therapy (Soetaert et al. 2020) and theragnostic (Zhao et al. 2020).

A study shows that an iron oxide-based nanoparticles can be used as a fast response humidity sensor, for a real-time monitoring and noncontact sensing, allowing the diagnosis of asthma, among other respiratory diseases (Dhariwal et al. 2023). Mucopenetration of solid lipid nanoparticles containing magnetosensitive iron oxide have been studied in order to treat COPD airways secretion (Castellani et al. 2022).

Magnetic polyethylene glycol-functionalized iron oxide nanocarriers have been used on magnetic resonance imaging as an asthma biomarker in the lung within the inflammatory sites (Al Faraj et al. 2016).

Aluminum oxide nanoparticles are interesting candidates to bioapplications due to their low cytotoxicity associated to the aluminum oxide inertness (Radziun et al. 2011). Other characteristics are its high mechanical strength and large surface area/volume ratio that allow them to be used as catalyzers, adsorbents, additives, and lubricants, among others (Ziva et al. 2021). Also, the therapeutic effects of alpha alumina nanoparticle have been studied for the treatment of allergic asthma, showing that it can be considered an effective nano-drug (Athari et al. 2016).

Zinc oxide nanoparticles are a wide band gap semiconductor at room temperature (Kumar et al. 2012). These nanoparticles are ideal to catalytic reaction processes (Rostami-Charati and Akbari 2018) and has many more possible applications, such as sunscreens and UV blockers, textile coating, antibacterial function, dental fillings, and food preservative, among others, thanks to their characteristics such as photocatalytic properties, better stability, better crystallinity, good thermal conductivity, light emission, etc. (Czyżowska and Barbasz 2022).

The biosynthesized zinc oxide nanoparticles using *Tinospora cordifolia*, a plant known to have medicinal effect over diarrhea, asthma, and skin diseases, are more effective as an antibacterial agent than the stem extract of the plant alone (Libin et al. 2023). The effect in asthma has not been studied.

Herb-based zinc oxide nanoparticles anti-inflammatory properties have been studied for the treatment of asthma (Vandana et al. 2023).

Zirconium oxide nanoparticles have been used as adsorbent for heavy and toxic metals in wastewater (Cui et al. 2012; Su et al. 2013). Novel studies have shown that zirconium oxide nanoparticles supported on carbon nanotubes can work as a nonenzymatic glucose sensor (Huo et al. 2023).

6.5 Common and New Synthetic Procedures

Frequently, three types of approaches can be found to obtain metal and metal oxide nanoparticles: the physical approach (it is also called top-down approach), the chemical approach or so-called bottom-up approach, and the new green or biological or even enzymatic approaches that are considered bottom-up methods.

6.5.1 Top-Down Methods

Generally, these methods use bulk materials to be fragmented, creating the nanostructured materials. In this category, we can found the electrospinning technique (Ostermann et al. 2011), mechanical milling (Gorrasi and Sorrentino 2015), laser ablation (Amendola and Meneghetti 2009), sputtering (Muñoz-García et al. 2009), electron explosion (Joh et al. 2013), sonication (Shi et al. 2012), pulsed discharge (Patil et al. 2023), arc discharge (Berkmans et al. 2014), and lithography (McGrath et al. 2021), among others (Baig et al. 2021).

6.5.2 Bottom-up Methods

Generally, these methods use atoms and molecules to be combined to create the nanomaterials.

6.5.2.1 Chemical Approach

In the chemical approaches, we can find the chemical vapor deposition (Dikusar et al. 2009), sol-gel process (Baig et al. 2021), coprecipitation (Das and Srivasatava 2016), inert gas condensation (Pérez-Tijerina et al. 2008), and hydrothermal methods (Patil et al. 2021).

6.5.2.2 Green/Biological Methods

In these approaches, a bioactive agent is used. Normally, plants, microbes, biowastes, algae, vegetable waste, fruit peel, and eggshell, between others, are used to obtain nanoparticles (Kumari et al. 2021; Malhotra and Alghuthaymi 2022).

6.5.2.3 Biomimetic or Enzymatic Methods

Biomimetic refers to a chemical process that resembles biological synthesis. In these approaches, proteins, enzymes, cells, viruses, and others can be used to synthesize nanoparticles (Ghosh et al. 2021; Hulkoti and Taranath 2014; Qi et al. 2013).

6.6 Recent Advances and Experimental Findings

Recent research endeavors have delved into the application of metal and metal oxide nanoparticles for the treatment of allergic airways diseases, showcasing promising outcomes in preclinical studies. From gold and silver nanoparticles to more specialized formulations, these nanoparticles exhibit potential in mitigating inflammation, improving airway function, and enhancing the overall treatment outcomes. Also, recent investigations regarding the utilization of metal oxide nanoparticles have yielded promising results (Di Gioacchino et al. 2022). From in vitro studies elucidating cellular responses to in vivo models showcasing the efficacy of metal oxide formulations, the experimental landscape is dynamic. Whether through the mitigation of inflammation, enhancement of lung function, or innovative drug delivery systems, these advances lay the groundwork for translating experimental findings into clinical applications (Abbasi et al. 2023; Deeba et al. 2021).

6.7 Conclusion

While the therapeutic potential of metal and metal oxide nanoparticles is promising, their application in the treatment of allergic airways diseases is not without challenges. Concerns related to biocompatibility, long-term safety, and potential off-target effects necessitate thorough consideration and further investigation. Addressing these challenges is pivotal for responsible development and eventual clinical translation of metal oxide nanoparticle-based therapies.

As our understanding of allergic airways diseases deepens, so too does the potential for innovative therapies and preventive strategies. From advancements in precision medicine to the exploration of novel immunomodulatory agents, the future holds promise for more targeted and effective interventions. As we stand at the crossroads of allergic airways diseases and metal and metal oxide nanoparticle therapeutics, the future unfolds with exciting possibilities. The synergy between nanotechnology and respiratory medicine holds the potential to revolutionize treatment strategies, offering personalized and targeted interventions. Finally, this chapter sets the stage for the exploration of these emerging frontiers, laying the groundwork for subsequent discussions on diagnostics, management strategies, and the societal impact of allergic airways diseases using metal or metal oxide nanoparticles as principal components of these innovative solutions.

References

Abbasi R, Shineh G, Mobaraki M, Doughty S, Tayebi L (2023) Structural parameters of nanoparticles affecting their toxicity for biomedical applications: a review. J Nanopart Res 25(3):43. https://doi.org/10.1007/s11051-023-05690-w

Abdelkader DH, Negm WA, Elekhnawy E, Eliwa D, Aldosari BN, Almurshedi AS (2022) Zinc oxide nanoparticles as potential delivery carrier: green synthesis by Aspergillus niger endophytic fungus, characterization, and in vitro/in vivo antibacterial activity. Pharmaceuticals 15(9):1057. https://doi.org/10.3390/ph15091057

Agarwal H, Nakara A, Shanmugam VK (2019) Anti-inflammatory mechanism of various metal and metal oxide nanoparticles synthesized using plant extracts: A review. Biomed Pharmacother 109:2561–2572. https://doi.org/10.1016/j.biopha.2018.11.116

Aiswarriya GR, Gayathri R, Priya Veeraraghavan V, Sankaran K, Prakash Francis A (2023) Green synthesis, characterization and biocompatibility study of quercetin-functionalized biogenic silver nanoparticles. Nano 18(07):2350055. https://doi.org/10.1142/S1793292023500558

Akash BA, Kanagaraj S, Sundaravadivelu S, Varalakshmi GS, Manikantan V, Pillai AS, Alexander A, Enoch IVMV (2024) Cytotoxicity of 5-fluorouracil- loaded, β-Cyclodextrin-tethered metallic copper nanoparticles on triple-negative breast cancer cells. J Mol Struct 1295:136530. https://doi.org/10.1016/j.molstruc.2023.136530

Al Faraj A, Shaik AS, Afzal S, Al-Muhsen S, Halwani R (2016) Specific targeting and noninvasive magnetic resonance imaging of an asthma biomarker in the lung using polyethylene glycol functionalized magnetic nanocarriers. Contrast Media Mol Imaging 11(3):172–183. https://doi.org/10.1002/cmmi.1678

Amendola V, Meneghetti M (2009) Laser ablation synthesis in solution and size manipulation of noble metal nanoparticles. Phys Chem Chem Phys 11(20):3805. https://doi.org/10.1039/b900654k

Athari SS, Pourpak Z, Folkerts G, Garssen J, Moin M, Adcock IM, Movassaghi M, Ardestani MS, Moazzeni SM, Mortaz E (2016) Conjugated alpha-alumina nanoparticle with vasoactive intestinal peptide as a Nano-drug in treatment of allergic asthma in mice. Eur J Pharmacol 791:811–820. https://doi.org/10.1016/j.ejphar.2016.10.014

Baig N, Kammakakam I, Falath W (2021) Nanomaterials: a review of synthesis methods, properties, recent progress, and challenges. Mater Adv 2(6):1821–1871. https://doi.org/10.1039/D0MA00807A

Batool M, Khurshid S, Daoush WM, Siddique SA, Nadeem T (2021) Green synthesis and biomedical applications of ZnO nanoparticles: role of PEGylated-ZnO nanoparticles as doxorubicin drug carrier against MDA-MB-231(TNBC) cells line. Crystals 11(4):344. https://doi.org/10.3390/cryst11040344

Berkmans AJ, Jagannatham M, Priyanka S, Haridoss P (2014) Synthesis of branched, nano channeled, ultrafine and nano carbon tubes from PET wastes using the arc discharge method. Waste Manag 34(11):2139–2145. https://doi.org/10.1016/j.wasman.2014.07.004

Bi J, Mo C, Li S, Huang M, Lin Y, Yuan P, Liu Z, Jia B, Xu S (2023) Immunotoxicity of metal and metal oxide nanoparticles: from toxic mechanisms to metabolism and outcomes. Biomater Sci 11(12):4151–4183. https://doi.org/10.1039/D3BM00271C

Brożek JL, Bousquet J, Agache I, Agarwal A, Bachert C, Bosnic-Anticevich S, Brignardello-Petersen R, Canonica GW, Casale T, Chavannes NH, Correia De Sousa J, Cruz AA, Cuello-Garcia CA, Demoly P, Dykewicz M, Etxeandia-Ikobaltzeta I, Florez ID, Fokkens W, Fonseca J et al (2017) Allergic Rhinitis and its Impact on Asthma (ARIA) guidelines—2016 revision. J Allergy Clin Immunol 140(4):950–958. https://doi.org/10.1016/j.jaci.2017.03.050

Casals E, Zeng M, Parra-Robert M, Fernández-Varo G, Morales-Ruiz M, Jiménez W, Puntes V, Casals G (2020) Cerium oxide nanoparticles: advances in biodistribution, toxicity, and preclinical exploration. Small 16(20):1907322. https://doi.org/10.1002/smll.201907322

Castellani S, Trapani A, Elisiana Carpagnano G, Cotoia A, Laselva O, Pia Foschino Barbaro M, Corbo F, Cinnella G, De Giglio E, Larobina D, Di Gioia S, Conese M (2022) Mucopenetration study of solid lipid nanoparticles containing magneto sensitive iron oxide. Eur J Pharm Biopharm 178:94–104. https://doi.org/10.1016/j.ejpb.2022.07.017

Chau TP, Veeraragavan GR, Narayanan M, Chinnathambi A, Alharbi SA, Subramani B, Brindhadevi K, Pimpimon T, Pikulkaew S (2022) Green synthesis of zirconium nanoparticles using Punica granatum (pomegranate) peel extract and their antimicrobial and antioxidant potency. Environ Res 209:112771. https://doi.org/10.1016/j.envres.2022.112771

Chen M, Shou Z, Jin X, Chen Y (2022) Emerging strategies in nanotechnology to treat respiratory tract infections: realizing current trends for future clinical perspectives. Drug Deliv 29(1):2442–2458. https://doi.org/10.1080/10717544.2022.2089294

CODELCO (2011) *Nuevas líneas de Metro utilizarán cobre bactericida*. https://www.codelco.com/sustentabilidad/noticias/2011/nuevas-lineas-de-metro-utilizaran-cobre-bactericida

Cremillieux Y, Bianchi A, Dufort S, Coll J-L, Lux F, Tillement O (2012) *Ultrafine nanoparticles as multimodal contrast agent* (Patent WO2013153197A1)

Cui H, Li Q, Gao S, Shang JK (2012) Strong adsorption of arsenic species by amorphous zirconium oxide nanoparticles. J Ind Eng Chem 18(4):1418–1427. https://doi.org/10.1016/j.jiec.2012.01.045

Czyżowska A, Barbasz A (2022) A review: zinc oxide nanoparticles—friends or enemies? Int J Environ Health Res 32(4):885–901. https://doi.org/10.1080/09603123.2020.1805415

Das S, Srivasatava VC (2016) Synthesis and characterization of ZnO–MgO nanocomposite by co-precipitation method. Smart Sci 4(4):190–195. https://doi.org/10.1080/23080477.2016.1260425

Deeba F, Umer F, Nainan D (2021) Applications of nanoparticles in treatment of respiratory disorders. Life Sci 3(1):8. https://doi.org/10.37185/LnS.1.1.186
Dhariwal N, Yadav P, Kumari M, Jain P, Sanger A, Kumar V, Thakur OP (2023) Iron oxide-based nanoparticles for fast-response humidity sensing, real-time respiration monitoring, and noncontact sensing. IEEE Sensors J 23(19):22217–22224. https://doi.org/10.1109/JSEN.2023.3306475
Di Gioacchino M, Di Giampaolo L, Mangifesta R, Gangemi S, Petrarca C (2022) Exposure to nanoparticles and occupational allergy. Curr Opin Allergy Clin Immunol 22(2):55–63. https://doi.org/10.1097/ACI.0000000000000818
Dikusar AI, Globa PG, Belevskii SS, Sidel'nikova SP (2009) On limiting rate of dimensional electrodeposition at meso- and nanomaterial manufacturing by template synthesis. Surf Eng Appl Electrochem 45(3):171–179. https://doi.org/10.3103/S1068375509030016
Ekinci Y, Solak HH, Löffler JF (2008) Plasmon resonances of aluminum nanoparticles and nanorods. J Appl Phys 104(8):083107. https://doi.org/10.1063/1.2999370
Elbakary R, Okasha E, Hassan Ragab A, Ragab M (2018) Histological effects of gold nanoparticles on the lung tissue of adult male albino rats. J Microsc Ultrastruct 6(2):116. https://doi.org/10.4103/JMAU.JMAU_25_18
El-Fakharany EM, El-Maradny YA, Ashry M, Abdel-Wahhab KG, Shabana ME, El-Gendi H (2023) Green synthesis, characterization, anti-SARS-CoV-2 entry, and replication of lactoferrin-coated zinc nanoparticles with halting lung fibrosis induced in adult male albino rats. Sci Rep 13(1):15921. https://doi.org/10.1038/s41598-023-42702-0
Eshed M, Pol S, Gedanken A, Balasubramanian M (2011) Zirconium nanoparticles prepared by the reduction of zirconium oxide using the RAPET method. Beilstein J Nanotechnol 2:198–203. https://doi.org/10.3762/bjnano.2.23
Faisal S, Jan H, Shah SA, Shah S, Khan A, Akbar MT, Rizwan M, Jan F, Wajidullah, Akhtar N, Khattak A, Syed S (2021) Green synthesis of zinc oxide (ZnO) nanoparticles using aqueous fruit extracts of *Myristica fragrans*: their characterizations and biological and environmental applications. ACS Omega 6(14):9709–9722. https://doi.org/10.1021/acsomega.1c00310
Geiser M, Quaile O, Wenk A, Wigge C, Eigeldinger-Berthou S, Hirn S, Schäffler M, Schleh C, Möller W, Mall MA, Kreyling WG (2013) Cellular uptake and localization of inhaled gold nanoparticles in lungs of mice with chronic obstructive pulmonary disease. Part Fibre Toxicol 10(1):19. https://doi.org/10.1186/1743-8977-10-19
Ghosh S, Ahmad R, Zeyaullah M, Khare SK (2021) Microbial Nano-factories: synthesis and biomedical applications. Front Chem 9:626834. https://doi.org/10.3389/fchem.2021.626834
Global Initiative for Asthma (2022) Global strategy for asthma management and prevention. http://www.ginasthma.org
Golnaraghi Ghomi AR, Mohammadi-Khanaposhti M, Vahidi H, Kobarfard F, Ameri Shah Reza M, Barabadi H (2019) Fungus-mediated extracellular biosynthesis and characterization of zirconium nanoparticles using standard Penicillium species and their preliminary bactericidal potential: a novel biological approach to nanoparticle synthesis. Iranian J Pharma Res 18:4. https://doi.org/10.22037/ijpr.2019.112382.13722
Gorrasi G, Sorrentino A (2015) Mechanical milling as a technology to produce structural and functional bio-nanocomposites. Green Chem 17(5):2610–2625. https://doi.org/10.1039/C5GC00029G
Granmayeh Rad A, Abbasi H, Afzali MH (2011) Gold nanoparticles: synthesising, characterizing and reviewing novel application in recent years. Phys Procedia 22:203–208. https://doi.org/10.1016/j.phpro.2011.11.032
Gudkov SV, Serov DA, Astashev ME, Semenova AA, Lisitsyn AB (2022) Ag2O nanoparticles as a candidate for antimicrobial compounds of the new generation. Pharmaceuticals 15(8):968. https://doi.org/10.3390/ph15080968
Han R, Xiao Y, Bai Q, Choi CHJ (2023) Self-therapeutic metal-based nanoparticles for treating inflammatory diseases. Acta Pharm Sin B 13(5):1847–1865. https://doi.org/10.1016/j.apsb.2022.07.009

Hulkoti NI, Taranath TC (2014) Biosynthesis of nanoparticles using microbes—A review. Colloids Surf B: Biointerfaces 121:474–483. https://doi.org/10.1016/j.colsurfb.2014.05.027

Huo C, Xiao J, Xiao S, Wang M, Qi P, Liu T, Hu Y (2020) Pre-treatment with zirconia nanoparticles reduces inflammation induced by the pathogenic H5N1 influenza virus. Int J Nanomedicine 15:661–674. https://doi.org/10.2147/IJN.S221667

Huo Y, Li R, Xiu S, Wang Y, Zhang L, Jin A, Quan B (2023) A high–performance non–enzymatic glucose sensor based au and ZrO2 nanoparticles supported on multi-walled carbon nanotubes. Diam Relat Mater 140:110455. https://doi.org/10.1016/j.diamond.2023.110455

Hussain S, Vanoirbeek JAJ, Haenen S, Haufroid V, Boland S, Marano F, Nemery B, Hoet PHM (2013) Prior lung inflammation impacts on body distribution of gold nanoparticles. Biomed Res Int 2013:1–6. https://doi.org/10.1155/2013/923475

Ilves M, Kinaret PAS, Ndika J, Karisola P, Marwah V, Fortino V, Fedutik Y, Correia M, Ehrlich N, Loeschner K, Besinis A, Vassallo J, Handy RD, Wolff H, Savolainen K, Greco D, Alenius H (2019) Surface PEGylation suppresses pulmonary effects of CuO in allergen-induced lung inflammation. Part Fibre Toxicol 16(1):28. https://doi.org/10.1186/s12989-019-0309-1

Iqbal S, Fakhar-e-Alam M, Akbar F, Shafiq M, Atif M, Amin N, Ismail M, Hanif A, Farooq WA (2019) Application of silver oxide nanoparticles for the treatment of cancer. J Mol Struct 1189:203–209. https://doi.org/10.1016/j.molstruc.2019.04.041

Jiang J, Pi J, Cai J (2018) The advancing of zinc oxide nanoparticles for biomedical applications. Bioinorg Chem Appl 2018:1–18. https://doi.org/10.1155/2018/1062562

Joh D-W, Jung T-K, Lee H-S, Kim D-H (2013) Synthesis of nanoparticles using electrical explosion of Ni wire in Pt solution. J Nanosci Nanotechnol 13(9):6092–6094. https://doi.org/10.1166/jnn.2013.7677

Khare T, Oak U, Shriram V, Verma SK, Kumar V (2019) Biologically synthesized nanomaterials and their antimicrobial potentials. In: Comprehensive analytical chemistry, vol 87. Elsevier, pp 263–289. https://doi.org/10.1016/bs.coac.2019.09.002

Kievit FM, Stephen ZR, Veiseh O, Arami H, Wang T, Lai VP, Park JO, Ellenbogen RG, Disis ML, Zhang M (2012) Targeting of primary breast cancers and metastases in a transgenic mouse model using rationally designed multifunctional SPIONs. ACS Nano 6(3):2591–2601. https://doi.org/10.1021/nn205070h

Kumar D, Jat SK, Khanna PK, Vijayan N, Banerjee S (2012) Synthesis, characterization, and studies of PVA/Co-doped ZnO nanocomposite films. Int J Green Nanotechnol 4(3):408–416. https://doi.org/10.1080/19430892.2012.738509

Kumari SC, Dhand V, Padma PN (2021) Green synthesis of metallic nanoparticles: A review. In: Nanomaterials. Elsevier, pp 259–281. https://doi.org/10.1016/B978-0-12-822401-4.00022-2

Lerner MI, Glazkova EA, Lozhkomoev AS, Svarovskaya NV, Bakina OV, Pervikov AV, Psakhie SG (2016) Synthesis of Al nanoparticles and Al/AlN composite nanoparticles by electrical explosion of aluminum wires in argon and nitrogen. Powder Technol 295:307–314. https://doi.org/10.1016/j.powtec.2016.04.005

Li Z, Bai W, Liu D, Han B, Liang Y, Qi J (2024) Preloaded oxygen vacancy conditioning Ni/TiO2 to enhance photocatalytic CO2 reduction. Sep Purif Technol 330:125250. https://doi.org/10.1016/j.seppur.2023.125250

Libin KV, Sisodiya S, Debnath M (2023) Biosynthesis and antimicrobial activities of Tinospora cordifolia zinc nanoparticles. Mater Today Proc 79:367–374. https://doi.org/10.1016/j.matpr.2022.12.159

Lin J, Dong L, Liu Y-M, Hu Y, Jiang C, Liu K, Liu L, Song Y-H, Sun M, Xiang X-C, Qu K, Lu Y, Wen L-P, Yu S-H (2023) Nickle-cobalt alloy nanocrystals inhibit activation of inflammasomes. Natl Sci Rev 10(8):nwad179. https://doi.org/10.1093/nsr/nwad179

Luan R, Ding D, Xue Q, Li H, Wang Y, Yang J (2023) Protective role of zinc in the pathogenesis of respiratory diseases. Eur J Clin Nutr 77:427–435. https://doi.org/10.1038/s41430-022-01191-6

Majzlik P, Strasky A, Adam V, Nemec M, Trnkova L, Zehnalek J, Hubalek J, Provaznik I, Kizek R (2011) Influence of zinc(II) and copper(II) ions on streptomyces bacteria revealed by electrochemistry. Int J Electrochem Sci 6(6):2171–2191. https://doi.org/10.1016/S1452-3981(23)18175-2

Malhotra SPK, Alghuthaymi MA (2022) Biomolecule-assisted biogenic synthesis of metallic nanoparticles. In: Agri-waste and microbes for production of sustainable nanomaterials. Elsevier, pp 139–163. https://doi.org/10.1016/B978-0-12-823575-1.00011-1

McGrath F, Qian J, Gwynne K, Kumah C, Daly D, Hrelescu C, Zhang X, O'Carroll DM, Louise Bradley A (2021) Structural, optical, and electrical properties of silver gratings prepared by nanoimprint lithography of nanoparticle ink. Appl Surf Sci 537:147892. https://doi.org/10.1016/j.apsusc.2020.147892

Mehrabi Nasab D, Taheri A, Athari SS (2023) Evaluation anti-inflammatory effect of conjugated gold nanoparticles with Cortistatin peptide as drug delivery to asthmatic lung tissue. Int J Pept Res Ther 29(1):16. https://doi.org/10.1007/s10989-022-10487-x

Minkner R, Xu J, Takemura K, Boonyakida J, Wätzig H, Park EY (2020) Ni-modified magnetic nanoparticles for affinity purification of His-tagged proteins from the complex matrix of the silkworm fat body. J Nanobiotechnol 18(1):159. https://doi.org/10.1186/s12951-020-00715-1

Mohamed A, Atta RR, Kotp AA, Abo El-Ela FI, Abd El-Raheem H, Farghali A, Alkhalifah DHM, Hozzein WN, Mahmoud R (2023) Green synthesis and characterization of iron oxide nanoparticles for the removal of heavy metals (Cd2+ and Ni2+) from aqueous solutions with antimicrobial investigation. Sci Rep 13(1):7227. https://doi.org/10.1038/s41598-023-31704-7

Mubaraki MA, Mustafa K, Fozia F, Aslam M, Ahmad I, Ahmad N (2023) Antimicrobial, anti-asthmatic and cytotoxic activities of silver nanoparticles synthesized by green method using Zingiber officinale extract. Crystals 13(2):333. https://doi.org/10.3390/cryst13020333

Muñoz-García J, Vázquez L, Cuerno R, Sánchez-García JA, Castro M, Gago R (2009) Self-organized surface Nanopatterning by ion beam sputtering. In: Wang ZM (ed) Toward functional nanomaterials. Springer US, pp 323–398. https://doi.org/10.1007/978-0-387-77717-7_10

Nasab DM, Taheri A, Athari SS (2023) Conjugation of cortistatin peptide with gold nanoparticles synthesized to investigate anti-inflammatory effects in allergic asthma. J Med Chem Sci 6(1). https://doi.org/10.26655/JMCHEMSCI.2023.1.3

Negrescu AM, Killian MS, Raghu SNV, Schmuki P, Mazare A, Cimpean A (2022) Metal oxide nanoparticles: review of synthesis, characterization and biological effects. J Funct Biomater 13(4):274. https://doi.org/10.3390/jfb13040274

Nemmar A, Yuvaraju P, Beegam S, Fahim MA, Ali BH (2017) Cerium oxide nanoparticles in lung acutely induce oxidative stress, inflammation, and DNA damage in various organs of mice. Oxidative Med Cell Longev 2017:1–12. https://doi.org/10.1155/2017/9639035

Obare SO, Bonner CDJ (2022) Assessing the environmental impact of anthropogenic nanoparticles. In: Separation science and technology, vol 15. Elsevier, pp 55–67. https://doi.org/10.1016/B978-0-323-90763-7.00007-X

Ostermann R, Cravillon J, Weidmann C, Wiebcke M, Smarsly BM (2011) Metal–organic framework nanofibers via electrospinning. Chem Commun 47(1):442–444. https://doi.org/10.1039/C0CC02271C

Park P, Kim JW, Park L, Kim K, Jung SS (2010) Attenuation of allergic airway inflammation and hyperresponsiveness in a murine model of asthma by silver nanoparticles. Int J Nanomedicine 505. https://doi.org/10.2147/IJN.S11664

Parmar M, Sanyal M (2024) Biosynthesis of gold nanoparticles using aqueous extract of Ricinus cummunis leaves to augment catalytic degradation of organic dyes and study of its antifungal and antibacterial activities. Particuology 87:87–98. https://doi.org/10.1016/j.partic.2023.07.018

Patil N, Bhaskar R, Vyavhare V, Dhadge R, Khaire V, Patil Y (2021) Overview on methods of synthesis of nanoparticles. Int J Curr Pharm Res:11–16. https://doi.org/10.22159/ijcpr.2021v13i2.41556

Patil S, Kulkarni R, Patil M, Malik VR (2023) Investigations on material removal and tool wear rate of silver nanoparticles coated copper electrodes for electric discharge machining. Adv Mater Process Technol:1–29. https://doi.org/10.1080/2374068X.2023.2194519

Paulkumar K, Jesi Reeta T, Emmanuel Joshua Jebasingh S, Mangalanagasundari S, Muthu K, Murugan K (2021) Potential utilization of zinc nanoparticles for wastewater treatment. In: Aquananotechnology. Elsevier, pp 437–466. https://doi.org/10.1016/B978-0-12-821141-0.00026-4

Pérez-Tijerina E, Gracia Pinilla M, Mejía-Rosales S, Ortiz-Méndez U, Torres A, José-Yacamán M (2008) Highly size-controlled synthesis of au/Pd nanoparticles by inert-gas condensation. Faraday Discuss 138:353–362. https://doi.org/10.1039/B705913M

Porto GD, Haupenthal DPDS, Souza PS, Silveira GDB, Nesi RT, Feuser PE, Possato JC, De Andrade VM, Pinho RA, Silveira PCL (2022) Effects of the intranasal application of gold nanoparticles on the pulmonary tissue after acute exposure to industrial cigarette smoke. J Biomed Mater Res B Appl Biomater 110(6):1234–1244. https://doi.org/10.1002/jbm.b.34994

Prasad A (2009) Zinc: role in immunity, oxidative stress and chronic inflammation. Curr Opin Clin Nutr Metab Care 12:646–652. https://doi.org/10.1097/MCO.0b013e3283312956

Qi P, Zhang D, Wan Y (2013) Sulfate-reducing bacteria detection based on the photocatalytic property of microbial synthesized ZnS nanoparticles. Anal Chim Acta 800:65–70. https://doi.org/10.1016/j.aca.2013.09.015

Radziun E, Dudkiewicz Wilczyńska J, Książek I, Nowak K, Anuszewska EL, Kunicki A, Olszyna A, Ząbkowski T (2011) Assessment of the cytotoxicity of aluminium oxide nanoparticles on selected mammalian cells. Toxicol In Vitro 25(8):1694–1700. https://doi.org/10.1016/j.tiv.2011.07.010

Rahimi E, Asefi F, Afzalinia A, Khezri S, Zare-Zardini H, Ghorani-Azam A, Es-haghi A, Yazdi MET (2023) Chitosan coated copper/silver oxide nanoparticles as carriers of breast anticancer drug: cyclin D1/P53 expressions and cytotoxicity studies. Inorg Chem Commun 158:111581. https://doi.org/10.1016/j.inoche.2023.111581

Rajadura M, Kulkarni P, Sevilmedu A, Saxena U (2018). *Magnetic nanoparticle formulations for targeted delivery of drugs to lungs for treatment of pulmonary diseases* (Patent WO2018150362)

Raliya R, Tarafdar JC (2013) ZnO nanoparticle biosynthesis and its effect on phosphorous-mobilizing enzyme secretion and gum contents in clusterbean (Cyamopsis tetragonoloba L.). Agric Res 2(1):48–57. https://doi.org/10.1007/s40003-012-0049-z

Rautela M, Kumar J (2023) Efficiency enhancement of GaAs nanowire array-based solar cell by plasmonic Al nanoparticles. Mater Today Commun 37:106984. https://doi.org/10.1016/j.mtcomm.2023.106984

Roach KA, Stefaniak AB, Roberts JR (2019) Metal nanomaterials: immune effects and implications of physicochemical properties on sensitization, elicitation, and exacerbation of allergic disease. J Immunotoxicol 16(1):87–124. https://doi.org/10.1080/1547691X.2019.1605553

Rostami-Charati F, Akbari R (2018) ZnO-nanoparticles as an efficient catalyst for the synthesis of functionalized benzenes: multicomponent reactions of sulfonoketenimides. Comb Chem High Throughput Screen 20(9):781–786. https://doi.org/10.2174/1386207320666171004163437

Saleh TA (2022) Properties of nanoadsorbents and adsorption mechanisms. In: Interface science and technology, vol 34. Elsevier, pp 233–263. https://doi.org/10.1016/B978-0-12-849876-7.00010-5

Seabra A, Durán N (2015) Nanotoxicology of metal oxide nanoparticles. Metals 5(2):934–975. https://doi.org/10.3390/met5020934

Sengul AB, Asmatulu E (2020) Toxicity of metal and metal oxide nanoparticles: A review. Environ Chem Lett 18(5):1659–1683. https://doi.org/10.1007/s10311-020-01033-6

Shenashen MA, El-Safty SA, Elshehy EA (2014) Synthesis, morphological control, and properties of silver nanoparticles in potential applications. Part Part Syst Charact 31(3):293–316. https://doi.org/10.1002/ppsc.201300181

Shi L, Fang X, Zhang Z, Zhou T, Jiang D, Wu H, Tang Z (2012) Preparation of nano-ZnO using sonication method and its antibacterial characteristics. Int J Food Sci Technol 47(9):1866–1871. https://doi.org/10.1111/j.1365-2621.2012.03043.x

Shume WM, Murthy HCA, Zereffa EA (2020) A review on synthesis and characterization of ag $_2$ O nanoparticles for photocatalytic applications. J Chem 2020:1–15. https://doi.org/10.1155/2020/5039479

Siddiqi KS, Ur Rahman A, Tajuddin, Husen A (2018) Properties of zinc oxide nanoparticles and their activity against microbes. Nanoscale Res Lett 13(1):141. https://doi.org/10.1186/s11671-018-2532-3

Singh AK (2023) Nanostructured coatings based on metallic nanoparticles as viral entry inhibitor to combat COVID-19. Sustain Mater Technol 35:e00544. https://doi.org/10.1016/j.susmat.2022.e00544

Sirelkhatim A, Mahmud S, Seeni A, Kaus NHM, Ann LC, Bakhori SKM, Hasan H, Mohamad D (2015) Review on zinc oxide nanoparticles: antibacterial activity and toxicity mechanism. Nano-Micro Lett 7(3):219–242. https://doi.org/10.1007/s40820-015-0040-x

Smallcombe CC, Harford TJ, Linfield DT, Lechuga S, Bokun V, Piedimonte G, Rezaee F (2020) Titanium dioxide nanoparticles exaggerate respiratory syncytial virus-induced airway epithelial barrier dysfunction. Am J Phys Lung Cell Mol Phys 319(3):L481–L496. https://doi.org/10.1152/ajplung.00104.2020

Soetaert F, Korangath P, Serantes D, Fiering S, Ivkov R (2020) Cancer therapy with iron oxide nanoparticles: agents of thermal and immune therapies. Adv Drug Deliv Rev 163–164:65–83. https://doi.org/10.1016/j.addr.2020.06.025

Su Y, Cui H, Li Q, Gao S, Shang JK (2013) Strong adsorption of phosphate by amorphous zirconium oxide nanoparticles. Water Res 47(14):5018–5026. https://doi.org/10.1016/j.watres.2013.05.044

Swinton D, McClay T, Yarbrough B, Tyree K (2019, 23/10) Incorporation of hybrid gold nanoparticles to investigate differences in plasma protein biomarkers of chronic obstructive pulmonary disease (COPD) in a population of smokers. 71st Southeastern Regional Meeting of the American Chemical Society, Savannah, GA, United States

Taghavizadeh Yazdi ME, Qayoomian M, Beigoli S, Boskabady MH (2023) Recent advances in nanoparticle applications in respiratory disorders: A review. Front Pharmacol 14:1059343. https://doi.org/10.3389/fphar.2023.1059343

Tortella GR, Pieretti JC, Rubilar O, Fernández-Baldo M, Benavides-Mendoza A, Diez MC, Seabra AB (2022) Silver, copper and copper oxide nanoparticles in the fight against human viruses: Progress and perspectives. Crit Rev Biotechnol 42(3):431–449. https://doi.org/10.1080/07388551.2021.1939260

Tsai S-M, Duran-Robles E, Goshia T, Mesina M, Garcia C, Young J, Sibal A, Chiu M-H, Chin W-C (2018) CeO2 nanoparticles attenuate airway mucus secretion induced by TiO2 nanoparticles. Sci Total Environ 631–632:262–269. https://doi.org/10.1016/j.scitotenv.2018.03.001

Umair M, Javed I, Rehman M, Madni A, Javeed A, Ghafoor A, Ashraf M (2016) Nanotoxicity of inert materials: the case of gold, silver and iron. J Pharm Pharm Sci 19(2):161. https://doi.org/10.18433/J31021

Urso A, Meloni F, Malatesta M, Latorre R, Damoci C, Crapanzano J, Pandolfi L, Giustra MD, Pearson M, Colombo M, Schilling K, Glabonjat RA, D'Ovidio F (2023) Endotracheal nebulization of gold nanoparticles for noninvasive pulmonary drug delivery. Nanomedicine 18(4):317–330. https://doi.org/10.2217/nnm-2022-0179

Vaishnavi A, Rajeshkumar S, Roy A, Lakshmi T (2020) Symplocos racemosa bark functionalized copper nanoparticles and its characterisation. J Complement Med Res 11(5):1–7

Valdiglesias V, Fernández-Bertólez N, Kiliç G, Costa C, Costa S, Fraga S, Bessa MJ, Pásaro E, Teixeira JP, Laffon B (2016) Are iron oxide nanoparticles safe? Current knowledge and future perspectives. J Trace Elem Med Biol 38:53–63. https://doi.org/10.1016/j.jtemb.2016.03.017

Vandana, Gupta S, Aggarwal S, Ashzad S (2023) Herb-based nano-formulations used as anti-inflammatory activity. Res Rev: A Journal of Pharmaceutical Science 14(1):42–52

Wang N, Wei C, Zhang Z, Liu T, Wang T (2020) Aluminum nanoparticles acting as a pulmonary vaccine adjuvant-delivery system (VADS) able to safely elicit robust systemic and mucosal immunity. J Inorg Organomet Polym Mater 30:4203–4217

Xiaoshang G, Murakonda GK, Ma Y (2021) Biofabrication of silver nanoparticles using Ostericum koreanum plant extract for treatment of allergic rhinitis in nursing care. Mater Res Express 8(1):015004. https://doi.org/10.1088/2053-1591/abd4b5

Yamashita N, Higaki M, Ishihara T, Mizushima Y (2006) *Therapeutic agent for bronchial asthma* (patent JP2008162904A)

Yaqoob AA, Umar K, Ibrahim MNM (2020) Silver nanoparticles: various methods of synthesis, size affecting factors and their potential applications–a review. Appl Nanosci 10(5):1369–1378. https://doi.org/10.1007/s13204-020-01318-w
Zhao S, Yu X, Qian Y, Chen W, Shen J (2020) Multifunctional magnetic iron oxide nanoparticles: an advanced platform for cancer theranostics. Theranostics 10(14):6278–6309. https://doi.org/10.7150/thno.42564
Zhao T, Zong X, Liu J, Chen J, Xu K, Wang X, Chen X, Yang W, Liu F, Yu M, Cheng F (2024) Functionalizing Cu nanoparticles with fluoric polymer to enhance C2+ product selectivity in membraned CO2 reduction. Appl Catal B Environ 340:123281. https://doi.org/10.1016/j.apcatb.2023.123281
Zhong W, Zhang X, Zeng Y, Lin D, Wu J (2021) Recent applications and strategies in nanotechnology for lung diseases. Nano Res 14(7):2067–2089. https://doi.org/10.1007/s12274-020-3180-3
Zhou L, Tan Y, Wang J, Xu W, Yuan Y, Cai W, Zhu S, Zhu J (2016) 3D self-assembly of aluminium nanoparticles for plasmon-enhanced solar desalination. Nat Photonics 10(6):393–398. https://doi.org/10.1038/nphoton.2016.75
Ziva AZ, Suryana YK, Kurniadianti YS, Nandiyanto ABD, Kurniawan T (2021) Recent progress on the production of aluminum oxide (Al2O3) nanoparticles: A review. Mech Eng Soc Industry 1(2):54–77. https://doi.org/10.31603/mesi.5493
Zuo X, Guo X, Gu Y, Zheng H, Zhou Z, Wang X, Jiang S, Wang G, Xu C, Wang F (2022) Recent advances in nanomaterials for asthma treatment. Int J Mol Sci 23(22):14427. https://doi.org/10.3390/ijms232214427

7 Nanomicelles for the Management of Allergic Airway Diseases

Moushumi Baidya, Satish Shilpi, and Shvetank Bhatt

Abstract

Asthma and allergic rhinitis are two examples of allergic airway disorders, which are common chronic respiratory diseases that have a major impact on patients' quality of life. The standard medications used for treatment may cause negative effects. As a result, developing novel and safe treatment options for allergic airway disorders is critical. Nanomicelles (NMs) have emerged as a viable treatment for allergic airway disorders. NMs are self-assembling spherical nanoparticles (NPs) with a hydrophobic core and a hydrophilic shell. They can encapsulate medications and target specific cells or tissues, resulting in increased efficacy and decreased toxicity. NMs can deliver continuous and tailored release of anti-inflammatory medications, such as glucocorticoids or cyclosporine, minimizing the likelihood of adverse responses. Furthermore, NMs can be designed to be stable in the presence of stimuli like pH, temperature, storage conditions, and light, allowing for controlled drug release at the desired spot. NMs have been proven in preclinical investigations to have strong anti-inflammatory and anti-allergic effects in allergic airway disease models. As a result, NMs have the potential to transform the treatment of allergic airway illnesses by circumventing the limits of traditional therapies and providing more targeted and efficient medication delivery.

M. Baidya
Bharat Pharmaceutical Technology, Agartala, West Tripura, India

S. Shilpi
Faculty of Pharmacy, School of Pharmaceuticals and Population Health Informatics, DIT University, Dehradun, Uttarakhand, India

S. Bhatt (✉)
School of Health Sciences and Technology, Dr. Vishwanath Karad MIT World Peace University, Pune, India

P. Prasher et al. (eds.), *Understanding Allergic Airway Diseases*,
https://doi.org/10.1007/978-981-97-1953-2_7

Keywords

Nanomicelles · Nanoparticles · Allergic airway diseases

7.1 Introduction

Micelles are amphiphilic colloidal structures with particle sizes ranging from 5 to 100 nm. Micelles are made up of molecules with competing water affinity. Micelles are formed when amphiphilic molecules join together at appropriate temperatures and concentrations. The critical micelle concentration is the concentration at which micelles begin to form and aggregation begins. The crucial temperature for micelle formation, according to this, is the temperature below which micellar molecules stay only as monomers. The aggregation number of a micelle is the total number of monomer molecules that comprise the micelle. The amphiphilic molecules cluster and form micelles as a result of the removal of hydrophobic micelle fragments from the aqueous environment and the formation of hydrogen bonds in water, which results in a decrease in the system's free energy. Micelles are commonly used as drug carriers by pharmaceutical corporations because they can transport lipophilic drugs inside their cores while interacting with polar molecules on their surface (Xu et al. 2013).

In recent years, NMs have received a lot of attention as prospective treatment agents for allergic airway disorders. Asthma and allergic rhinitis are allergic airway illnesses characterized by chronic inflammation of the airways in response to allergen exposure. Traditional therapies frequently include the use of corticosteroids and bronchodilators, which can have negative side effects when used for an extended period of time (Yokoyama et al. 1992).

NMs are colloidal particles that range in size from 10 to 100 nanometers and have a hydrophilic outer surface and a hydrophobic interior core (Hsu et al. 2008). Because of their distinct structure, they can encapsulate and transport hydrophobic medicines such as corticosteroids or anti-inflammatory medications, which are routinely employed in the treatment of allergic airway illnesses.

Because of their microscopic size, NMs can penetrate deep into the airway tissues, boosting drug delivery to the target site and therapy efficacy. They can also shield the drug from degradation and metabolic processes, reducing early drug release and extending the therapeutic impact (Gao et al. 2006).

The capacity of NMs to target specific cell types or receptors involved in the inflammatory response is one of the most significant advantages of employing NMs for allergic airway disease therapy. Surface changes to NMs allow them to selectively bind to receptors produced on immune cells such as mast cells or eosinophils, which play an important role in allergic reactions (Kuo et al. 2008). This tailored administration improves overall treatment outcomes by increasing medication concentration at the site of inflammation and lowering systemic side effects.

NMs can also circumvent some of the constraints of traditional medication delivery techniques. For example, by encapsulating hydrophobic pharmaceuticals, they can help overcome poor solubility concerns, allowing for the use of a broader range of therapeutic treatments. Furthermore, NMs can avoid mucociliary clearance in the airways, which restricts the bioavailability of medicines administered by traditional routes (Yoncheva et al. 2012).

Targeted drug delivery utilizes NMs, which improves medicine bioavailability and allows for deeper tissue penetration (Lee et al. 2003a). Treatment of chronic illnesses, infectious diseases, and cancer, which is becoming more prevalent, all require sustained pharmaceutical release prodrug, new polymers, layer-by-layer micellar assembly on solid supports, reverse micelle creation, drug polymer conjugate micelle manufacturing, and the creation of polymer films that generate micelles in vivo (Bromberg et al. 2008).

The micelles' small size gives them an advantage over larger systems for drug targeting (Lee et al. 2003b). Because of its unique qualities like a small size, high surface area to volume ratio, and capacity to encapsulate a variety of therapeutic agents such as proteins, peptides, and small molecules, the use of NMs as a therapeutic delivery system is garnering interest. The purpose of employing these NMs is to improve therapeutic agent delivery and efficacy while lowering the risk of unwanted side effects (Tang et al. 2009).

Chronic inflammation in the airways characterizes allergic airway disease such as asthma, allergic rhinitis, and COPD. The major goal of therapy for above diseases is to control the chronic inflammation and minimize symptoms including wheezing, shortness of breath, and coughing. Existing medications, such as corticosteroids taken via inhalation or oral routes, can cause side effects such as an increased risk of infection, bruising, and osteoporosis, especially when used for an extended period of time (Smart et al. 2005).

NMs, due to their small size, can penetrate deep into the lung tissue and deliver the encapsulated therapeutics directly to the site of inflammation. This offers the potential to reduce the overall dose of the therapeutic agent required and improve its efficacy while minimizing side effects. Furthermore, NMs can be designed to release the therapeutic agent over an extended period of time, resulting in sustained anti-inflammatory effects (Ponchel et al. 1998).

Overall, the use of NMs for the management of allergic airway disease presents an exciting potential for the development of safe and effective therapies that can improve the quality of life of individuals affected by this condition. However, further research in this area is critical to fully understand their safety and efficacy in clinical settings (Crater et al. 2010). Ideal properties of micelles have been shown in Fig. 7.1.

7.2 Sources of NMs

NMs can be synthesized from a variety of amphiphilic (dual hydrophilic and hydrophobic) molecules, typically consisting of a hydrophilic "head" group and a hydrophobic "tail" group. These molecules self-assemble in aqueous environments to form nanoscale structures with a hydrophobic core and a hydrophilic shell (Lin

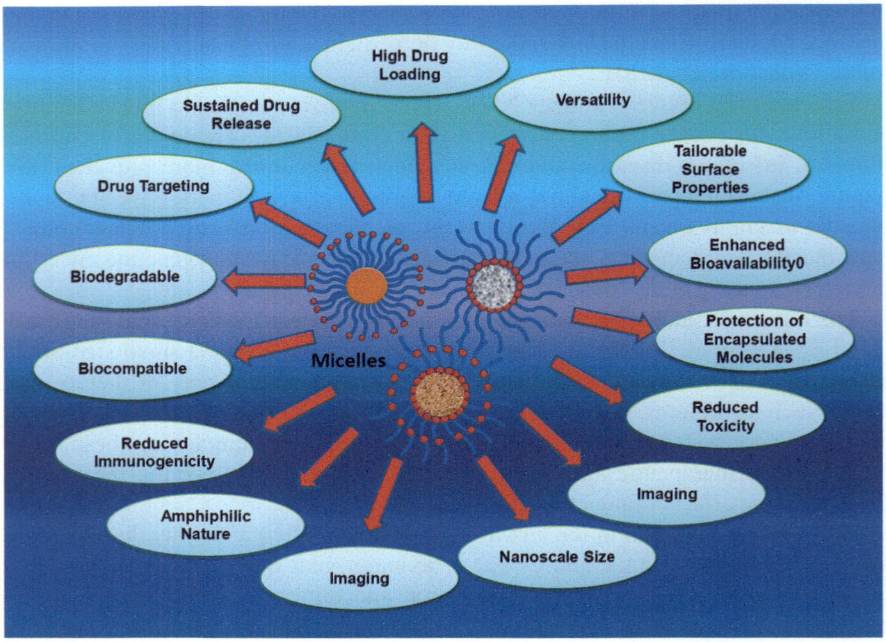

Fig. 7.1 Ideal properties of micelles

et al. 2010). Here are some common sources and types of molecules used for creating NMs:

Surfactants: Surfactant molecules, such as phospholipids and fatty acids, are commonly used to form NMs. Phospholipids, like phosphatidylcholine and phosphatidylglycerol, are natural constituents of cell membranes and are often used in drug delivery systems due to their biocompatibility. Examples include liposomes and lipid-based micelles.

Block Copolymers: Block copolymers consist of two or more distinct polymer blocks with differing chemical properties. Pluronics (poloxamers) and polyethylene glycol (PEG)-block-poly (propylene glycol) (PEG-b-PPG) are commonly used block copolymers for NMs formation. They can provide stability and prolonged circulation time in the bloodstream.

Dendrimers: Dendrimers are highly branched, three-dimensional macromolecules with well-defined structures. They can be engineered to encapsulate drugs or other bioactive molecules within their cavities, making them suitable for drug delivery applications.

Cyclodextrins: Cyclodextrins are cyclic oligosaccharides with a hydrophilic exterior and a hydrophobic interior cavity. They can form inclusion complexes with hydrophobic drugs, improving their solubility and stability.

Solid Lipid NPs (SLNs) and Nanostructured Lipid Carriers (NLCs): These lipid-based NPs utilize solid lipids or a mixture of solid and liquid lipids to form the core of the nanostructure. They offer controlled drug release and improved drug stability.

Polymeric Micelles: Synthetic polymers, such as poly (lactic-co-glycolic acid) (PLGA), poly (caprolactone) (PCL), and poly (ethylene glycol) (PEG), can self-assemble into micellar structures. These polymeric micelles can encapsulate hydrophobic drugs and exhibit tunable properties.

Nucleic Acid Micelles: Nucleic acids, such as DNA and RNA, can also be used to create micellar structures for gene delivery and other therapeutic applications.

Protein-Based Micelles: Proteins and peptides with amphiphilic properties can be engineered to form micelles. These protein-based micelles can be used for drug delivery and other biotechnological applications.

It is important to note that the choice of micelle-forming molecule depends on the specific application, drug payload, desired release profile, and biocompatibility. Researchers often tailor the structure and properties of NMs to optimize their performance for targeted drug delivery and therapeutic outcomes (Sammalkorpi et al. 2009).

7.3 NMs Preparation

NMs are formed as a result of the self-assembly of amphiphilic molecules in aqueous fluids. Amphiphilic compounds have both polar or hydrophilic and nonpolar or hydrophobic components. When a hydrophilic or polar solvent is added, these particles form, with the hydrophobic component oriented away from the solvent and forming the interior, and the hydrophilic polar component functioning as the exterior and rotating toward the solvent (Alvarez-Lorenzo et al. 2009). The hydrophilic or polar component is inside the structure because it is hydrophobic and cannot interact with the aqueous solvent. Regular micelles are amphiphilic molecular aggregates in this orientation (Kedar et al. 2010). In a nonpolar or hydrophobic fluid, the amphiphilic molecules are arranged in an opposite direction, with the hydrophobic portion on the outside and the hydrophilic portion on the inside. These kinds of micelles are referred to as reverse micelles (Chen et al. 2011).

Surface-active substances like surfactants and synthetic block copolymers are used to make NMs. Monomers that are amphiphilic might be ionic, nonionic, or zwitterionic. Depending on their charge, ionic surfactants can be positive (cation) or negative (anion). Dodecyl trimethyl ammonium bromide is a cationic surfactant, whereas sodium dodecyl is an anionic surfactant (Sant et al. 2005). Zwitterionic surfactants, such dioctanoyl phosphatidyl choline, have both positive and negative charges while nonionic surfactants, like n-dodecyl tetra (ethylene oxide), are neutral. The linear diblock (A-B), triblock (A-B-A), pentablock (A-A-B-A-A), and branched kinds of polymeric micelles can all be arranged (Xie et al. 2005). In block copolymer configurations, A and B can be any polymer, including but not limited to poly (ethylene glycol), poly (L-lysine), poly (D,L-lactic acid), polypropylene oxide, polyglycolic acid, poly (aspartic acid), poly (glutamic acid), poly (L-lysine), and poly (histidine) (Meng et al. 2003).

7.4 Nanomicelles (NMs) Characteristics

NMs are unique nanostructures with distinct properties that make them well suited for various biomedical and pharmaceutical applications, including drug delivery. Here are some key properties of NMs (Kabanov and Alakhov 2000):

- **Amphiphilic Nature:** NMs are formed by amphiphilic molecules, which have both hydrophilic and hydrophobic parts. This allows them to self-assemble into stable structures in aqueous solutions, with hydrophobic cores shielded by hydrophilic shells.
- **Nanoscale Size:** NMs typically have sizes ranging from 10 to 100 nanometers, making them suitable for interactions at the cellular and subcellular levels. Their small size allows for efficient uptake by cells and potential penetration through barriers like mucus layers.
- **High Drug-Loading Capacity:** The hydrophobic core of NMs provides an environment conducive to encapsulating hydrophobic drugs or biomolecules. This high drug-loading capacity can improve the solubility and stability of poorly water-soluble compounds.
- **Sustained and Controlled Drug Release:** NMs enable controlled and sustained drug release over time. The gradual release of drugs from the hydrophobic core can lead to prolonged therapeutic effects and reduced dosing frequency.
- **Targeted Drug Delivery:** Surface modification of NMs with targeting ligands, such as antibodies or peptides, allows for site-specific drug delivery. This property enhances drug accumulation at the desired target while minimizing off-target effects.
- **Enhanced Bioavailability:** NMs can improve the bioavailability of drugs by enhancing their solubility and absorption in the body, leading to better therapeutic outcomes.
- **Protection of Encapsulated Molecules:** NMs can shield encapsulated drugs or molecules from degradation by enzymes, pH changes, or other harsh conditions in the body's environment.
- **Biocompatibility and Reduced Toxicity:** Many materials used to form NMs, such as phospholipids and biodegradable polymers, are biocompatible and have low toxicity, reducing the risk of adverse effects.
- **Versatility:** NMs can be engineered for a wide range of applications, from drug delivery to gene therapy, due to their ability to encapsulate various types of therapeutic agents, including small molecules, proteins, nucleic acids, and more.
- **Tailorable Surface Properties:** The surface properties of NMs, including charge and hydrophilicity, can be modified to optimize interactions with biological systems and improve their stability and circulation time in the bloodstream.

- **Enhanced Bioavailability:** NMs can improve the bioavailability of drugs by enhancing their solubility and absorption in the body, leading to better therapeutic outcomes.
- **Protection of Encapsulated Molecules:** NMs can shield encapsulated drugs or molecules from degradation by enzymes, pH changes, or other harsh conditions in the body's environment.
- **Biocompatibility and Reduced Toxicity:** Many materials used to form NMs, such as phospholipids and biodegradable polymers, are biocompatible and have low toxicity, reducing the risk of adverse effects.
- **Potential for Combination Therapies:** NMs can encapsulate multiple drugs or bioactive molecules, allowing for combination therapy and synergistic effects in treating complex diseases.
- **Imaging Capabilities:** NMs can be engineered to carry imaging agents, enabling visualization and monitoring of drug distribution and therapeutic effects.
- **Ease of Formulation:** NMs are relatively easy to formulate, and their composition and properties can be adjusted through simple modifications to tailor them to specific applications.
- **Reduced Immunogenicity:** Some NMs formulations, particularly those based on natural lipids or polymers, tend to have lower immunogenicity compared to other NPs platforms.

It is important to note that the properties of NMs can vary based on factors such as the type of amphiphilic molecules used, the formulation process, and the specific application. Researchers continue to explore and optimize these properties to develop innovative and effective drug delivery systems and therapies.

7.5 Types of NMs

NMs are versatile nanostructures formed by self-assembly of amphiphilic molecules in an aqueous environment. They have a hydrophilic shell and a hydrophobic core, making them suitable for encapsulating hydrophobic drugs, enhancing drug solubility, and improving drug delivery. There are various types of NMs based on the composition of the amphiphilic molecules used (Cholkar et al. 2015). Here are some common types:

7.5.1 Polymeric Micelles

- These micelles are composed of synthetic polymers, often block copolymers that self-assemble in water to form micellar structures.
- Examples of polymers used include polyethylene glycol (PEG), poly-(caprolactone) (PCL), poly (lactic-co-glycolic acid) (PLGA), and poly (ethylene oxide)-block-poly(propylene oxide)-block-poly(ethylene oxide) (PEO-PPO-PEO) triblock copolymers (Pluronics).

- Polymeric micelles offer tunable properties and stability and can be functionalized for targeted drug delivery.

7.5.2 Lipid-Based Micelles

- Lipid-based micelles are formed from amphiphilic lipids, often phospholipids or fatty acids.
- Phospholipids, such as phosphatidylcholine and phosphatidylglycerol, are commonly used to create liposomes and other lipid-based micelles.
- These micelles can encapsulate hydrophobic drugs within their lipid bilayers.

7.5.3 Block Copolymer Micelles

- Block copolymers with distinct hydrophilic and hydrophobic segments can self-assemble into micelles.
- Examples include Pluronics PEG-poly(propylene glycol) and PEG-poly(propylene oxide)-PEG.
- Block copolymer micelles can improve drug solubility and provide controlled drug release.

7.5.4 Dendritic Micelles

- Dendritic micelles are formed from dendrimers, which are highly branched, three-dimensional polymers with well-defined structures.
- These micelles can encapsulate drugs within their cavities and offer precise control over drug loading and release.

7.5.5 Nucleic Acid Micelles

- Nucleic acids, such as DNA and RNA, can self-assemble into micellar structures suitable for gene delivery.

7.5.6 Cyclodextrin Micelles

- Drugs that are hydrophobic can form inclusion complexes with cyclic oligosaccharides called cyclodextrins.
- Cyclodextrin-based micelles enhance drug solubility and stability.

7.5.7 Protein-Based Micelles

- Proteins or peptides with amphiphilic properties can be engineered to form micelles.
- Protein-based micelles can be used for drug delivery and other biomedical applications.

7.5.8 Mixed Micelles

- Mixed micelles are composed of a combination of amphiphilic molecules, such as lipids and polymers, to take advantage of the benefits offered by each component.

Each type of NMs have its own advantages and limitations, and the choice of micelle type depends on the specific application, drug payload, desired release profile, and compatibility with biological systems. Researchers continue to explore and develop various types of NMs for targeted drug delivery and improved treatment strategies in conditions like allergic airway diseases.

7.6 Methods of Drug Delivery Using NMs

NMs, which range in size from 5 to 100 nm, are colloidal structures made of amphiphilic monomers with a short hydrophobic head and a long hydrophilic tail I believe that it is better suitable for medication distribution because of its small size and the structural stability that its nanostructure offers. Drugs may be contained inside a micelle, which would then operate as a vehicle to transport the drugs to the desired tissue. As was already mentioned, because it is nanoscale, it may easily pass through any capillary or vascular system and can quickly reach its target. The amount of micelle moieties has a major impact on drug release, causing slower and more precise release. Doxorubicin and paclitaxel are two effective examples of chemotherapy-carrying micelles that are now approved for clinical trials (Cholkar et al. 2014). The next sections go over the common techniques for loading medicines inside micelles.

7.6.1 Oil-in-Water Emulsion Techniques

Submicron emulsions are emulsions with nanoscopic droplet sizes, typically between 20 and 200 nm, and are made up of two immiscible liquids where one liquid is dispersed as droplets in the other. The oil droplets in a nano emulsion act as a reservoir for hydrophobic medicines. Saturated and unsaturated fatty acids, fatty acid esters, and soybean oils are the most widely used oil molecules.

The stabilization of the nano emulsions relies heavily on the surfactant molecules. The nonionic or amphoteric surfactants poloxamer, lecithin, and tween 80 are frequently utilized. To increase the stability of nano emulsions and regulate droplet size, mixtures of various surfactants have also been used (Savić et al. 2006).

7.6.2 Water-in-Oil-in-Water Emulsion Techniques

At room temperature, freshly prepared W/O/W emulsions are examined under an optical microscope. Samples were carefully placed on microscope slides and covered with a coverslip to avoid damaging the emulsion structures. The microstructures of the W/O/W emulsions were studied using an oil immersion objective (100× magnification) and suitable light intensity (Ahn et al. 2015).

7.6.3 Direct Dialysis

In order to remove the organic solvent from a solution of polymer and drug in a water-miscible organic solvent, such as dimethylformamide, the dialysis method involves performing dialysis for several hours against an excess of water while adding small amounts of water while stirring (Trinh et al. 2017).

7.6.4 Cosolvent Evaporation

A suitable quantity of any drug, such as simvastatin (SIM), is dissolved in 100 ml of methanol and a suitable quantity of hydroxypropyl methylcellulose [HPMC (K3LV)] is dissolved in a suitable volume of distilled water. Both solutions are then mixed to create a clear solution, which is then evaporated at a specific pressure and temperature in a rotary evaporator for 30 minutes.

7.6.5 Spray Drying

Typically, a spray drier is used to evaporate the solvent. The solutions are made by dissolving the right amount of the medicine in the right amount of methanol and the right amount of the surfactant in the right amount of distilled water and then combining the two solutions to create a clear solution. With the feed pump running at 10 ml/min and aspiration at 45%, the solvent is evaporated at 110 °C inlet and 60 °C outlet. In 20–30 min, the medication mixture is ready for spray drying.

Among the methods stated above, O/W, direct dialysis, and cosolvent evaporation are well compatible for encapsulating hydrophobic medicines, but W/O/W is frequently favored for encapsulating more hydrophilic molecules (Wang and Zhang 2012). Because the micelles are solid enough at the nanoscale, the drug will not be released until all of the NMs have gathered in the intended spot, and also if intravenous delivery is employed.

7.7 Nanomicelle-Mediated Drug Loading and Release

In general, three major strategies for loading medicines into polymer micelle cores are reviewed in the sections that follow (Wu et al. 2010).

7.7.1 Chemical Conjugation

A drug is chemically linked to the center of a cell in order for it to be released in its active form inside the cell. Assembling a copolymer block with the appropriate pH or enzyme sensitivity, so that it can be divided by a linker. Variations in the durability of the drug conjugate and the type of polymer drug linkage. The rate of drug release and, as a result, the efficacy of the prodrug can be affected by linking. For instance, in a recent study, PEO-bpoly(aspartate hydrazone doxorubicin) pH-sensitive polymer micelles were developed. Doxorubicin was attached to the hydrophobic segments using acid-sensitive hydrazone linkers, which are stable at extracellular pH 7.4 but breakdown and release the free drug in endosomes and lysosomes at acidic pH 5.0 to 6.0 (Bae et al. 2005). Doxorubicin was amide-bonded to the poly (aspartic acid) chain of the PEObpoly (aspartic acid) block copolymer. The total eradication of solid tumors implanted in mice was proven by changing the composition of the block copolymer and the concentration of conjugated doxorubicin. The anticancer activity was later discovered to be attributable to doxorubicin, which was physically present within the micellar core. This discovery led to the application of PEObpoly (aspartate doxorubicin) conjugates as nanocontainers for doxorubicin that was physically confined (Singh and Lillard Jr. 2009).

7.7.2 Physical Entrapment

In general, micelle-forming polymer-drug conjugates are preferred over block copolymer micelles with physical drug inclusion or solubilization, especially for hydrophobic medicinal molecules. For each type of medication, specific block copolymers must be designed because many polymers and medicinal molecules lack the reactive functional groups required for chemical conjugation. By contrast, a variety of drugs can be physically incorporated into the micelles by altering the shape of the core-forming portion. Additionally, the molecular characteristics of a homologous copolymer series (molecular weight, composition, and presence of functional groups for active targeting) can be tailored to maximize a drug's efficiency in a given drug delivery scenario. Our group first put forth this concept in the late 1980s under the moniker "micellar micro container," but it is now more frequently referred to as a "micellar nanocontainer." An increase in neuroleptic activity was observed when haloperidol was encapsulated in Pluronic block copolymer micelles and the micelles were administered to the brain using insulin or antibodies that are specific to the brain

7.7.3 Polyionic Complexation

Charged helpful operators can group together to form square copolymer micelles via electrostatic interactions with an oppositely charged ionic component of the polymer. This approach, which Kabanov and Kataoka independently proposed in 1995, is now widely employed to link various polynucleic acids into square ionomer structures in order to create nonviral quality conveyance frameworks. The creation of stable square ionomer edifices is influenced by ionic square lengths, charge thickness, and ionic quality of the arrangement, which, in turn, regulates the amount of medication that can be combined inside the micelles. Such square ionomer micelles offer a singular opportunity to control the activated arrival of the dynamic restorative agent due to their pH and salt affectability. Square ionomer structures can also be interested in polyion exchange reactions, which are thought to symbolize the entry of DNA and a helping specialist into a functional structure inside of cells. There are several thorough studies that focus on square ionomer micelles as drug and quality delivery frameworks can be located in the writing.

7.8 Advantages of NMs in Drug Delivery System

The nanomicelle-based drug delivery system has some advantages over the conventional drug delivery system due to its size and structural make-p. Here are a few advantages:

(a) Hydrophobic pharmaceuticals attach to the hydrophobic centers of nanomicelles because they are amphiphilic molecules, creating a clear aqueous solution that greatly increases the solubility of lipophilic medications (Bae and Kataoka 2009). For instance, ordinary efavirenz (non-nanomicelle), which is poorly soluble in water, would have only had a solubility of 4 micro g/ml, but antiretroviral efavirenz has a solubility of 34 mg/ml. As a result, hydrophobic medications are captured by nanomicelles, improving their solubility (Rapoport 2007).

(b) The active targeting capabilities of nanomicelles allow for target-specific delivery with the fewest possible adverse effects. They can be targeted particularly because of their conjugation with the target moieties. There may be interactions between certain surface receptors, transporter proteins, applied signal proteins, or phage fusion proteins (Gong et al. 2012). Ahn et al. created an antibody fragment linked with platinum-loaded nanomicelles to target pancreatic cancer on tumor xenografts. Ahn et al. developed an antibody fragment coupled with platinum-loaded NMs.

(c) When specific triggers, such as pH or temperature, are present, drugs are released from NMs, resulting in improved selectivity and less toxicity. Because tumor cells, lysosomes, and endosomes are acidic by nature, pharmaceuticals that are normally imprisoned inside NMs (pH 7.2) are released in the area of the body that is being targeted by the acidic situation. The stimulation provided by this pH gradient permits the medication to be released precisely where it is needed (Oerlemans et al. 2010). For instance, because to the hydrazine linker's acid sensitivity, Bae et al. developed pH-sensitive polyethylene glycol (PEG)-poly aspartate hydrazine-loaded NMs with Adriamycin that release at low pH.

(d) NMs are created using hydrophobic, biodegradable polymeric NPs. By keeping medications where they are needed, these biodegradable NPs act as a neighborhood drug depot. This depot acts as the source of constant delivery of the therapeutic agent or medication contained in the NMs at the ill site, such as solid tumors (Civiale et al. 2009).

(e) Due to their size, NMs (>1 m) are more suited for the administration of drugs intravenously. The body's tiniest capillary measures 5–6 m in diameter. NMs deliver appropriate drug distribution into the bloodstream while preventing embolism because they are smaller than the body's tiniest capillaries (Patel et al. 2013).

7.9 Nanomicelle-Mediated Drug Delivery for Several Therapeutic Treatment

7.9.1 In Cancer Treatment

Due to their significant physical and biological advantages over alternative nanocarriers, polymeric micelles produced from amphiphilic block copolymers have attracted a lot of interest for the delivery of certain cancer medicines. They have an intrinsic sheath effect and a hydrophilic shell to provide colloidal stability, as well as a hydrophobic core for effective drug loading of medications that are poorly water soluble. Their sizes range from 20 to 100 nm. The targeted and regulated administration of hydrophobic anticancer medicines like paclitaxel (PTX) and doxorubicin (DOX) is also made possible by biodegradable polymeric micelles. The anticancer drugs' water solubility is increased by these nanocarriers, which can also prolong drug circulation times, passively target tumor tissues using the EPR effect, increase bioavailability, have excellent biocompatibility, and degrade in vivo into nontoxic byproducts that can be absorbed and excreted further from the human body. Due to their propensity to biodegrade, micelles are a useful tool for regulating medication release and preventing long-term toxicity from the drug's buildup in human tissues. The treatment of breast, ovarian, and lung cancers with Genexol PM, a PTX formulation based on biodegradable poly(ethylene glycol)-b-poly (D,L-lactide) (PEG-PLA) copolymer micelles, has been approved in countries like Japan, South Korea, the UK, and the USA. The traditional treatments for cancer include radiation, chemotherapy, and surgical removal of the tumor. Chemotherapeutic chemicals' cytotoxicity and nonspecificity make them a prime candidate for nonspecific targeting of both cancer and healthy cells.

As a result, they have negative side effects. Due to their low molecular weight, polymeric micelles have proven to be effective drug carriers that aid in indiscriminate biodistribution to both normal and tumor tissues (Vadlapudi and Mitra 2013).

Additionally, the majority of anticancer medications are water insoluble, which causes poor absorption and low bioavailability; therefore, a carrier that transports the medication to the tumor is needed.

7.9.2 In Eye Treatment

Nanomicellar compositions have allowed for far more technological advancement in the delivery of drugs through the eyes. Their high drug encapsulation capacity, ease of production, small size, and capacity to create aqueous solutions through hydrophilic nanomicellar corona are just a few of their benefits. Drugs are more bioavailable in ocular tissues thanks to micellar formulations (Fireman et al. 2011). For instance, PHEAC (16) and PEGylated PHEAC (16) copolymers were used to create dexamethasone-loaded NMs that were delivered to the eye's anterior region. Currently, scientists are looking at the posterior part of the eye drug. For the topical administration of hydrophobic medicines, this mode of medication delivery appears to be a viable pharmaceutical carrier (Lee et al. 2010). With this technique, which is also very patient friendly, age-related visual disorders such as macular degeneration, diabetic retinopathy, diabetic macular edema, and posterior uveitis can all be efficiently treated (Makhmalzade and Chavoshy 2018).

7.9.3 In Skin Treatment

Drugs are unable to penetrate the skin barrier easily because both hydrophobic and hydrophilic substances are extremely refractory to the skin. The new method of drug delivery involves using nano-sized carriers for skin penetration. Pharmaceutical substances are enclosed in nanocarriers to enable functions including penetrating hair follicles and interacting with skin's lipid to transport.

As possible therapeutic agents for the treatment of allergic airway disorders, NMs have drawn a lot of attention in recent years. Chronic airway inflammation brought on by allergen exposure is a hallmark of allergic airway disorders like asthma and allergic rhinitis. Traditional therapies frequently use corticosteroids and bronchodilators, which can have negative side effects if used for an extended period of time.

Colloidal particles known as NMs have a hydrophilic outer surface and a hydrophobic inner core. Their size ranges from 10 to 100 nanometers. Due to their special shape, they may transport hydrophobic medications like corticosteroids or anti-inflammatory medicines, which are frequently used to treat allergic airway illnesses.

The ability of micro micelles to penetrate deeply into the tissues of the airways improves drug delivery to the target site and therapeutic effectiveness. Additionally, they can shield the drug from metabolizing and degrading, delaying drug release and extending the therapeutic impact.

Targeting particular cell types or receptors involved in the inflammatory response is one of the key benefits of utilizing NMs for the therapy of allergic airway illness.

The micro micelles' surfaces can be altered to make it possible for them to selectively attach to receptors expressed on immune cells, including mast cells and eosinophils, which are key players in allergic reactions. By increasing drug concentration at the site of inflammation, this tailored delivery decreases systemic side effects and enhances overall therapy outcomes.

Furthermore, micro micelles have the potential to overcome some of the limitations of traditional drug delivery systems. For example, by encapsulating hydrophobic pharmaceuticals, they can help overcome poor solubility concerns, allowing for the use of a broader range of therapeutic treatments. Furthermore, micro micelles can avoid mucociliary clearance in the airways, which reduces the bioavailability of medications administered by traditional routes.

It is vital to remember that the unique application, therapeutic payload, intended release profile, and biocompatibility all influence the choice of micelle-forming chemical. The shape and properties of NMs are frequently tailored by researchers to maximize their performance for targeted drug delivery and therapeutic outcomes.

7.10 Management of Allergic Airway Disease

The management of allergic airway diseases, such as asthma and allergic rhinitis, using NMs is an evolving field of research with the potential to offer targeted and effective therapeutic approaches (Javadzadeh and Bahari 2017). While there is ongoing research, here is how NMs could potentially be utilized for the management of allergic airway diseases:

1. Targeted Drug Delivery: NMs can be engineered to deliver anti-inflammatory drugs, bronchodilators, or immunomodulatory agents directly to the inflamed airway tissues. Functionalization of the micelles' surface with ligands that bind to specific receptors overexpressed in the inflamed tissues enables precise targeting and reduces systemic side effects.
2. Encapsulation of Therapeutic Agents: Hydrophobic drugs, which are often used to manage allergic airway diseases, can be encapsulated within the hydrophobic core of NMs. This enhances drug solubility and stability, ensuring efficient delivery to the target site.
3. Prolonged Drug Release: NMs offer controlled and sustained drug release. This property can maintain therapeutic drug levels in the airways over an extended period, reducing the frequency of dosing and providing consistent relief.
4. Combination Therapy: NMs allow for the encapsulation of multiple therapeutic agents in a single system. This enables combination therapy, where drugs with different mechanisms of action can be administered together to achieve enhanced therapeutic outcomes.
5. Reduced Systemic Side Effects: By targeting the drug delivery specifically to the inflamed airways, NMs can minimize systemic exposure to the drugs, thereby reducing the risk of systemic side effects associated with long-term medication use.

6. Immunomodulation: NMs can carry immunomodulatory agents that help regulate the immune response underlying allergic airway diseases. This can lead to a balanced immune reaction and reduced inflammation.
7. Mucus Penetration: NMs can potentially penetrate mucus barriers in the airways, ensuring efficient drug delivery to the target cells beneath the mucus layer.
8. Imaging and Monitoring: Some NMs formulations can incorporate imaging agents, allowing healthcare providers to monitor drug distribution and evaluate treatment effectiveness.
9. Enhanced Drug Stability: Encapsulation of drugs within the hydrophobic core of NMs can protect them from degradation and inactivation, leading to improved drug stability during storage and administration.
10. Minimized First-Pass Metabolism: NMs can potentially bypass first-pass metabolism in the liver, allowing a higher fraction of the administered drug to reach the target airway tissues.
11. Prevention of Drug Resistance: Controlled and sustained drug release from NMs might help prevent the development of drug resistance often associated with frequent or high-dose drug administration.
12. Localized Action: NMs provide the opportunity for localized drug action within the airways, which can be particularly beneficial for reducing inflammation and bronchoconstriction directly where they occur.
13. Personalized Treatment: The flexibility of NMs formulations may allow for personalized treatment strategies, tailoring drug combinations and dosages based on individual patient profiles.
14. Theranostic Applications: Some NMs formulations can incorporate imaging agents that allow for both therapeutic delivery and real-time monitoring of drug distribution and response, enabling a theranostic approach.
15. Inhalable Formulations: NMs-based therapies can potentially be formulated for inhalation, providing a noninvasive route of administration directly to the respiratory system.
16. Local Immune Modulation: NMs carrying immunomodulatory agents can directly interact with immune cells in the airways, promoting a balanced immune response and potentially reducing allergic inflammation.
17. Reduced Dosage Frequency: The sustained release properties of NMs could potentially lead to less frequent dosing compared to conventional treatments, improving patient compliance.
18. Overcoming Blood-Air Barrier: NMs can be designed to overcome the barriers posed by the respiratory epithelium and reach deeper layers of the airway, ensuring effective drug delivery.
19. Pediatric and Geriatric Applications: NMs might offer a more adaptable and patient-friendly approach for children and the elderly, groups that can face challenges with conventional drug delivery methods.
20. Biodegradability: Many NMs formulations are designed with biodegradable materials, potentially reducing long-term environmental impact.

21. Patient Compliance: The potential for reduced dosing frequency due to sustained drug release can improve patient compliance, as patients may need to take medications less often.
22. Challenges and Future Directions: While the concept of using NMs for allergic airway disease management is promising, there are challenges to overcome. These include optimizing the formulation for stability, achieving efficient mucus penetration, ensuring biocompatibility and safety, and conducting thorough preclinical and clinical studies to demonstrate efficacy and safety (Bose et al. 2021).

7.11 Conclusions

NMs for the management of allergic airway diseases have been summed up in this chapter. We have also outlined the sources, types, properties of NMs, NMs-mediated drug loading and release, and management of allergic airway disease.

It is important to note that while the concept of using NMs for allergic airway disease management holds promise, there are challenges to address. These include optimizing the formulation to ensure stability, developing effective targeting strategies, verifying safety, and conducting rigorous preclinical and clinical studies to demonstrate efficacy. Regulatory approvals from health authorities are also essential before such treatments can be used in clinical practice.

References

Ahn J et al (2015) Antibody fragment-conjugated polymeric micelles incorporating platinum drugs for targeted therapy of pancreatic cancer. Biomaterials 39:23–30

Alvarez-Lorenzo C et al (2009) Light-sensitive intelligent drug delivery systems. Photochem Photobiol 85(4):848–860

Bae Y, Kataoka K (2009) Intelligent polymeric micelles from functional poly (ethylene glycol)-poly (amino acid) block copolymers. Adv Drug Deliv Rev 61(10):768–784

Bae Y et al (2005) Preparation and biological characterization of polymeric micelle drug carriers with intracellular pH-triggered drug release property: tumour permeability, controlled subcellular drug distribution, and enhanced in vivo antitumor efficacy. Bioconjug Chem 16(1):122–130

Bose A, Roy Burman D, Sikdar B, Patra P (2021) Nanomicelles: types, properties and applications in drug delivery. IET Nanobiotechnol 15:19–27

Bromberg L et al (2008) Polymeric micelles in oral chemotherapy. J Control Release 128(2):99–112

Chen H et al (2011) Nanonization strategies for poorly water-soluble drugs. Drug Discov Today 16(7–8):354–360

Cholkar K et al (2014) Optimization of dexamethasone mixed nanomicellar formulation. AAPS PharmSciTech 15(6):1454–1467

Cholkar K et al (2015) Nanomicellar topical aqueous drop formulation of rapamycin for back-of-the-eye delivery. AAPS Pharm Sci Tech 16(3):610–622

Civiale C et al (2009) Polyhydroxyethylaspartamide-based micelles for ocular drug delivery. Int J Pharm 378(1–2):177–186

Crater JS et al (2010) Barrier properties of gastrointestinal mucus to nanoparticles transport. Macromol Biosci 10(12):1473–1483

Fireman S et al (2011) A look at emerging delivery systems for topical drug products. Dermatol Ther 24(5):477–488
Gao WP et al (2006) Controlling vesicle formation via interpolymer hydrogen-bonding complexation between poly (ethylene oxide)-block polybutadiene and poly (acrylic acid) in solution. Macromolecules 39(14):4894–4898
Gong J et al (2012) Polymeric micelles drug delivery system in oncology. J Control Release 159(3):312–323
Hsu CH et al (2008) Self-assembly behavior of A-B diblock and C-D random copolymer mixtures in the solution state through mediated hydrogen bonding. Langmuir 24(15):7727–7734
Javadzadeh Y, Bahari LA (2017) Therapeutic nanostructures for dermal and transdermal drug delivery. In: Nano-and Microscale Drug Delivery Systems, pp. 131–146
Kabanov AV, Alakhov VY (2000) Micelles of amphiphilic block copolymers as vehicles for drug delivery. In: Amphiphilic block copolymers, pp, 347–376
Kedar U et al (2010) Advances in polymeric micelles for drug delivery and tumour targeting. Nanomed Nanotechnol Biol Med 6(6):714–729
Kuo S-W et al (2008) Supramolecular micellization of diblock copolymer mixtures mediated by hydrogen bonding for the observation of separated coil and chain aggregation in common solvents. Macromol Rapid Commun 29(3):229–233
Lee ES et al (2003a) Poly (L-histidine)-PEG block copolymer micelles and pH-induced destabilisation. J Control Release 90(3):363–374
Lee ES et al (2003b) Polymeric micelle for tumour pH land folate-mediated targeting. J Control Release 91(1–2):103–113
Lee RW, Shenoy DB, Sheel R (2010) Micellar nanoparticles: applications for topical and passive transdermal drug delivery. In: Handbook of noninvasive drug delivery systems. Elsevier, pp 37–58
Lin HM et al (2010) Light-sensitive intelligent drug delivery systems of coumarin-modified mesoporous bioactive glass. Acta Biomater 6(8):3256–3263
Makhmalzade BS, Chavoshy F (2018) Polymeric micelles as cutaneous drug delivery system in normal skin and dermatological disorders. J Adv Pharm Technol Res 9(1):2
Meng FT et al (2003) W/O/W double emulsion technique using ethyl acetate as organic solvent: effects of its diffusion rate on the characteristics of microparticles. J Control Release 91(3):407–416
Oerlemans C et al (2010) Polymeric micelles in anticancer therapy: targeting, imaging and triggered release. Pharm Res 27(12):2569–2589
Patel A et al (2013) Ocular drug delivery systems: an overview. World J Pharmacol 2(2):47–64
Ponchel G et al (1998) Specific and non-specific bioadhesive particulate systems for oral delivery to the gastrointestinal tract. Adv Drug Deliv Rev 34(2–3):191–219
Rapoport N (2007) Physical stimuli-responsive polymeric micelles for anticancer drug delivery. Prog Polym Sci 32(8–9):962–990
Sammalkorpi M et al (2009) Ionic surfactant aggregates in saline solutions: sodium dodecyl sulphate (SDS) in the presence of excess sodium chloride (NaCl) or calcium chloride (CaCl2). J Phy Chem B 113(17):5863–5870
Sant VP et al (2005) Enhancement of oral bioavailability of poorly water soluble drugs by poly (ethylene glycol)-block-poly (alkyl acrylate-comethacrylic acid) self-assemblies. J Control Release 104(2):289–300
Savić R et al (2006) Block copolymer micelles as delivery vehicles of hydrophobic drugs: micelle-cell interactions. J Drug Target 14(6):343–355
Singh R, Lillard JW Jr (2009) Nanoparticle-based targeted drug delivery. Exp Mol Pathol 86(3):215–223
Smart JD et al (2005) The basics and underlying mechanisms of mucoadhesion. Adv Drug Deliv Rev 57(11):1556–1568
Tang BC et al (2009) Biodegradable polymer nanoparticles that rapidly penetrate the human mucus barrier. Proc Natl Acad Sci USA 106(46):19268–19273

Trinh HM et al (2017) Nanomicelles in diagnosis and drug delivery. In: Emerging nanotechnologies for diagnostics, drug delivery and medical devices. Elsevier, pp 45–58
Vadlapudi AD, Mitra AK (2013) Nanomicelles: an emerging platform for drug delivery to the eye. Ther Deliv 4(1):1–3
Wang XQ, Zhang Q (2012) pH-sensitive polymeric nanoparticles to improve oral bioavailability of peptide/protein drugs and poorly water-soluble drugs. Eur J Pharm Biopharm 82(2):219–229
Wu XL et al (2010) Tumour-targeting peptide conjugated pH-responsive micelles as a potential drug carrier for cancer therapy. Bioconjug Chem 21(2):208–213
Xie J et al (2005) Self-assembled biodegradable nanoparticles developed by direct dialysis for the delivery of paclitaxel. Pharma Res 22(12):2079–2090
Xu W et al (2013) Polymeric micelles, a promising drug delivery system to enhance bioavailability of poorly water-soluble drugs. J Drug Deliv 15:340315
Yokoyama M et al (1992) Block copolymers as drug carriers. Crit Rev Ther Drug Carrier Syst 9(3–4):213–248
Yoncheva K et al (2012) Stabilized micelles as delivery vehicles for paclitaxel. Int J Pharm 436(1–2):253–264

Carbon Nanotubes (CNTs): A Novel Therapeutic Approach in the Management of Allergic Airway Diseases

8

Pooja Tiwary, Krishil Oswal, and Dileep Kumar

Abstract

Globally, allergic airway diseases caused immense threats with increased mortality rates. Current pharmacological agents have resulted in poor patient survival due to decreased bioavailability. Nanoparticle discoveries improved the therapeutic efficacy allowing for extensive functionalization. Carbon nanotubes hold promising effects for bringing revolution worldwide in drug delivery systems. They offer the potential for improved therapeutic efficacy while at the same time, they are also associated with potential risks. Nanotechnology provides significant benefits including therapy for respiratory diseases such as asthma, COPD, and pulmonary fibrosis. However, there is large evidence showing that carbon nanotubes exacerbate lung disease in rodents with pre-existing lung disease if used in drug delivery and are known to cause adverse effects such as pulmonary fibrosis. CNTs have also opened the gateway for the detection, delivery, and treatment of a multitude of pulmonary diseases. This chapter is based on how we can overcome the hurdles for growing pathologic events and improve therapeutic efficacy. It will also describe the interaction of nanomaterials in the biological system that will be essential in overcoming toxicity thereby allowing the safe use of carbon nanotubes in the management of allergic airway disease.

Keywords

Asthma · CNT · Nanotechnology · Nanomaterials · Respiratory · Toxicity

P. Tiwary · K. Oswal · D. Kumar (✉)
Poona College of Pharmacy, Bharati Vidyapeeth (Deemed to be) University, Pune, Maharashtra, India

P. Prasher et al. (eds.), *Understanding Allergic Airway Diseases*,
https://doi.org/10.1007/978-981-97-1953-2_8

8.1 Introduction

Allergic diseases are chronic inflammatory conditions caused by the body's abnormal immune response to foreign particles, known as allergens (León 2023). This response involves both innate and adaptive immunity and can result in a variety of symptoms. While immunoglobulin E (IgE), CD4+ T cells, macrophages, epithelial cells, and natural killer cells are all involved in the inflammatory process, many other factors contribute to the development of allergic airway disease (Hough and Deshane 2019). These diseases can include asthma, chronic rhinosinusitis, and allergic rhinitis, which are commonly found in children. Extensive research has been conducted to understand the underlying causes of these conditions.

The incidence of allergic diseases has shown substantial rise in recent years. It is pertinent to mention that allergic rhinitis (AR) is a prevalent condition in children, with percentage ranging from 25% to 40% (Nur Husna et al. 2022). Global urbanization is the primary cause of the prevalence of AR, which can be attributed to various risk factors. This is due to the alarming increase in particulate matter and pollutants that further exacerbate AR symptoms. Based on research, it is evident that smoking does not greatly impact the severity of nasal symptoms. However, chronic rhinitis is certainly affected by it. In addition, pediatric allergic rhinitis is more significantly influenced by maternal smoking. It is important to note that the use of e-cigarettes and heated tobacco products increases the risk of AR in adolescents (Chung et al. 2020). The treatment guidelines for AR currently mandate the use of inhaled corticosteroids like mometasone furoate (for those over 3 years old) and triamcinolone acetonide (for those over 4 years old), alongside leukotriene receptor antagonists (such as montelukast and zafirlukast) to prevent allergic inflammation and alleviate symptoms like mucous production and nasal congestion (Peters-Golden et al. 2006; Zhang et al. 2022). There is controversial debate surrounding the combination of H1 antagonists with first-line medications of AR. Despite meta-analysis studies, the results continue to conflict with some indicating better relief for nighttime symptoms while others suggesting greater effectiveness for daytime symptoms (Feng et al. 2021; Liu et al. 2018; Seresirikachorn et al. 2019).

According to the National Asthma and Education and Prevention Program Expert Panel Report 3 (NAEPPR3), asthma is an enduring inflammatory ailment that targets the airways of individuals who smoke and suffer from occupational allergies (National Asthma Education and Prevention Program 2007). It must be noted that those with these conditions undergo a significant surge in mast cells, T lymphocytes, macrophages, and epithelial cells, leading to inflammation. This inflammation manifests itself clinically as repeated episodes of coughing, sneezing, breathlessness, and dyspnea in those who are susceptible (Stern et al. 2020). The treatment of asthmatic patients usually involves corticosteroids and approved monoclonal antibodies such as omalizumab and benralizumab. It is important to note, however, that these medications can lead to immunosuppression, which can significantly increase the risk of infection in patients.

The drug delivery nanotechnology platform is comprised of diverse nanostructures, such as dendrimers, polymeric nanoparticles, fullerenes, and carbon nanotubes, all of which possess distinctive properties (De Jong and Borm 2008). It is crucial to acknowledge that using carbon nanotubes (CNTs) could potentially enhance the efficacy of treatments for respiratory conditions, including asthma, COPD, allergic rhinitis, and pulmonary fibrosis. However, it is equally important to recognize the potential hazards associated with their application (Card et al. 2008). It possesses a tube or fiber-like structure that makes it an invaluable tool for drug delivery. Their unique configuration enables significant functionalization, which can offer tremendous therapeutic benefits (Bonner 2011). Altering them can effectively enhance the pharmacokinetic profile of drugs, leading to reduced dosage requirements for treatment. Modified CNTs are particularly advantageous in delivering drugs to affected tissues and monitoring their accumulation and clearance through development of imaging molecules. It is important to recognize that single-walled CNTs (SWCNTs) have much greater ability to be absorbed by the body and remain in tissues after being inhaled through an aerosol. Their ability to be functionalized in numerous ways and transport various agents, such as drugs, imaging agents, and targeting agents, is also noteworthy. Furthermore, their diverse physical and chemical properties enable both covalent and noncovalent functionalization, making them highly adaptable for a wide range of applications (Prato et al. 2008). While carbon nanotubes have been found to have valuable properties for drug delivery, it's important to also consider the potential risks. Studies conducted on rodents and cell cultures have shown that human exposure can exacerbate lung conditions, elicit immune reactions, and result in pulmonary fibrosis. These findings emphasize the importance of carefully weighing the benefits and drawbacks of utilizing CNTs for drug delivery. This chapter provides comprehensive analysis of the advantages and drawbacks associated with incorporating CNTs into the management of respiratory conditions. Additionally, we will assess and tackle the significant challenges that must be surmounted to ensure their safe and reliable implementation in medical care.

8.2 Risk vs. Benefit of CNT in Allergic Airway Disease

CNTs are widely used in many industries because of their exceptional electronic, mechanical, and optical properties. However, their effect on human health and well-being is still being studied, particularly their impact on pulmonary function. Various in vitro and in vivo studies have shown that it could lead to adverse effects on pulmonary toxicity, fibrosis, and inflammation. This is a significant concern since they hold great potential for producing innovative products, and human exposure through inhalation and dermal contact during handling and manufacturing is a major contributor to dermal and pulmonary contraction (Luanpitpong et al. 2014).

Additionally, they are strongly associated with allergic asthma and can cause inflammation in the respiratory tract, which exacerbates asthma and causes oxidative stress. This further leads to the release of pro-inflammatory mediators, causing increased inflammation and disruption of the airway epithelial barrier. As a result, irritants and allergens can penetrate the tracheal tubes, leading to increased inflammation and airway hyperresponsiveness (Deng et al. 2023). Animal models have shown that exposure to CNTs can aggravate symptoms and airway inflammation.

Studies have investigated the impact of SWCNTs with various impurities in metal on its toxicity. The intratracheal route of administration to the mice lungs at a dose concentration of 0.1–0.5 mg produced granuloma associated with lung inflammation in a dose-dependent manner (Shvedova et al. 2008). Furthermore, a short-term inhalation study consecutively for 4 days for 5 h/day of 5 mg/m^3 in mice disclosed chronic lung inflammation, subsequent buildup of granulomas, and persistent interstitial fibrosis (Shvedova et al. 2008).

Studies showed the activity of protease-activated receptor (PAR 2) in the exacerbation of house dust mite (HDM)-induced lung disease by multi-walled CNTs (MWCNTs). This reported the pro-fibrotic and pro-inflammatory effects; however, the complete mechanism is not fully understood. The result shows that PAR 2 mediates airway fibrosis and is not involved in eosinophilic lung inflammation (Lee et al. 2023). The study used wild-type (WT) male C57BL6 mice and PAR 2 KO mice and investigated the exposure of HDM and MWCNT which resulted in increased bronchioalveolar lavage fluid (BALF) total proteins and LDL levels in both mice. However, the WT mice displayed more airway fibrosis than the PAR 2 KO as the pro-fibrotic mediator AIG 1 was downregulated but the pro-inflammatory cytokines were increased. These findings suggest PAR 2 could be used as a potential therapeutic target while CNT as a tool of management in allergic airway diseases. Treating the disease by targeting PAR 2 could potentially reduce the risk of airway fibrosis which is a common complication associated with MWCNT.

CNTs show potential for drug delivery, but researchers must analyze their effects on multiple organs before they can be safely used for medical purposes (Maruyama et al. 2015). Studies have proposed several mechanisms for cellular toxicity induced by CNTs, with endocytosis being the major one (Jain et al. 2012; Witzmann and Monteiro-Riviere 2017). CNTs can penetrate the cellular lipid bilayer, leading to the generation of reactive oxygen species (ROS) and causing cellular toxicity (Wani et al. 2011). This excess free radical generation can cause the oxidation of DNA, lipids, and proteins in the cells, leading to inflammation (Wani et al. 2011). Inhalation via the respiratory channel is the most commonly reported mode for CNT delivery (Mohanta et al. 2019), and physicochemical properties play a crucial role in pulmonary toxicity after inhalation. SWCNTs, in particular, get deposited in the intratracheal lining of the lungs, leading to transient inflammation and cytotoxicity (Polimeni et al. 2015; Warheit et al. 2004). When functionalized carboxyl SWCNTs were injected intravenously (IV), they produced pulmonary embolization and granuloma over a period of 90 days (Qin et al. 2017). Long-term IV injection of CNTs can lead to the accumulation of CNTs in the respiratory tract, leading to the formation of tumors. MWCNTs have also shown toxicity in the lungs, resulting in granuloma formation (Nerl et al. 2011).

Despite the potential toxicities, studies have shown that nano-based medicine holds great potential in treating respiratory diseases such as asthma, COPD, and lung infections (Yazdi et al. 2023). Compared to conventional drug treatments, it has proven to be more effective due to its high surface-to-volume ratio. This enables it to attach multiple ligands to its surface and provide targeted therapy (Adel et al. 2022; Taghavizadeh Yazdi et al. 2021). Due to its increased chemico-biological stability, it can deliver drugs in hydrophobic and hydrophilic forms through inhalation (Deng et al. 2021; Luo et al. 2021). Furthermore, aerosol drug delivery is one of the most effective methods for treating pulmonary disorders as the deposition spot in the respiratory tract is determined by physicochemical properties (Bianco et al. 2021; Vartak et al. 2021; Yazdi et al. 2023). Nanomedicine, such as CNTs, has been found to be particularly effective as it can be functionalized for targeted therapy and designed to target specific cells and tissues in the body (Mousavi-Kouhi et al. 2021). With all these advantages, it's no surprise that nano-based medicine is a promising avenue for treating respiratory diseases.

Moreover, it's important to note that carbon nanomaterials possess increased biocompatibility, which means they cause less harm and are less likely to be rejected by the immune system. However, the potential risks and benefits of using these materials are still highly controversial. It's worth mentioning that macroscopic carbon materials lack a suitable bandgap, which makes it difficult for them to act as efficient fluorescent materials and therefore unsuitable for the treatment of respiratory diseases (Liu et al. 2020). Notwithstanding these limitations, interdisciplinary fields are utilizing new carbon-based nanomaterials for medical and pharmacological applications. Diverse carbon-based nanomaterials are being developed with substantial properties to overcome the limitations and toxicity of conventional CNTs, allowing them to be used in diagnostic applications (Chauhan et al. 2020). Carbon nanotubes and carbon nanodiamonds, for instance, have been functionalized to improve their biocompatibility, making them more suitable for biomedical applications (Chauhan et al. 2020; Maiti et al. 2019). The functionalized CNTs possess the ability to deliver drug molecules into living cells. Their natural shape allows them to penetrate biological membranes without causing damage (Panczyk et al. 2016). Studies have extensively researched the use of polymeric and lipid nanomedicine for the establishment of carbon nanotubes (CNTs) in the lung delivery system (Chan et al. 2023). Clinical trials have recommended the use of an inhaler with a developed delivery system to overcome toxicities and provide targeted therapy with fewer side effects and increased effectiveness (Forest and Pourchez 2022). Other nanoparticles, such as mesoporous silica and solid lipid nanoparticles, can also be functionalized with CNTs to create inhaled powders for respiratory drug delivery (Chan et al. 2023). The spray-drying method can be used with thermolabile materials to improve mucociliary clearance and transport the drug to the respiratory epithelium (Pramanik et al. 2021; Sham et al. 2004). However, further research is necessary to understand the drug transport mechanism and potential side effects.

Nanomedicines have demonstrated anti-inflammatory effects in conditions such as asthma that is characterized by chronic airway inflammation, reversible airway obstruction, and bronchial hyperresponsiveness (Yang et al. 2019). In vitro and in vivo studies have reported the potential of nanomedicines as a treatment option for this respiratory disorder. The poly(lactic acid) nanoparticle, for instance, has the ability to encapsulate both hydrophilic and hydrophobic drugs. It can load two potential drugs, theophylline and budesonide, using a process called double emulsion solvent diffusion. The drug release was observed with the help of Franz-type diffusion cells, and the airway epithelial cell line used was 16HBE14o-. The result was a percentage fraction of 75% and 48% for theophylline and budesonide, respectively, indicating the effectiveness of the drug-loaded nanoparticle in delivering the drugs to the respiratory tract (Buhecha et al. 2019). Research on rodents has demonstrated that CNTs can worsen allergic airway inflammation when combined with allergens like ovalbumin. The combination of ovalbumin allergen and multi-walled carbon nanotubes (MWCNTs) has been found to increase airway fibrosis in mice, indicating that CNTs pose a significant threat to people with allergic lung inflammatory diseases such as asthma.

Liposomal nanotubes have been found to be an effective measure for anti-asthmatics in in vivo studies, with the ability to prolong the effect compared to solutions without liposomes (Hussain et al. 2011). In addition, the use of quantum dots (QDs) in HDM-induced allergic airway disease has been proven to be efficacious in improving cytokine production and regulating the immune response in respiratory diseases (Scoville et al. 2019). Furthermore, study on a murine model of diisocyanate-induced asthma found that a low dose of nanomedicine, including gold nanoparticles and TiO2, was effective in improving pulmonary function (Hussain et al. 2011). This strongly suggests that nano-based medicine can be utilized to enhance pulmonary function.

It is imperative to thoroughly evaluate the potential risks and benefits before implementing nanotechnology in the treatment of pulmonary disease. The recommended dose of carbon nanotubes for respiratory illnesses is 0.4–20 μg/ml for a duration of 24 h, but despite the numerous studies in favor of nanomedicine for allergic airway disease, evidence of its toxicity cannot be ignored. To establish the use of CNTs as a treatment option, multiple in vitro and in vivo studies as well as clinical trials are necessary. It is crucial to accurately determine the duration of CNTs to optimize their therapeutic potential. Thus, conducting pharmacokinetic and pharmacodynamic studies of CNTs in respiratory system disorders is essential. Such studies will aid in determining the dosing regimen and administration of CNTs.

8.3 Conclusion

This chapter provides a comprehensive analysis of the impact of CNTs, covering their toxicity, efficacy, drug safety, and drug delivery based on both in vitro and in vivo studies. Although CNTs have shown promising results in reducing the number of inflammatory cells and markers and improving oxidative stress and pulmonary function in allergic asthma and other respiratory conditions, certain studies have highlighted their toxicity. Therefore, it is crucial to determine the most effective doses for specific respiratory disorders to ensure maximum therapeutic benefits and minimal side effects. A thorough evaluation of CNTs is essential in determining their efficacy and safety in treating allergic airway diseases, which can only be achieved through well-designed clinical trials. While nanomedicine has proven to be more effective due to its physicochemical properties, more clinical and experimental data is necessary to evaluate the effect of CNTs. To ensure safe drug delivery, it is necessary to make CNTs more biodegradable, as recent studies have shown that enzymes like human neutrophil myeloperoxidase or horseradish peroxidase can break down CNTs and reduce their toxic effects. However, the development of safe nanotechnology-based therapies, including those based on CNTs, remains a challenge due to the lack of regulatory guidelines and safety testing protocols specific to nanotechnology. Nevertheless, with more experience and better data, effective nanotechnology-based therapies for respiratory diseases are expected to be developed. It is important to note that public knowledge regarding the benefits of CNTs for respiratory disorders should only be encouraged after strong evidence of their advantages over the risks is obtained. Systematic meta-analyses must be conducted to assess the results of research on CNTs, so they can be applied for clinical use.

Acknowledgments Not applicable.

Compliance with Ethical Standards

Disclosure of Potential Conflict of Interest Authors declare no conflict of interest.

Research Involving Human Participants/Animals Not applicable.

Informed Consent Not applicable.

References

Adel G, Javad M, Amiri MS, Mohammad M, Alireza H, Aliakbar H, Fahimeh N, Mohabat N, Mohsen Q, Yazdi MET (2022) Resveratrol-mediated gold-nanoceria synthesis as green nanomedicine for phytotherapy of hepatocellular carcinoma. Front Biosci 27(8):1–6

Bianco F, Salomone F, Milesi I, Murgia X, Bonelli S, Pasini E, Dellacà R, Ventura ML, Pillow J (2021) Aerosol drug delivery to spontaneously-breathing preterm neonates: lessons learned. Respir Res 22:1–31

Bonner JC (2011) Carbon nanotubes as delivery systems for respiratory disease: do the dangers outweigh the potential benefits? Expert Rev Respir Med 5(6):779–787

Buhecha MD, Lansley AB, Somavarapu S, Pannala AS (2019) Development and characterization of PLA nanoparticles for pulmonary drug delivery: co-encapsulation of theophylline and budesonide, a hydrophilic and lipophilic drug. J Drug Deliv Sci Technol 53:101128

Card JW, Zeldin DC, Bonner JC, Nestmann ER (2008) Pulmonary applications and toxicity of engineered nanoparticles. Am J Phys Lung Cell Mol Phys 295(3):L400–L411

Chan HW, Chow S, Zhang X, Zhao Y, Tong HHY, Chow SF (2023) Inhalable nanoparticle-based dry powder formulations for respiratory diseases: challenges and strategies for translational research. AAPS Pharm Sci Tech 24(4):98

Chauhan S, Jain N, Nagaich U (2020) Nanodiamonds with powerful ability for drug delivery and biomedical applications: recent updates on in vivo study and patents. J Pharm Anal 10(1):1–12

Chung SJ, Kim B, Oh JH, Shim J, Chang Y, Cho S, Yang M (2020) Novel tobacco products including electronic cigarette and heated tobacco products increase risk of allergic rhinitis and asthma in adolescents: analysis of Korean youth survey. Allergy 75(7):1640–1648

De Jong WH, Borm PJA (2008) Drug delivery and nanoparticles: applications and hazards. Int J Nanomedicine 3(2):133–149

Deng Z, Kalin GT, Shi D, Kalinichenko VV (2021) Nanoparticle delivery systems with cell-specific targeting for pulmonary diseases. Am J Respir Cell Mol Biol 64(3):292–307

Deng R, Zhu Y, Wu X, Wang M (2023) Toxicity and mechanisms of engineered nanoparticles in animals with established allergic asthma. Int J Nanomedicine Volume 18:3489–3508

Feng Y, Meng Y-P, Dong Y-Y, Qiu C-Y, Cheng L (2021) Management of allergic rhinitis with leukotriene receptor antagonists versus selective H1-antihistamines: a meta-analysis of current evidence. Allergy, Asthma Clin Immunol 17(1):1–12

Forest V, Pourchez J (2022) Nano-delivery to the lung-by inhalation or other routes and why nano when micro is largely sufficient? Adv Drug Deliv Rev 183:114173

Hough KP, Deshane JS (2019) Exosomes in allergic airway diseases. Curr Allergy Asthma Rep 19:1–8

Hussain S, Vanoirbeek JAJ, Luyts K, De Vooght V, Verbeken E, Thomassen LCJ, Martens JA, Dinsdale D, Boland S, Marano F (2011) Lung exposure to nanoparticles modulates an asthmatic response in a mouse model. Eur Respir J 37(2):299–309

Jain S, Singh SR, Pillai S (2012) Toxicity issues related to biomedical applications of carbon nanotubes. J Nanomed Nanotechol 3(140):2

Lee HY, You DJ, Taylor-Just A, Tisch LJ, Bartone RD, Atkins HM, Ralph LM, Antoniak S, Bonner JC (2023) Role of the protease-activated receptor-2 (PAR2) in the exacerbation of house dust mite-induced murine allergic lung disease by multi-walled carbon nanotubes. Part Fibre Toxicol 20(1):32

León B (2023) Understanding the development of Th2 cell-driven allergic airway disease in early life. Front Allergy 3:1080153

Liu G, Zhou X, Chen J, Liu F (2018) Oral antihistamines alone vs in combination with leukotriene receptor antagonists for allergic rhinitis: a meta-analysis. Otolaryngol–Head Neck Surg 158(3):450–458

Liu J, Li R, Yang B (2020) Carbon dots: a new type of carbon-based nanomaterial with wide applications. ACS Cent Sci 6(12):2179–2195

Luanpitpong S, Wang L, Rojanasakul Y (2014) The effects of carbon nanotubes on lung and dermal cellular behaviors. Nanomedicine 9(6):895–912

Luo M-X, Hua S, Shang Q-Y (2021) Application of nanotechnology in drug delivery systems for respiratory diseases. Mol Med Rep 23(5):1–17

Maiti D, Tong X, Mou X, Yang K (2019) Carbon-based nanomaterials for biomedical applications: a recent study. Front Pharmacol 9:1401

Maruyama K, Haniu H, Saito N, Matsuda Y, Tsukahara T, Kobayashi S, Tanaka M, Aoki K, Takanashi S, Okamoto M (2015) Endocytosis of multiwalled carbon nanotubes in bronchial epithelial and mesothelial cells. BioMed Res Int 2015:793186

Mohanta D, Patnaik S, Sood S, Das N (2019) Carbon nanotubes: evaluation of toxicity at biointerfaces. J Pharm Anal 9(5):293–300

Mousavi-Kouhi SM, Beyk-Khormizi A, Amiri MS, Mashreghi M, Yazdi MET (2021) Silver-zinc oxide nanocomposite: from synthesis to antimicrobial and anticancer properties. Ceram Int 47(15):21490–21497

National Asthma Education and Prevention Program (2007) Expert Panel Report 3 (EPR-3): guidelines for the diagnosis and management of asthma-summary report 2007. J Allergy Clin Immunol 120(5 Suppl):S94–S138

Nerl HC, Cheng C, Goode AE, Bergin SD, Lich B, Gass M, Porter AE (2011) Imaging methods for determining uptake and toxicity of carbon nanotubes in vitro and in vivo. Nanomedicine 6(5):849–865

Nur Husna SM, Tan H-TT, Md Shukri N, Mohd Ashari NS, Wong KK (2022) Allergic rhinitis: a clinical and pathophysiological overview. Front Med 9:874114

Panczyk T, Wolski P, Lajtar L (2016) Coadsorption of doxorubicin and selected dyes on carbon nanotubes. Theoretical investigation of potential application as a pH-controlled drug delivery system. Langmuir 32(19):4719–4728

Peters-Golden M, Gleason MM, Togias A (2006) Cysteinyl leukotrienes: multi-functional mediators in allergic rhinitis. Clin Exp Allergy 36(6):689–703

Polimeni M, Gulino GR, Gazzano E, Kopecka J, Marucco A, Fenoglio I, Cesano F, Campagnolo L, Magrini A, Pietroiusti A (2015) Multi-walled carbon nanotubes directly induce epithelial-mesenchymal transition in human bronchial epithelial cells via the TGF-β-mediated Akt/GSK-3β/SNAIL-1 signalling pathway. Part Fibre Toxicol 13:1–19

Pramanik S, Mohanto S, Manne R, Rajendran RR, Deepak A, Edapully SJ, Patil T, Katari O (2021) Nanoparticle-based drug delivery system: the magic bullet for the treatment of chronic pulmonary diseases. Mol Pharm 18(10):3671–3718

Prato M, Kostarelos K, Bianco A (2008) Functionalized carbon nanotubes in drug design and discovery. Acc Chem Res 41(1):60–68

Qin Y, Li S, Zhao G, Fu X, Xie X, Huang Y, Cheng X, Wei J, Liu H, Lai Z (2017) Long-term intravenous administration of carboxylated single-walled carbon nanotubes induces persistent accumulation in the lungs and pulmonary fibrosis via the nuclear factor-kappa B pathway. Int J Nanomedicine 12:263–277

Scoville DK, Nolin JD, Ogden HL, An D, Afsharinejad Z, Johnson BW, Bammler TK, Gao X, Frevert CW, Altemeier WA (2019) Quantum dots and mouse strain influence house dust mite-induced allergic airway disease. Toxicol Appl Pharmacol 368:55–62

Seresirikachorn K, Chitsuthipakorn W, Kanjanawasee D, Khattiyawittayakun L, Snidvongs K (2019) Leukotriene receptor antagonist addition to H1-antihistamine is effective for treating allergic rhinitis: a systematic review and meta-analysis. Am J Rhinol Allergy 33(5):591–600

Sham JO-H, Zhang Y, Finlay WH, Roa WH, Löbenberg R (2004) Formulation and characterization of spray-dried powders containing nanoparticles for aerosol delivery to the lung. Int J Pharm 269(2):457–467

Shvedova AA, Kisin E, Murray AR, Johnson VJ, Gorelik O, Arepalli S, Hubbs AF, Mercer RR, Keohavong P, Sussman N (2008) Inhalation vs. aspiration of single-walled carbon nanotubes in C57BL/6 mice: inflammation, fibrosis, oxidative stress, and mutagenesis. Am J Physiol Lung Cell Mol Physiol 295(4):L552–L565

Stern J, Pier J, Litonjua AA (2020) Asthma epidemiology and risk factors. Semin Immunopathol 42:5–15

Taghavizadeh Yazdi ME, Nourbakhsh F, Mashreghi M, Mousavi SH (2021) Ultrasound-based synthesis of ZnO Ag2 O3 nanocomposite: characterization and evaluation of its antimicrobial and anticancer properties. Res Chem Intermed 47:1285–1296

Vartak R, Patil SM, Saraswat A, Patki M, Kunda NK, Patel K (2021) Aerosolized nanoliposomal carrier of remdesivir: an effective alternative for COVID-19 treatment in vitro. Nanomedicine 16(14):1187–1202

Wani MY, Hashim MA, Nabi F, Malik MA (2011) Nanotoxicity: dimensional and morphological concerns. Adv Phys Chem 2011:1

Warheit DB, Laurence BR, Reed KL, Roach DH, Reynolds GAM, Webb TR (2004) Comparative pulmonary toxicity assessment of single-wall carbon nanotubes in rats. Toxicol Sci 77(1):117–125

Witzmann FA, Monteiro-Riviere NA (2017) Multi-walled carbon nanotube exposure alters protein expression in human keratinocytes. In: Nanomedicine in Cancer. Jenny Stanford Publishing, pp 461–485

Yang B, Choi H, Kim S-H, Yoon HJ, Lee H (2019) How will nanotechnology lead to better control of asthma? Ann Trans Med 7(20):515

Yazdi MET, Qayoomian M, Beigoli S, Boskabady MH (2023) Recent advances in nanoparticle applications in respiratory disorders: a review. Front Pharmacol 14:1059343

Zhang M, Ni J-Z, Cheng L (2022) Safety of intranasal corticosteroids for allergic rhinitis in children. Expert Opin Drug Saf 21(7):931–938

Liposomes for the Management of Allergic Airway Diseases

9

C. Sarath Chandran, Krishnameera Sajayan, P. K. Anjana,
K. K. Swathy, Alan Raj,
and Amna Mohammed Mohamed AlSahi

Abstract

To treat allergic airway illnesses, lipid-based vesicles called liposomes can be employed as drug delivery vehicles. Asthma, allergic rhinitis, and allergic bronchopulmonary aspergillosis are a few examples of these disorders. For treating allergic obstructive pulmonary illnesses, liposomes provided several benefits, including improved drug stability, prolonged drug release, targeted drug delivery inside the airways, etc. The application of liposomes in allergic airway illnesses has been investigated in a number of research. For the treatment of asthma, liposomal corticosteroids have been studied. These liposomes may reduce the necessary dose and lessen systemic side effects by enhancing the drug's pulmonary deposition and anti-inflammatory effects. Liposomes that contain allergens or immunomodulatory substances have also been investigated as a potential immunotherapy strategy for allergic airway disorders. By promoting immunological tolerance and immune response modulation, these liposomes can lessen allergic inflammation and symptoms. Antibodies or ligands that precisely target receptors or cells involved in allergic airway inflammation can be attached to the surface of liposomes. This method of targeted drug delivery increases the concentration of the drug at the site of inflammation, lowering the dosage required and reducing systemic exposure. To treat allergic airway illnesses, lipo-

C. Sarath Chandran (✉) · K. Sajayan · P. K. Anjana · K. K. Swathy
College of Pharmaceutical Sciences, Govt. Medical College, Kannur, India

A. Raj
Department of Pharmaceutical Biotechnology, Manipal College of Pharmaceutical Sciences, Manipal Academy of Higher Education, Manipal, India
e-mail: alan raj@learner.manipal.edu

A. M. M. AlSahi
Pharmacy Department, Emirates Health Service, Dubai, UAE
e-mail: amna.alsahi@ehs.gov.ae

P. Prasher et al. (eds.), *Understanding Allergic Airway Diseases*,
https://doi.org/10.1007/978-981-97-1953-2_9

somes have been used to deliver anti-inflammatory drugs like curcumin or resveratrol. These liposomal compositions deliver drugs precisely at the inflamed airways enabling localized anti-inflammatory benefits and possibly minimizing systemic side effects. In preclinical trials, these liposomal formulations showed anti-inflammatory efficacy and enhanced lung function. Liposomal formulations may improve the efficacy of already effective treatments and provide new options for the treatment of allergic airway disorders.

Keywords

Liposome · Allergic, Airway disease · Targeted delivery · Pulmonary

9.1 Introduction

Almost six decades ago, Bangham and co-workers discovered vesicular systems known as liposomes (Bangham et al. 1965). Since then, they have garnered extensive attention in the field of drug delivery at the nanoscale. Their excellent biocompatibility, degradability, and capacity for targeted administration as carriers, along with their ease of modification and minimal toxicity, have contributed to their prominence. Liposomes are spherical and contain one or more layers of phospholipids (Matos 2023). Hence, their fundamental structural element is phospholipids. Phospholipid arrangement usually occurs in bilayers, i.e. two layers of phospholipid molecules arranged with their hydrophilic heads facing outwards and the hydrophobic tails oriented inwards, thus forming a lipidic membrane. Generally, liposomes also contain cholesterol molecules, which help to regulate the membrane's fluidity and maintain the lipid bilayer's stability (Fig. 9.1) (Arpita et al. 2020). Liposomes can encapsulate a wide range of drugs, both hydrophilic and hydrophobic.

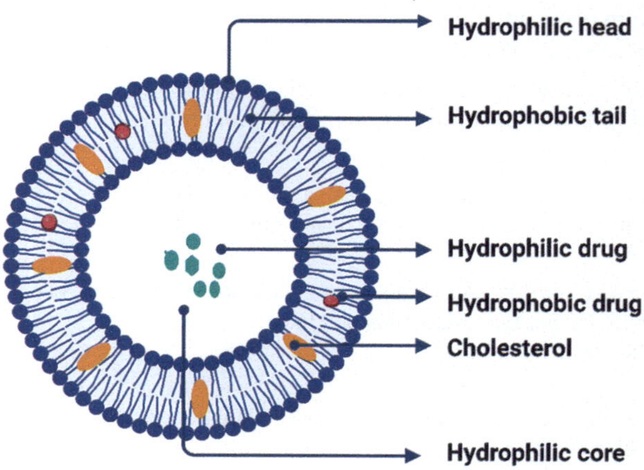

Fig. 9.1 Structure of liposome

Water-soluble drugs are trapped within the aqueous core of liposomes, while fat-soluble drugs are accommodated within the phospholipid bilayer.

The mechanism underlying liposomal drug delivery entails the encapsulation of drugs within the liposome's structure. This encapsulation not only shields the drugs from the harsh biological environment but also enables their controlled release, leading to delayed clearance from circulation (Matos 2023). The primary objective of utilizing liposomes as a drug delivery system is to regulate the drug's release, enhance its activity both in vivo and in vitro, decrease its toxicity, and ultimately improve its therapeutic effectiveness. Liposomes possess adaptable physicochemical and biophysical characteristics, enabling them to address the limitations associated with traditional free-drug therapies (Kim and Jeong 2021). They have been the subject of extensive research and application in various fields, including cancer treatment, immunology, genetic engineering, and the management of infections (Mulla et al. 2019).

9.1.1 Mechanism of Drug Delivery by Liposomes

Liposomes utilize several mechanisms to efficiently deliver drugs. One of the major mechanisms is endocytosis. Through endocytosis, they attach to cell surfaces via receptors, ligands, or antibodies, getting engulfed by the cell membrane. Once inside, liposomes release their drug payload, preventing initial metabolic breakdown. Additionally, their structural similarity to human cell membranes enables content transfer, allowing liposomal phospholipids to interact with cell membrane lipids and transfer the drug substance. Other mechanisms followed by liposomes are fusion and adsorption. Fusion occurs as liposome phospholipids merge with cell membrane phospholipids, forming a single membrane and facilitating drug transfer. Through adsorption, liposomes adhere to the cell membrane, followed by drug release and cellular uptake. These versatile mechanisms safeguard drugs from degradation, enhance bioavailability, and make liposome-based drug delivery systems indispensable tools in pharmaceuticals, improving efficacy and mitigating side effects (Chudasma et al. 2023).

9.1.2 Classification of Liposomes

Liposomes are typically categorized according to their structural characteristics, such as being unilamellar, multilamellar, small-sized, giant-sized, or medium-sized. Additionally, they are classified into different groups based on the technique used for their creation, as well as their specific composition and the purposes they serve (Duong et al. 2023). A detailed classification of liposomes is given in Table 9.1.

Table 9.1 Different classifications of liposomes

Classification of liposomes	Structure based
	Unilamellar vesicles (ULV) Multivesicular vesicles (UVV) Small unilamellar vesicles (SUV) Large unilamellar vesicles (LUV) Medium unilamellar vesicles (MUV) Giant unilamellar vesicles (GUL) Multilamellar vesicles (MLV) Oligolamellar vesicles (OLV)
	Based on the method of preparation
	Frozen and thawed MLV Stable multilamellar vesicles Dehydration and rehydration method MLV made by reverse-phase evaporation SUV/OLV prepared by the reverse-phase evaporation Vesicles prepared by extrusion technique
	Based on application and composition
	Conventional liposome Cationic liposome pH-sensitive liposome Long circulatory (stealth) liposome Fusogenic liposome Immuno-liposome

9.1.3 Preparation of Liposomes

Liposomes are straightforward to prepare and relatively stable. As discussed earlier they consist of a hydrophilic core encased by a single or multiple phospholipid bilayer shells, and the methods used for their formation and fabrication play a crucial role in developing stable formulations (Šturm and Poklar Ulrih 2021). Liposomes can be developed through various methods. Conventional techniques are simple and don't demand intricate equipment, but they have limitations in efficiently encapsulating active pharmaceutical ingredients (APIs) and in terms of scaling up production processes. Conversely, microfluidic approaches offer precise control, high reproducibility, and scalability for industrial production, making them well-suited for encapsulating diverse APIs (Lombardo and Kiselev 2022). Recent advancements in liposome preparation methods have enhanced their predictability and scalability, particularly for industrial-scale production. Some of the techniques that are available for the preparation of liposomes are given in the flow chart below (Fig. 9.2), such as solvent dispersion, mechanical dispersion, and detergent removal methods. Mechanical dispersion encompasses processes like sonication and lipid film hydration. Conversely, the solvent dispersion method includes approaches like reverse-phase evaporation, ethanol/ether injection, and double emulsion (Arpita et al. 2020).

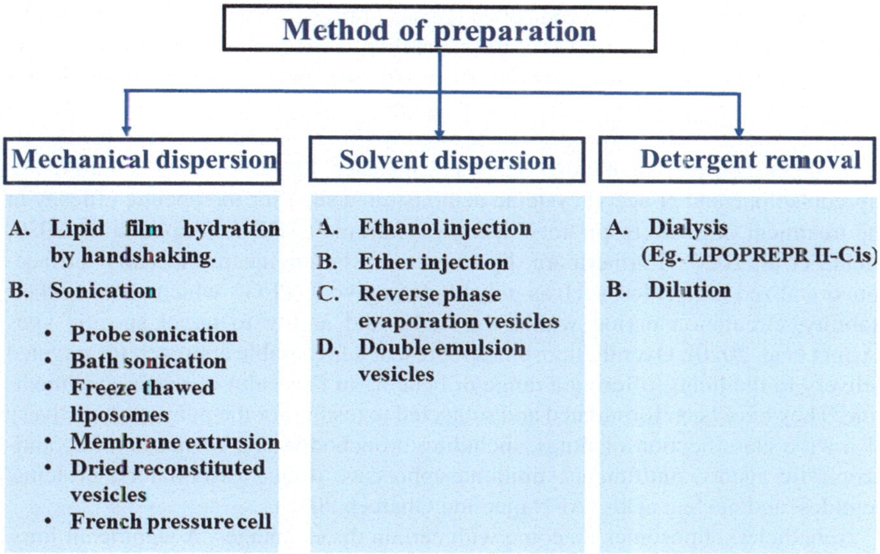

Fig. 9.2 Methods of preparation of liposomes

9.1.4 Characterization of Liposomes

Following the preparation of liposomes, general characterization studies are typically conducted to assess several key properties. These include particle size, morphology, zeta potential, and encapsulation efficiency. To determine these properties, various analytical techniques are employed, for example, dynamic light scattering (DLS) is used for particle size analysis; high-performance liquid chromatography (HPLC) and UV-visible spectroscopy are generally used for drug encapsulation efficiency measurements; etc. (Rahman et al. 2022; Ravindar et al. 2022). These comprehensive analyses help in understanding and optimizing the performance of liposomal drug delivery systems. It is worth noting that studying the morphology of liposomes is crucial because lipids play a major role in the formation and shape of biological vesicular structures. Understanding the morphology of liposomes helps researchers gain insights into the fundamental processes that govern the structure and behaviour of these lipid-based vesicles, which can have implications for aerosol administrations (Hotani 1984).

9.1.5 Liposomes as the Delivery System for the Pulmonary Region: Advantages and Disadvantages

Liposomes offer numerous advantages in the context of the delivery of drugs to the lungs. Their capability to selectively accumulate in lung tissue results in increased effectiveness and reduced adverse effects. Utilizing liposomes through pulmonary administration allows for the controlled release of drugs, enhancing patient compliance. Moreover, liposomes possess the capability to encapsulate drugs with diverse properties and can be customized for active targeting purposes. Their composition closely mimics that of lung surfactants, promoting compatibility with the lung environment. Additionally, liposomes can be employed in combination with other agents to augment their therapeutic outcomes. For instance, liposomes loaded with methylprednisolone and N-acetyl cysteine demonstrated superior therapeutic efficacy in the treatment of acute respiratory distress syndrome (ARDS) (Arpita et al. 2020; Mehta et al. 2020). Furthermore, liposomes can be engineered to carry surface-functionalized molecules, such as polyethylene glycol (PEG), which enhance their stability, circulation period within the body, and ability to target specific sites (Arpita et al. 2020). Overall, liposomes represent a favourable approach for targeted delivery to the lungs, offering a range of benefits in the realm of respiratory medicine. They have been formulated and subjected to testing for the pulmonary delivery of a wide classification of drugs, including bronchodilators, corticosteroids, antineoplastic agents, antifungals, immunosuppressive drugs, vasodilators, proteins/peptides, and nucleic acids (Al-Najjar and Ghareeb 2020).

Nonetheless, liposomes do come with certain disadvantages. A significant limitation is their stability, encompassing both physical and chemical aspects, which can impede their broader adoption. Furthermore, liposomes encounter challenges related to their in vivo aerodynamic behaviour and aerosolization performance, which can influence their efficacy in pulmonary drug delivery. Regulatory hurdles also exist, including the need to establish new chemistry, manufacturing, and control protocols and address upcoming challenges from the Center for Devices and Radiological Health. Additionally, robust toxicological data are scarce, necessitating a thorough examination of pulmonary pharmacokinetics for a more comprehensive understanding (Fig. 9.3) (Antimisiaris et al. 2021; Mehta et al. 2020).

9.1.6 Liposomal Dosage Forms for Pulmonary Delivery

For delivery to the lungs, liposomal products are generally aerosolized using a nebulizer from an aqueous dispersion. The creation of effective liposomal vehicles for inhalation relies on two major factors: the composition of the formulation and the design of the nebulizer. The primary objective is to mitigate the potentially harmful impact of shearing forces on the stability of liposomes while also optimizing their delivery to the deep regions of the lungs (Elhissi 2017). Hence, it's essential to maintain certain characteristics of liposomal formulations, including lamellarity, size of the liposomes, drug encapsulation, and the rate of in vitro drug release, even after aerosolization. Usually in the market, there are three types of nebulizers: pneumatic/jet, ultrasonic, or vibration mesh. Nebulizers work by atomizing the aqueous solution formulation into small droplets using methods such as piezoelectric vibrations, compressed air, or oscillation discs, depending on the type of nebulizer (ultrasonic, pneumatic/jet, or vibration mesh, respectively). Liposomes face a greater amount of shear forces at the liquid-air interface during the atomization process,

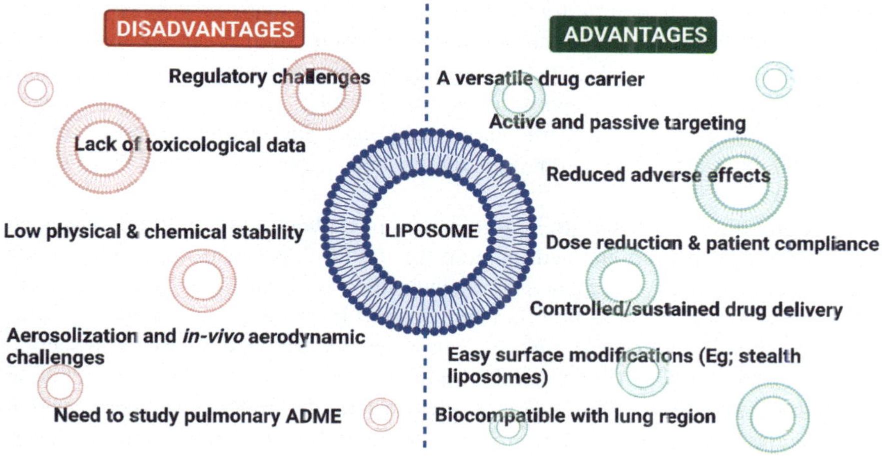

Fig. 9.3 Advantages and disadvantages of using liposomes in airway diseases

which could potentially destroy the liposome membranes resulting in a change in their lamellarity, particle size, and drug encapsulation (Cipolla et al. 2013; Smyth and Hickey 2011).

It's noteworthy that multilamellar liposomes (microns in size) tend to reduce their size when subjected to aerosolization by a pneumatic nebulizer, whereas unilamellar liposomes tend to increase in size under similar conditions. Liposomes possessing a higher gel-to-liquid transition temperature (T_c) generally exhibit greater membrane stability during nebulization or when introduced in bronchoalveolar lavage fluid. The incorporation of cholesterol and phospholipids with high Tc values, particularly those containing saturated and/or lengthy fatty acyl chains, can increase the Tc of liposomes and, consequently, its laminate stiffness (Cipolla et al. 2013; Smyth and Hickey 2011).

Liposomes can also be administered via dry powder inhalers (DPI) from dry powder formulations and, to a lesser extent, through pressurized metered dose inhalers (pMDI) from propellant formulations. When liposomes are in a powdered form, they are commonly referred to as liposomes, lipospheres, or proliposomes. Liposomes developed as dry powders often adopt an MLV state, which has better stability thermodynamically (Cipolla et al. 2013; Smyth and Hickey 2011; Willis et al. 2012). Drying the vesicles through methods like spray drying or freeze-drying enhances their stability and improves their shelf life during storage (Willis et al. 2012).

9.1.7 Phospholipids for Pulmonary Drug Delivery Liposomes

Since liposomes are commonly composed of compounds that share structural similarities with substances found in the lungs, such as cholesterol and phospholipids, they are considered to be safe for the delivery of drugs intended for airway diseases (Unida et al. 2016). Several types of phospholipids are used for the preparation of liposomes such as 1,2-dioleoyl-sn-glycero-3-phosphocholine (DOPC) and 1,2-dimyristoyl-sn-glycero-3-phosphocholine (DMPC). However, studies suggest that phospholipids similar to endogenous pulmonary surfactants are more suitable for drug delivery into the pulmonary region than exogenous surfactants (Mehta et al. 2020).

Phospholipid 1,2-dipalmitoyl-sn-glycero-3-phosphocholine (DPPC) has been vastly explored in pulmonary drug delivery. DPPC has been considered a safe (GRAS) excipient by the regulatory agency, the Food and Drug Administration (FDA). It is a type of phospholipid characterized by having two completely saturated C16:0 palmitoyl chains connected through ester linkages to first and second carbons of the glycerol backbone. Additionally, it features a phosphatidylcholine moiety bound to third carbon of the central structural component glycerol, which serves as the hydrophilic head group (Wauthoz and Amighi 2014). Liposomes primarily made of DPPC exhibit poor recognition by macrophages and thereby reduced phagocytosis and lower uptake by type II alveolar epithelial cells and macrophages of the alveolar region in comparison to liposomes with a micellar-like composition (Poelma et al. 2004). Another benefit of DPPC is its ability to create liposomes containing densely packed solid layers. As a result, this leads to a decreased surface tension at the interface between alveolar air and liquid, effectively emulating the function of lung surfactants. This emulation helps diminish the effort required for breathing and maintains the stability of alveolar deflation and inflation, preventing alveoli from collapsing or excessively expanding at the end of exhalation (Notter 2000). DPPC also have higher T_c which further prevents membrane stability complications during nebulization.

9.2 Application of Liposomes in Allergic Airway Diseases

The biocompatibility of phospholipids with the lung region has attracted many formulation scientists to conduct research on liposomal systems for the delivery of drugs to the pulmonary region. One of the most troubled diseases of humankind is the allergic airway disease, and there has been advancement in novel drug delivery systems using liposomes to tackle this condition. This chapter delves into various recent studies that explore the utilization of liposomes as drug carriers for the management of allergic airway diseases, shedding light on promising developments in this field (Table 9.2).

Liposomes offer distinct advantages when it comes to surface modifications, which have opened up opportunities for various targeted approaches, including both active and passive targeting strategies. Active targeting methods with liposomes involve the incorporation of modified targeting peptides, ligands such as antibodies, or oligosaccharides. By doing so, liposomes can be tailored to deliver their cargo to specific tissues or cells in the body. For example, liposomes can be equipped with targeting peptides to enhance delivery to specific locations. Additionally, liposomes

Table 9.2 Developed liposomal formulation for the management of allergic airway diseases

Active pharmaceutical ingredient	Method	Key ingredients	Indication
Salbutamol sulphate	Spray drying and sonication	Soya phosphatidylcholine and cholesterol	Asthma (Al-Najar and Ghareeb 2020)
Salbutamol sulphate	Thin film hydration and spray drying	Soya phosphatidylcholine and cholesterol	Asthma (Honmane et al. 2019)
Salbutamol sulphate	Spray drying	DPPC and cholesterol	Asthma (Daman et al. 2014)
DOPC	Thin film hydration	DOPC and cholesterol	Chronic respiratory diseases (Komalla et al. 2020)
MBD2 siRNA	Vortexing and ultracentrifugation	Lipidoids, cholesterol, DSPC, and mPEG-DMG	Allergic airway inflammation in asthma (Wu et al. 2022)
SOCS3 protein	Vortexing and serial extrusion	Dioleoyl phosphocholine and dioleoyl phosphoglycerol	Allergic airway inflammation in asthma (Draijer et al. 2020)
Budesonide	Thin film hydration	Phosphatidylglycerol-phosphatidylcholine-PEG, distearoylphosphatidylethanolamine, and cholesterol	Asthma (Konduri et al. 2003)
DP allergen vaccine	Thin film hydration	Lipoid-S-100 phosphatidylcholine, soybean lecithin, and DDAB	Allergic rhinitis and asthma (Chaisri et al. 2017)

DPPC 1,2-dipalmitoyl-sn-glycero-3-phosphocholine, *DOPC* 1,2-dioleoyl-sn-glycero-3--phosphocholine, *DSPC* distearoylphosphatidylcholine, *MBD2* methyl-CpG-binding domain protein 2, *mPEG-DMG* 1,2-dimyristoyl-rac-glycero-3-methoxypolyethylene glycol-2000, *SOCS3* suppressor of cytokine signalling 3, *PEG* polyethylene glycol, *DDAB* didecyldioctadecylammonium bromide, *DP Dermatophagoides pteronyssinus*

can be modified with ligands like antibodies or oligosaccharides, enabling precise targeting of dendritic cells (DCs). This targeted approach allows for the delivery of antigens and immunostimulatory molecules to DCs, initiating antigen-specific immune responses. These customized liposomes hold promise for the development of vaccines against allergic airway diseases (Wagener et al. 2020).

On the other hand, passive targeting with liposomes doesn't rely on active targeting mechanisms but takes advantage of liposomes' natural properties to deliver drugs to specific cells or tissues. The ability to modify liposomal membranes has also addressed the issue of liposome instability in aqueous solutions. Coating the liposomal membranes can lead to the creation of dried products like proliposomes, ensuring stability during storage and transportation. Figure 9.4 provides an overview of some of the approaches observed in liposomal formulations for managing allergic airway diseases, showcasing the versatility and potential of these delivery systems.

To address the issue of liposome instability in aqueous formulations, a study was conducted involving the development of proliposomes containing salbutamol sulphate, a bronchodilator commonly used in asthma treatment. Proliposomes are stable precursor particles of liposomes that can be easily converted into liposomes when hydrated. These structures consist of a lipid bilayer surrounded by a carbohydrate layer, offering advantages such as enhanced stability and ease of handling compared to conventional liposomes.

The study employed the spray drying technique to prepare proliposomes, using various core carriers, including sucrose, mannitol, trehalose, and lactose, which

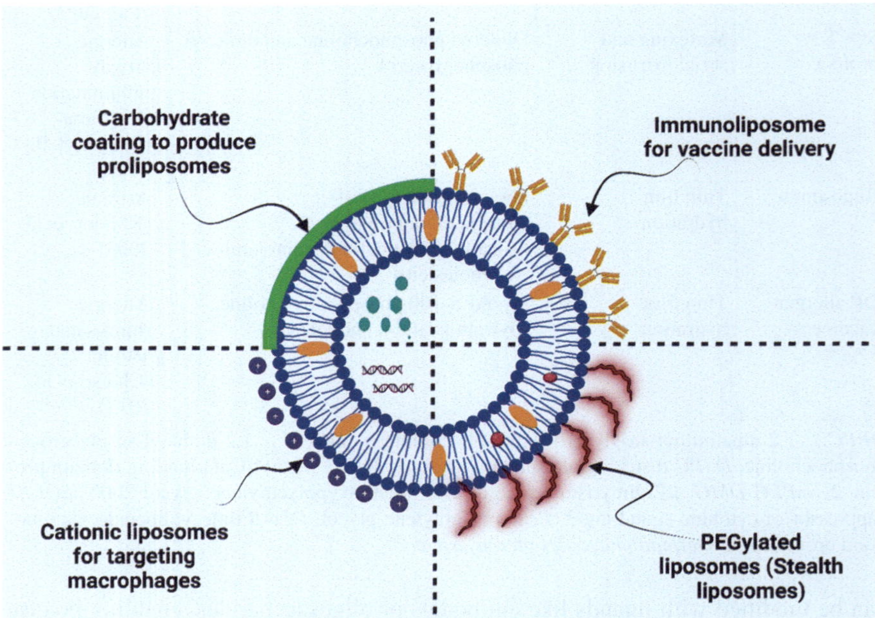

Fig. 9.4 Different approaches utilized for active and passive targeting of liposomes for the management of allergic airway diseases

were combined with an ethanolic lipid solution containing salbutamol sulphate at a specific phospholipid to carrier ratio. Several parameters, including particle size distribution, polydispersity, zeta potential, entrapment capacity, and morphology, were evaluated to assess the quality of the prepared proliposomes.

Furthermore, the evaluation of the liposomes was carried out by hydrating the proliposomes in either simulated lung fluid or deionized water. The choice of hydration media had a notable impact on the properties of the liposomes, particularly affecting size, polydispersity, zeta potential, and entrapment efficiency, with simulated lung fluid showing more significant effects than deionized water. However, liposome morphology remained relatively stable regardless of the dissolution medium.

The study also revealed that manual dispersion could influence the morphology and aggregation of liposomes generated from carrier-based proliposomes. Additionally, specific types of copolymers and the inclusion of particular lipid components were identified as potential factors causing morphological changes in liposomes. Overall, this research shed light on various aspects of proliposomes, highlighting the importance of core carrier selection, the influence of hydration media, and the potential factors impacting liposome morphology (Al-Najjar and Ghareeb 2020).

To enhance the oral bioavailability of salbutamol sulphate, researchers developed a nanoliposomal DPI formulation for pulmonary delivery. The optimized liposomal formulation exhibited several favourable characteristics, including a small particle size (160–174.2 nm), high drug entrapment, and extended drug release. The in vitro aerosol conduct of the liposomal formulation was assessed using the Andersen Cascade Impactor. The results revealed a sustained drug release of 90% of salbutamol sulphate for 14 h. The drug release pattern followed Higuchi's controlled release model. This suggests that the developed formulation has the potential for efficient and controlled release of drugs for the treatment of asthma (Honmane et al. 2019).

Allergic asthma is characterized by increased mucus production, heightened airway sensitivity, and inflammation around the bronchial tubes. Macrophages, important cells in maintaining lung health, are implicated in asthma's development. These macrophages can adopt different roles, becoming either M1 or M2 types in response to various triggers. In allergic asthma patients, M2 macrophages are prevalent in the airways, generating inflammatory molecules that worsen the condition. Targeting M2 macrophages holds promise for asthma treatment. Liposomes, loaded with siRNA, have been utilized for effective drug delivery, including targeted delivery to macrophages utilizing cationic liposomes. Methyl-CpG-binding domain 2 (MBD2) is linked to polarizing macrophage towards the M2 phenotype. Using MBD2 siRNA-loaded liposomes, a feasible therapeutic approach for combating asthma was developed.

Liposomes encapsulating MBD2 siRNA were formulated and administered intratracheally to a mouse model that has allergic airway inflammation. The inflammation was triggered by ovalbumin (OVA). The results of the study showed that these liposomes successfully reduced the expression of MBD2 and prevented allergic airway inflammation in mice. This was evidenced by the reduced airway inflammation and mucus production. The study also demonstrated that MBD2 siRNA-loaded liposomes effectively targeted macrophages, with 40% of the liposomes being taken up by macrophages in lung tissue. These findings suggest that MBD2 could be a promising target for asthma treatment, and RNAi-based therapy using MBD2 siRNA containing liposomes holds potential as a therapeutic approach (Wu et al. 2022).

The lungs rely on controlled inflammatory responses to maintain gas exchange while encountering inhaled allergens, toxins, and microbes. Alveolar macrophages (AMs), immune cells residing in the lungs, have a suppressive role in allergic airway inflammation. AMs use paracrine communication, including the transfer of extracellular vesicles (EVs), to regulate lung inflammation. These EVs from AMs contain a suppressor of cytokine signalling 3 (SOCS3), which inhibits inflammatory signalling in alveolar and airway epithelial cells. When the delivery of SOCS3 from AMs to ECs is impaired during allergic inflammation, it can lead to increased inflammatory responses. The intrapulmonary administration of liposomes with SOCS3 as the sole cargo was found to have the potential to limit inflammation in an allergic mouse model. Synthetic multilamellar vesicles were developed using a 1:1 ratio of dioleoyl phosphocholine and dioleoyl phosphoglycerol. Pretreatment of the lungs with SOCS3 liposomes via intrapulmonary administration effectively decreased the recruitment of inflammatory cells and the production of cytokines in the lungs of mice subjected to OVA challenge. These liposomes containing SOCS3 can rectify the problem of impaired SOCS3 secretion, thus mitigating the inhibition of inflammatory signalling pathway activation in epithelial cells (Draijer et al. 2020).

In a study aimed at improving asthma treatment, researchers investigated the use of budesonide, an inhaled corticosteroid. These drugs are commonly prescribed to mitigate the inflammatory responses. However, challenges like side effects and patient noncompliance have limited their effectiveness. To address this, the study explored the use of sterically stabilized (stealth) liposomes as carriers for budesonide.

The study employed OVA-sensitized C57/Black 6 mice as an experimental model to simulate asthma. The mice received various treatment regimens, including aerosolized budesonide loaded in stealth liposomes, budesonide without liposomes (administered daily or weekly), aerosolized budesonide loaded in empty stealth liposomes, and conventional liposomes administered weekly. For 4 weeks, the researchers conducted histopathological examinations of lung tissues and measured markers of inflammation, including marginal blood eosinophil counts, eosinophil peroxidase activity, and overall serum IgE levels.

The findings revealed that once-in-a-week intervention with budesonide loaded in PEGylated liposomes was equally effective as daily budesonide intervention in dropping the number of inflammatory markers in the asthma model. This innovative approach offers a promising alternative to the conventional daily budesonide regimen. Notably, it has the potential to enhance treatment compliance, reduce side effects, and improve the overall management of asthma (Konduri et al. 2003). House dust mites (HDM), scientifically known as *Dermatophagoides farinae* (Df) and *Dermatophagoides pteronyssinus* (Dp), are known to generate macromolecules and proteins which act as allergens and sensitize humans to hypersensitivities such as allergic rhinitis and asthma on a global scale. Among these allergens, category 1 (Der f 1 and Der p 1) and category 2 (Der f 2 and Der p 2) allergens are the super-abundant and immunodominant components found in HDM.

Allergen-specific immunotherapy (AIT) offers the possibility of curing allergies, resulting in long-term symptom alleviation, reduced dependence on medication, and the prevention of new sensitizations. AIT triggers the development of regulatory B cells (Bregs) and T cells (Tregs) that give rise to immunosuppressive cytokines (TGF-b, IL-10, and IL-35), which play a critical role in managing hypersensitivities. Traditional immunotherapy for HDM allergies using crude mite extracts has limitations in terms of effectiveness and can lead to adverse effects. As a result, efforts have been made to reduce side effects by utilizing recombinant allergens or hypoallergenic derivatives. Hence, an innovative approach to immunotherapy for HDM allergies, involving the use of a refined allergen vaccine loaded with liposomes, was developed. The researchers conducted experiments using a mouse model of Dp allergy to evaluate the therapeutic efficacy of intranasally administered vaccine-loaded liposomes, including those made from single and combined Dp major allergic agents (L-Der p 1, L-Der p 2), as well as crude extract of Dp (L-CE). The findings demonstrated that liposomal vaccines containing single refined major allergic agents were more therapeutically effective than the vaccine made from the crude extract in managing Dp allergy in the mouse model. These single allergens-loaded liposomes induced immunosuppressive cytokine responses, including TGF-b, IL-10, and IL-35, which are vital for managing allergic diseases. This innovative approach holds promise for improving the effectiveness and safety of immunotherapy for HDM allergies (Chaisri et al. 2017).

9.2.1 AmBisome®

Various medical institutions, in collaboration with pharmaceutical industries, have investigated the use of inhaled liposomal amphotericin B, known as AmBisome®, for treating various pulmonary diseases. Amphotericin B exerts its antifungal effect by creating pores or channels in the biological membranes of fungi, primarily utilizing the sterol components present in the fungal membrane. This process compromises the structural integrity of the fungal membrane, ultimately resulting in the demise of the fungal cell. It is worth highlighting that, alongside its antifungal activity, amphotericin B can also exhibit significant toxicity towards mammalian cells.

Amphotericin B is also available as an amphotericin B deoxycholate formulation. Its therapeutic utility arises from a favourable balance between its effectiveness and toxicity, resulting in a therapeutic window. This is due to the fact that fungal membranes primarily employ ergosterol as a membrane sterol. On the other hand, mammalian membranes utilize cholesterol. Importantly, amphotericin B exhibits a binding preference for ergosterol that is approximately ten times higher than for cholesterol. This preference allows for a therapeutic range in which amphotericin B is effective against fungal infections while minimizing harm to mammalian cells. However, it's essential to recognize that this therapeutic window is limited, typically allowing for doses of 1 mg/kg. The deoxycholate preparation of amphotericin B carries significant risks, including fatal cardiac or cardiorespiratory arrest in cases of overdose, fever, and other adverse events related to infusion. Additionally, it can affect the renal function of the kidney. These events of toxicity are a direct or indirect result of amphotericin B's interactions with host cell membranes (Brüggemann et al. 2022).

A liposomal formulation of amphotericin B was developed to tackle this problem. Moore et al. adapted liposome preparation technology from Vestar Inc., a California-based company, to create AmBisome®. Vaster-engineered liposomes have specific characteristics: they were less than 100 nm in diameter size, had single bilayer membranes, possessed a neutral charge, and featured a lipid bilayer composition of distearoylphosphatidylcholine and cholesterol in a molar ratio 2:1. This particular formulation, combined with high shear production techniques, resulted in liposomes that remained relatively solid at normal human body temperature and exhibited prolonged stability after injection. When the APIs are stably incorporated into these liposomes, they tend to experience significantly reduced clearance through the typical clearance pathways (such as liver and kidney clearance and macrophage uptake). As a result, these liposomes can extend their presence in the bloodstream and achieve enhanced distribution throughout various tissues in the body. Moore et al. introduced an additional phospholipid, distearoylphosphatidylglycerol (DSPG), into the liposome formulation. This addition served to impart a negative charge to the liposome particles. Furthermore, the drug was incorporated into the phospholipid bilayer of the liposomes, leveraging the drug's inherent propensity to bind to cholesterol within the membrane. Thus, AmBisome® formulation has a particle size of approximately 100 nm and is made up of amphotericin B, DSPG, soy phosphatidylcholine (hydrogenated), and cholesterol in a specific molar ratio of 0.4:0.8:2:1. The interaction between cholesterol and amphotericin B is particularly crucial for the stability of the formulation. This modification likely contributed to the effectiveness and properties of the resulting liposomal formulation (Brüggemann et al. 2022).

AmBisome® exhibits a higher affinity for the fungal cell wall compared to the mammalian cell membrane, likely attributed to the ergosterol present within the fungal cell wall. The viscoelastic properties of the cell wall enable AmBisome® to easily penetrate it. Once within the cell wall, AmBisome® attaches to the fungal cell membrane, due to oppositely charged surfaces of the liposome and the fungal membrane. Guided by the existence of ergosterol, and considering amphotericin B's preference for binding to ergosterol over cholesterol, the drug gets delivered by the liposome to the fungal membrane, where it undergoes subsequent degradation (Walker et al. 2018). The mechanistic concept for the delivery of amphotericin B is demonstrated in Fig. 9.5.

Clinical trials were performed on AmBisome® to study its therapeutic efficiency and safety in patients suffering from invasive pulmonary aspergillosis and allergic bronchopulmonary aspergillosis (ABPA). A phase II clinical trial (NCT02273661) focused on determining the potential to maintain the remission of ABPA by

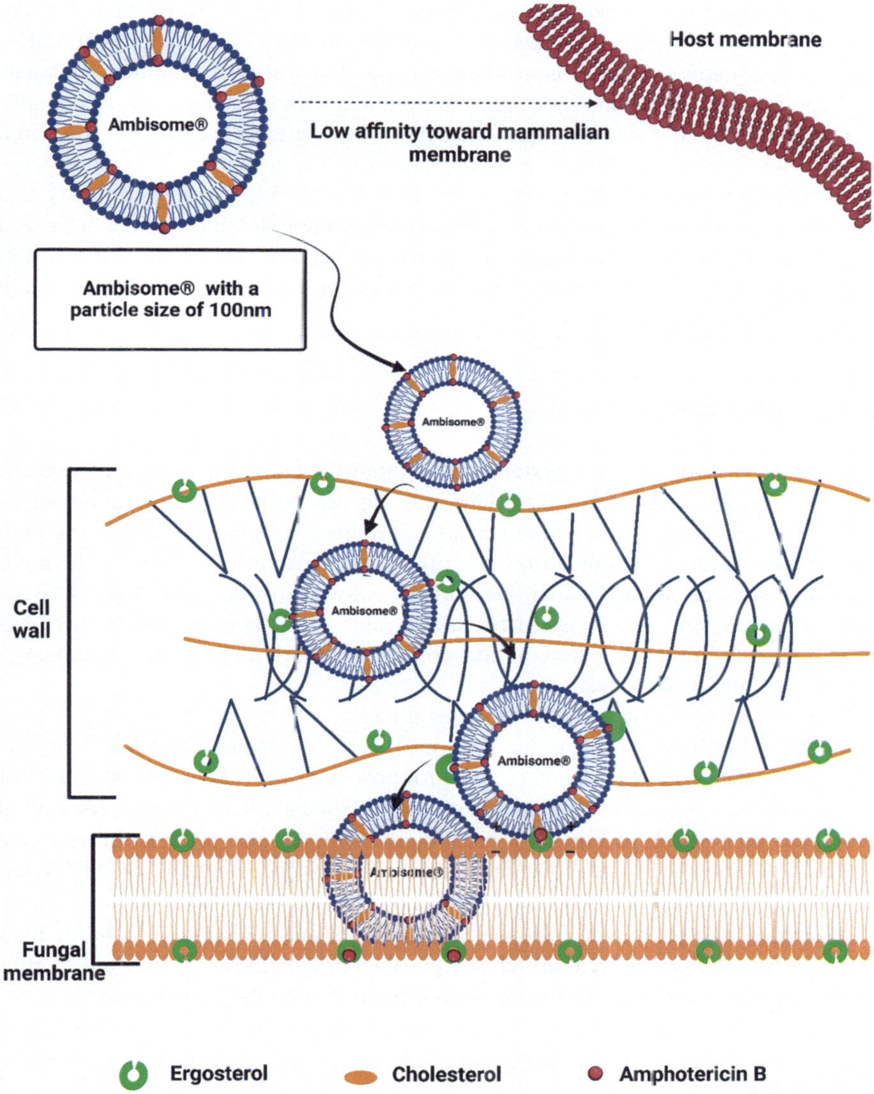

Fig. 9.5 Mechanistic concept for the delivery of amphotericin B by AmBisome®

AmBisome® when the nebulization was prolonged. The researchers conducted a randomized, single-blind clinical trial involving 30 centres. The participants were individuals with controlled ABPA following a 4-month ambush intervention involving itraconazole and corticosteroids. These patients were randomly allocated to receive either nebulized AmBisome® or a placebo for a half-year duration. The initial outcome measure was the development of the first severe clinical exacerbation within 1 year after randomization. The study enrolled 174 patients with ABPA, with 139 of them achieving control after the initial 4-month treatment and thus undergoing randomization. The results showed that nebulized active drugs containing liposomes did not significantly reduce the risk of critical exacerbation compared to the placebo group. However, some constructive secondary outcomes were observed, including a longer time to the first severe exacerbation clinically and reductions in overall *Aspergillus* precipitins and immunoglobulin E in the nebulized liposomal amphotericin B group. These findings suggest opportunities for further research to explore the useful benefits and constraints of this treatment approach in ABPA (Godet et al. 2022).

9.3 Conclusion

Liposomes, as drug delivery systems, hold significant promise in the treatment of pulmonary diseases, primarily due to the biocompatibility of their structural components with the lung region. These versatile systems are capable of delivering both hydrophilic and hydrophobic drugs effectively. Furthermore, the ability to manipulate the surfaces of liposomal membranes allows for targeted drug delivery approaches. Despite being in use for six decades since their discovery, there has been relatively limited research on liposomes, particularly in the context of managing allergic airway diseases.

Among allergic respiratory disorders, allergic asthma has received the most attention in liposome-based studies. Various types of liposomes, including cationic liposomes, PEGylated liposomes, and proliposomes, have been investigated for their potential in asthma treatment. Additionally, there have been attempts to develop intranasal liposomal vaccines for treating conditions like allergic rhinitis and asthma. Notably, only one liposome formulation for the treatment of ABPA has progressed to clinical trials. Overall, liposomes exhibit substantial potential in the treatment of allergic airway diseases. Researchers are encouraged to focus their efforts on exploring this area further, as liposomal drug delivery systems could offer more effective and targeted therapies for individuals with these conditions.

References

Al-Najjar BY, Ghareeb MM (2020) Preparation of inhalable salbutamol proliposome using different core carriers and investigate the impact of simulated lung fluid use in evaluation. Sys Rev Pharm 11:629–639

Antimisiaris SG, Marazioti A, Kannavou M, Natsaridis E, Gkartziou F, Kogkos G, Mourtas S (2021) Overcoming barriers by local drug delivery with liposomes. Adv Drug Deliv Rev 174:53–86. https://doi.org/10.1016/j.addr.2021.01.019

Arpita S, Kumar SV, Amresh G (2020) Liposome–a review. IJIHD V:1–6. https://doi.org/10.46956/ijihd.vi.116

Bangham AD, Standish MM, Watkins JC (1965) Diffusion of univalent ions across the lamellae of swollen phospholipids. J Mol Biol 13(1):238–252. https://doi.org/10.1016/S0022-2836(65)80093-6

Brüggemann RJ, Jensen GM, Lass-Flörl C (2022) Liposomal amphotericin B—the past. J Antimicrob Chemother 77(Supplement_2):ii3–ii10. https://doi.org/10.1093/jac/dkac351

Chaisri U, Tungtrongchitr A, Indrawattana N, Meechan P, Phurttikul W, Tasaniyananda N, Saelim N, Chaicumpa W, Sookrung N (2017) Immunotherapeutic efficacy of liposome-encapsulated refined allergen vaccines against *Dermatophagoides pteronyssinus* allergy. PLoS One 12(11):e0188627. https://doi.org/10.1371/journal.pone.0188627

Chudasma MP, Shah SA, Qureshi MHN, Shah N, Shah D, Trivedi R, Shah VH (2023) Brief insight on nanovesicular liposomes as drug-delivery carriers for medical applications. JERP. https://doi.org/10.14218/JERP.2022.00086

Cipolla D, Gonda I, Chan H-K (2013) Liposomal formulations for inhalation. Ther Deliv 4(8):1047–1072. https://doi.org/10.4155/tde.13.71

Daman Z, Gilani K, Rouholamini Najafabadi A, Eftekhari HR, Barghi MA (2014) Formulation of inhalable lipid-based salbutamol sulfate microparticles by spray drying technique. DARU J Pharm Sci 22(1):50. https://doi.org/10.1186/2008-2231-22-50

Draijer C, Speth JM, Penke LRK, Zaslona Z, Bazzill JD, Lugogo N, Huang YJ, Moon JJ, Peters-Golden M (2020) Resident alveolar macrophage-derived vesicular SOCS3 dampens allergic airway inflammation. FASEB J 34(3):4718–4731. https://doi.org/10.1096/fj.201903089R

Duong V-A, Nguyen T-T-L, Maeng H-J (2023) Recent advances in intranasal liposomes for drug, gene, and vaccine delivery. Pharmaceutics 15(1):207. https://doi.org/10.3390/pharmaceutics15010207

Elhissi A (2017) Liposomes for pulmonary drug delivery: the role of formulation and inhalation device design. Curr Pharm Des 23(3):362–372

Godet C, Couturaud F, Marchand-Adam S, Pison C, Gagnadoux F, Blanchard E, Taillé C, Philippe B, Hirschi S, Andréjak C, Bourdin A, Chenivesse C, Dominique S, Bassinet L, Murris-Espin M, Rivière F, Garcia G, Caillaud D, Blanc F-X et al (2022) Nebulised liposomal amphotericin-B as maintenance therapy in allergic bronchopulmonary aspergillosis: a randomised, multicentre trial. Eur Respir J 59(6):2102218. https://doi.org/10.1183/13993003.02218-2021

Honmane S, Hajare A, More H, Osmani RAM, Salunkhe S (2019) Lung delivery of nanoliposomal salbutamol sulfate dry powder inhalation for facilitated asthma therapy. J Liposome Res 29(4):332–342. https://doi.org/10.1080/08982104.2018.1531022

Hotani H (1984) Transformation pathways of liposomes. J Mol Biol 178(1):113–120. https://doi.org/10.1016/0022-2836(84)90234-1

Kim E-M, Jeong H-J (2021) Liposomes: biomedical applications. Chonnam Med J 57(1):27. https://doi.org/10.4068/cmj.2021.57.1.27

Komalla V, Allam VSRR, Kwok PCL, Sheikholeslami B, Owen L, Jaffe A, Waters SA, Mohammad S, Oliver BG, Chen H, Haghi M (2020) A phospholipid-based formulation for the treatment of airway inflammation in chronic respiratory diseases. Eur J Pharm Biopharm 157:47–58. https://doi.org/10.1016/j.ejpb.2020.09.017

Konduri KS, Nandedkar S, Düzgünes N, Suzara V, Artwohl J, Bunte R, Gangadharam PRJ (2003) Efficacy of liposomal budesonide in experimental asthma. J Allergy Clin Immunol 111(2):321–327. https://doi.org/10.1067/mai.2003.104

Lombardo D, Kiselev MA (2022) Methods of liposomes preparation: formation and control factors of versatile nanocarriers for biomedical and nanomedicine application. Pharmaceutics 14(3):543. https://doi.org/10.3390/pharmaceutics14030543

Matos CM (2023) Liposomes: the brave Old World. Int J Mol Sci 24(5):4343. https://doi.org/10.3390/ijms24054343

Mehta PP, Ghoshal D, Pawar AP, Kadam SS, Dhapte-Pawar VS (2020) Recent advances in inhalable liposomes for treatment of pulmonary diseases: concept to clinical stance. J Drug Deliv Sci Technol 56:101509. https://doi.org/10.1016/j.jddst.2020.101509

Mulla TS, Thorat MS, Rayate Y, Nitalikar M (2019) Liposome as a drug carrier. Asian J Res Pharm Sci 9(2):141. https://doi.org/10.5958/2231-5659.2019.00021.3

Notter RH (2000) Lung surfactants: basic science and clinical applications. CRC Press
Poelma DL, Ju MR, Bakker SC, Zimmermann LJ, Lachmann BF, Van Iwaarden JF (2004) A common pathway for the uptake of surfactant lipids by alveolar cells. Am J Respir Cell Mol Biol 30(5):751–758. https://doi.org/10.1165/rcmb.2003-0127OC
Rahman M, Singh JG, Afzal O, Altamimi ASA, Alrobaian M, Haneef J, Barkat MA, Almalki WH, Handa M, Shukla R, Ullah NMN, S., Kumar, V., & Beg, S. (2022) Preparation, characterization, and evaluation of curcumin–graphene oxide complex-loaded liposomes against *Staphylococcus aureus* in topical disease. ACS Omega 7(48):43499–43509. https://doi.org/10.1021/acsomega.2c03940
Ravindar C, Reddy ST, Sivaramakrishna D, Damera DP, Swamy MJ (2022) Base-triggerable lauryl sarcosinate–dodecyl sulfate catanionic liposomes: structure, biophysical characterization, and drug entrapment/release studies. Soft Matter 18(40):7814–7826. https://doi.org/10.1039/D2SM00965J
Smyth HDC, Hickey AJ (2011) Controlled pulmonary drug delivery. Springer Science & Business Media
Šturm L, Poklar Ulrih N (2021) Basic methods for preparation of liposomes and studying their interactions with different compounds, with the emphasis on polyphenols. Int J Mol Sci 22(12):6547. https://doi.org/10.3390/ijms22126547
Unida S, Ito Y, Onodera R, Tahara K, Takeuchi H (2016) Inhalation properties of water-soluble drug loaded liposomes atomized by nebulizer. Asian J Pharm Sci 11(1):205–206. https://doi.org/10.1016/j.ajps.2015.11.034
Wagener K, Bros M, Krumb M, Langhanki J, Pektor S, Worm M, Schinnerer M, Montermann E, Miederer M, Frey H, Opatz T, Rösch F (2020) Targeting of immune cells with trimannosylated liposomes. Adv Ther 3(6):1900185. https://doi.org/10.1002/adtp.201900185
Walker L, Sood P, Lenardon MD, Milne G, Olson J, Jensen G, Wolf J, Casadevall A, Adler-Moore J, Gow NAR (2018) The viscoelastic properties of the fungal cell wall allow traffic of AmBisome as intact liposome vesicles. mBio 9(1):e02383–e02317. https://doi.org/10.1128/mBio.02383-17
Wauthoz N, Amighi K (2014) Phospholipids in pulmonary drug delivery. Eur J Lipid Sci Technol 116(9):1114–1128. https://doi.org/10.1002/ejlt.201300368
Willis L, Hayes D, Mansour HM (2012) Therapeutic liposomal dry powder inhalation aerosols for targeted lung delivery. Lung 190(3):251–262. https://doi.org/10.1007/s00408-011-9360-x
Wu G-R, Zhou M, Wang Y, Zhou Q, Zhang L, He L, Zhang S, Yu Q, Xu Y, Zhao J, Xiong W, Wang C-Y (2022) Blockade of Mbd2 by siRNA-loaded liposomes protects mice against OVA-induced allergic airway inflammation via repressing M2 macrophage production. Front Immunol 13:930103. https://doi.org/10.3389/fimmu.2022.930103

Nanoparticles Solutions for Allergic Airway Disorders: Current Trends and Future Prospects

10

Sanchit Mondal, Sheetal Maan, Indu Rani, Anushree Jatrana, Ankita, Anil Kumar, Shulbhi Verma, Parteek Prasher, and Harish Mudila

Abstract

The present work explores the world of nanoparticles as a potential solution, for addressing allergic airway diseases. The ailments, such as asthma and allergic rhinitis, have an adverse impact on health, which calls for innovative approaches to their treatment. Polymeric nanoparticles possess qualities such as appreciable size, extraordinary surface properties, and controlled drug release capabilities that present a chance to revolutionize how we treat these conditions. This discussion delves into the benefits of using nanoparticles, for targeted drug delivery to the system. By allowing the administration of inflammatory drugs, bronchodilators, and immunomodulators directly to affected airways, polymeric nanoparticles minimize unwanted side effects in the body while increasing the effectiveness of treatment. Additionally, the versatility of materials allows for tailoring nanoparticle properties to suit patient needs and specific disease characteristics. In the coming years, sustained research efforts and interdisciplinary collaboration will be crucial in advancing the development of nanoparticles for treating airway diseases. As we venture into this realm of science, our aspiration is that these innovative approaches will ultimately improve the quality of life for indi-

S. Mondal · S. Maan · I. Rani · A. Jatrana
Department of Chemistry, Chaudhary Charan Singh Haryana Agricultural University, Hisar, Haryana, India

Ankita · A. Kumar · H. Mudila (✉)
Department of Chemistry, Lovely Professional University, Phagwara, Punjab, India

S. Verma
Department of Biotechnology, College of Basic Science and Humanities, SD Agricultural University, Dantiwada, Gujarat, India

P. Prasher
Department of Chemistry, University of Petroleum & Energy Studies, Dehradun, India

P. Prasher et al. (eds.), *Understanding Allergic Airway Diseases*, https://doi.org/10.1007/978-981-97-1953-2_10

viduals struggling with allergic airway diseases by offering them newfound hope and relief on their journey, toward better health and well-being.

Keywords

Nanoparticles · Solutions · Allergic airway disorders · Asthma · Allergic rhinitis · Treatment · Therapy · Nanomedicine · Drug delivery · Respiratory diseases

10.1 Introduction

Health issues are becoming more prevalent as the world's population and pollution levels increase. The reality is that population growth is exponentially growing worldwide. The population of the entire planet was 7.8 billion in 2020. It is anticipated that 9.7 billion people will live there by 2050. Pollution in the environment is getting worse. Numerous variables, including industrial pollution, automobile exhaust, and agricultural runoff, are responsible for this. The likelihood of experiencing health issues increases along with pollution levels. Respiratory issues, heart disease, cancer, and birth deformities are a few health issues that are correlated with pollution. One of the major classes of disease is allergic airway disease. Allergic airway disease is one of the emerging subjects of study in the field of health. A set of long-term inflammatory illnesses known as allergic airway diseases can affect the nose, sinuses, lungs, and other airways. Allergens are substances that are typically safe but can produce an allergic reaction in persons who are allergic to them, and many diseases are brought on by an overactive immune response to these allergens.

Asthma, chronic rhinosinusitis (CRS), and allergic rhinitis (hay fever) are the most common allergic airway illnesses. Allergens like pollen, dust mites, and mold can cause allergic rhinitis, an inflammation of the nose's lining. Itchy eyes, a runny nose, and sneezing are all signs of allergic rhinitis. Inflammation and airway constriction are two characteristics of the chronic lung illness asthma. Wheezing, shortness of breath, chest tightness, and coughing are all possible signs of asthma. CRS is a chronic sinus irritation that is frequently brought on by allergens. Postnasal drip, face discomfort, and nasal congestion are all signs of CRS. There are numerous drugs that can be used to treat allergic airway diseases. These drugs ease symptoms and enhance breathing in various ways. Some of these include leukotriene receptor antagonists, antihistamines, corticosteroids, decongestants, and omalizumab (Xolair). Despite their benefits, these medications also have some negative side effects, such as sleepiness, dry mouth, dizziness, sleeplessness, anxiety, headaches, weight gain, mood swings, high blood pressure, nausea, vomiting, and anaphylaxis. Polymeric nanoparticles are utilized as drug delivery agents to reduce the negative effects of conventional drugs. These polymeric nanoparticles can carry pharmaceuticals and be utilized to deliver medications to particular affected tissues or cells.

Polymeric nanoparticles are tiny, natural, or synthetic polymer-based particles. Large molecules known as polymers are formed from monomers, which are repeating units. These polymers generate polymeric nanoparticles when they are disintegrated into nanoscale particles. Due to their special qualities and adaptability, these nanoparticles have a wide range of applications in industries like drug delivery, nanomedicine, materials research, and more. Natural polymeric nanoparticles are tiny, biocompatible particles that are made of naturally occurring polymers such as chitosan, lignin, starch, gelatin, etc. Synthetic polymeric nanoparticles are nanoscale particles created in labs using artificial or synthetic polymers. Different chemical and physical techniques are used to produce these nanoparticles. Common synthetic polymers used to create polymeric nanoparticles include poly(lactic-co-glycolic acid) (PLGA), polyethylene glycol (PEG), polyvinyl alcohol (PVA), and many others (Fig. 10.1).

10.2 Types of Polymeric Nanoparticles

There are many different types of polymeric nanoparticles available, each with specific advantages and qualities for administering drugs to the airways (Fig. 10.2). The exact medicine, the airway size that is being targeted, and the desired drug release profile all influence the choice of nanoparticle type. The following are some popular polymeric nanoparticle types utilized to deliver medications to the airways:

10.2.1 Polymeric Nanospheres

Spherical nanoparticles produced from various polymers are known as polymeric nanospheres. Due to their numerous attributes, they are frequently utilized in medication delivery, nanomedicine, and materials science. The diameter of polymeric nanospheres typically ranges from 10 to 1000 nm. They can be designed to have a defined structure and uniform size (Cheng et al. 2015). Polymeric nanospheres can have their surfaces altered to facilitate targeted medication delivery. The surface of the nanoparticle can be coated with ligands, antibodies, or peptides that are peculiar to certain cells or tissues, assuring medication delivery to the intended region while limiting off-target effects (Alavi and Webster 2021).

10.2.2 Polymeric Nanocapsules

These capsules have a shell around a core and are used to deliver a variety of medications, nutrients, and other active ingredients. The active ingredient to be administered is contained in the core of a nanocapsule, which can be either solid or liquid. Various substances, such as medicines, vitamins, essential oils, or other bioactive molecules, can be encapsulated within this core. A polymeric or lipid-based covering called the shell envelops and protects the core. This shell serves to safeguard the chemical it is encasing, manage its release, and ensure its stability. As long as the shell material is biocompatible and biodegradable, it can be used in biomedical applications (Mahmoud et al. 2023).

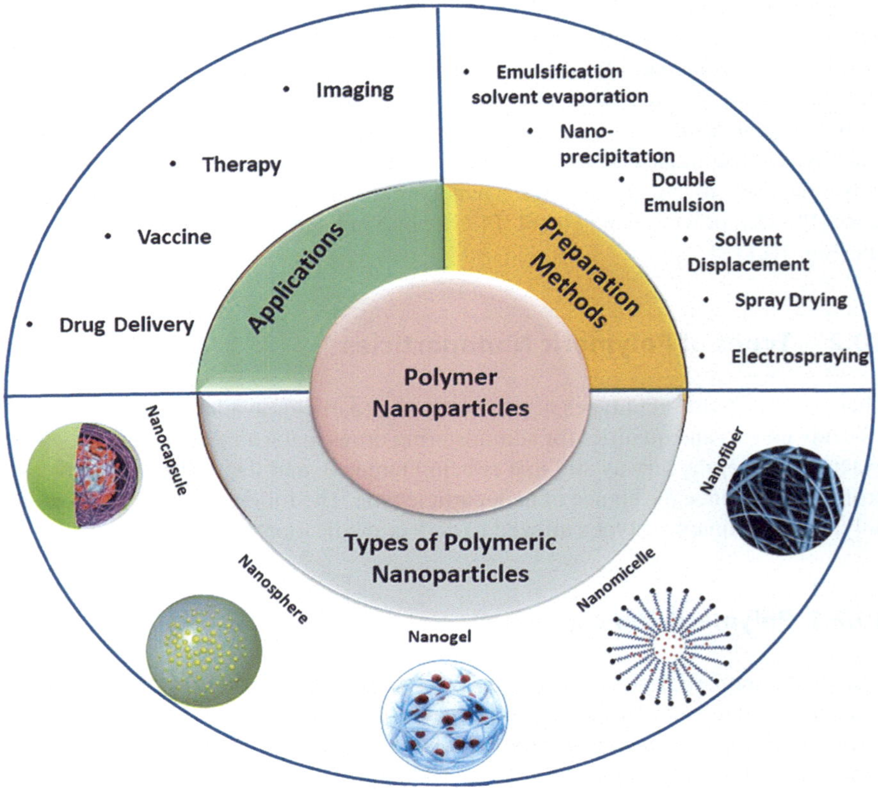

Fig. 10.1 Demonstration of polymeric nanomaterial, their types, preparation method, and applications

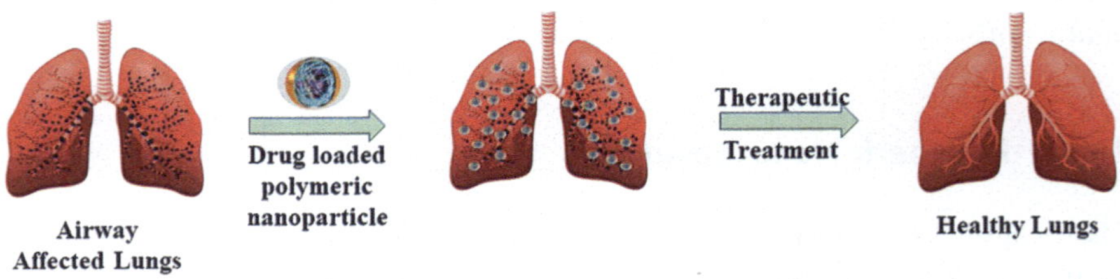

Fig. 10.2 Treatment of airway affected lungs by polymeric nanoparticles

10.2.3 Polymeric Micelles

Self-assembling nanoparticles known as polymeric micelles are created in an aqueous solution from amphiphilic block copolymers. The formation of polymeric micelles relies on the utilization of block copolymers, which are made up of two or more unique polymer segments with different hydrophilic and hydrophobic characteristics. For polymeric micelles, the most popular block copolymers are polyethylene glycol, polystyrene, polypropylene, poly(lactic-co-glycolic acid), etc. Block copolymer segments that are hydrophobic (water-repellent) make up the center of a polymeric micelle. In order to reduce contact with the surrounding aquatic environment, these hydrophobic segments group together. The block copolymer segments that are hydrophilic (attract water) make up the micelle's outer shell. The interaction of this shell with the water stabilizes the micelle in the aqueous media (Yokoyama 2014).

10.2.4 Polymeric Nanogels

A large amount of water or other aqueous solutions can be absorbed and retained by nanoscale hydrogel particles or networks, known as nanogels. They are renowned for having a three-dimensional, cross-linked structure that gives them special qualities and allows for a wide range of uses. A three-dimensional network of polymer chains that have been cross-linked to create a stable structure is often the building block of nanogels. They can absorb and hold onto water or other aqueous solutions inside their matrix owing to the network structure. Nanogels are highly hydrophilic, meaning they have a strong affinity for water (Kandil and Merkel 2019).

10.2.5 Polymeric Nanofibers

Nanofibers are incredibly thin fibers with typical sizes in the nanometer range, or fewer than 1000 nm (1 μm), on average. Due to their distinctive characteristics and extensive applications, these ultrafine fibers have attracted considerable attention from a variety of fields. There are various methods for the synthesis of nanofibers, i.e., template synthesis, phase separation, self-assembly, and electrospinning. One of the most typical processes for creating nanofibers is electrospinning. It includes drawing a polymer solution or melting it into a thin jet that goes to a collector where it solidifies into nanofibers with the help of an electric field. Nanofibers have a number of distinctive qualities, including a high surface area to volume ratio, porosity, flexibility, and customizable mechanical and chemical properties. They are appropriate for a variety of applications because of their qualities (Bhat 2015).

10.2.6 Lipid-Polymer Hybrid Nanoparticles

Lipid-polymer hybrid nanoparticles (LPNs), which combine the characteristics of lipids and polymers, are nanoscale drug delivery devices. LPNs are composed of a lipid core and a polymer shell, forming a core-shell structure. The shell is normally made of a biocompatible and biodegradable polymer, whereas the core is typically made of lipids or compounds that resemble lipids. Lipid and polymer strengths are combined in this two-component structure. Drugs that are hydrophobic can be encapsulated by lipid cores within their lipid bilayers, which will shield them from deterioration and improve their solubility (Mandal et al. 2013).

10.3 Advantages of Polymeric Nanoparticles

In a variety of disciplines, including medication delivery, materials science, and nanotechnology, polymeric nanoparticles have several benefits. They are a useful tool for a variety of applications because of their advantages (Fig. 10.1). The following are a few of the main benefits (Jarai et al. 2020) of polymeric nanoparticles:

10.3.1 Drug Delivery

Poorly water-soluble pharmaceuticals can become more soluble with the use of polymeric nanoparticles, increasing their bioavailability and potency. Additionally, they offer exact control over the kinetics of drug release, enabling sustained and regulated drug administration, which can lower the need for dosing. By allowing the attachment of ligands, antibodies, or peptides for targeted drug delivery, the surface changes reduce off-target effects and increase therapeutic efficacy. By limiting systemic exposure, targeted delivery can lower the likelihood of adverse medication reactions brought on by conventional drug formulations. Polymeric nanoparticles can prolong the shelf life and therapeutic potency of labile pharmaceuticals by shielding them from deterioration during storage and in vivo, such as proteins or nucleic acids.

10.3.2 Biocompatibility

Many of the polymers utilized to create polymeric nanoparticles are safe and biocompatible for usage in biological systems, lowering the possibility of negative reactions. Nanoparticles made from biodegradable polymers can gradually degrade into nontoxic byproducts, eliminating the requirement for extraction or removal.

10.3.3 Versatility

By choosing suitable polymers and fabrication techniques, the characteristics of polymeric nanoparticles, such as surface charge, drug release kinetics, and size, can be precisely customized to meet the needs of various applications. Combination therapy and diagnostic applications are made possible by the simultaneous transport of various medicinal or imaging substances via polymeric nanoparticles.

10.3.4 Stability

Drugs are more stable during storage and transportation when protected from environmental elements such as light, heat, and moisture by polymeric nanoparticles. Labile substances' shelf lives can be extended via encapsulation, which lowers waste and costs.

10.3.5 Reduced Toxicity

By regulating their release profile, polymeric nanoparticles can be employed to mitigate the toxicity of some medications while retaining therapeutic efficacy.

10.3.6 Diagnostic Applications

Imaging agents can be added to polymeric nanoparticles to improve their utility in diagnostic imaging methods such as MRI, CT scans, and fluorescence imaging.

10.4 Preparation of Polymeric Nanoparticles

Numerous processes are used to create polymeric nanoparticles, each one adapted to a particular application and the required qualities of the final product. The following are some typical techniques for creating polymeric nanoparticles:

10.4.1 Nanoprecipitation

The process of generating polymeric nanoparticles via nanoprecipitation, often referred to as solvent diffusion, is straightforward and popular. It is appropriate for encasing hydrophobic substances or medications. In this process, the respective polymer is solubilized in the water-miscible organic solvent (methanol, acetone, etc.). This organic solution is added into the aqueous solution through a constant rate flow and under continuous stirring. As the origin phase diffuses into the aqueous phase, the polymeric nanoparticles separate out spontaneously. Continuing stirring is provided to evaporate the organic solvent and to obtain a nanoparticle suspension (Almoustafa et al. 2017).

Gadad et al. (2012) created nanosuspensions of poly(lactic-co-glycolic acid) through the nanoprecipitation method. They were then compared to commercially available products for their performance in terms of zeta potential, particle size, surface morphology, drug entrapment effectiveness, in vitro release, and ex vivo transcorneal permeability. The cup plate method was used to examine the microbiological effectiveness against *Staphylococcus aureus* and *Pseudomonas aeruginosa*. For MF4 (drug-to-polymer ratio 1:0.4), spherical uniform particles (202.5 nm) with a polydispersity index of 0.226 and a negative zeta potential (−25.45 mV) were produced. The effectiveness of drug entrapment for MF4 was determined to be 83.1%. After 24 h, the cumulative percent medication release for formulation MF4 was 86.1%, demonstrating a sustained impact in treating bacterial conjunctivitis and preventing the need for repeated dosages. In ex vivo transcorneal permeation investigations, MFX-loaded PLGA nanoparticles (MF4) showed a much more advanced drug penetration ability than the commercially sold eye drops and also demonstrated superior antibacterial efficacy versus the marketed formulation. The findings suggest that moxifloxacin-loaded PLGA nanosuspension may be a useful drug delivery method for sustained release in ocular use.

10.4.2 Emulsification-Solvent Evaporation

Both hydrophobic and hydrophilic medicines are encapsulated in polymeric nanoparticles using the emulsification-solvent evaporation technique. In this method, an organic phase is obtained by dissolving the medication and the polymer in an organic solvent. To create an oil-in-water (O/W) emulsion, emulsify the organic phase in an aqueous phase that contains a surfactant. Evaporate the organic solvent, under reduced pressure, to harden the polymeric nanoparticles. The nanoparticles can be collected by ultrafiltration or centrifugation (Szczęch and Szczepanowicz 2020).

Babos et al. (2018) trapped doxorubicin and sorafenib (which have very different chemical properties) by the same polymeric carriers, namely, poly(d,l-lactide-co-glycolide) (PLGA) and polyethylene glycol-poly(d,l-lactide-co-glycolide) (PEG-PLGA), respectively, through emulsion solvent evaporation method. The nanopharmaceuticals' usual mean sizes were 142 and 177 nm, respectively. Doxorubicin was successfully encapsulated by PLGA and PEG-PLGA polymers with 52% and 69% efficiency, respectively, while sorafenib was successfully encapsulated with 55% and 88% efficiency. Doxorubicin was found to have sustained drug delivery under biorelevant circumstances, but sorafenib was released fast from the PLGA-doxorubicin-sorafenib and PEG-doxorubicin-sorafenib nanotherapeutics.

10.4.3 Double Emulsion

Drugs that are hydrophilic are encapsulated within polymeric nanoparticles using the double emulsion technique. In this technique, water-in-oil (W/O) emulsion is prepared by dissolving the drug in water and emulsifying it in an organic phase containing polymer. This solution is transferred into a second aqueous phase which contains surfactant, to create a water-in-oil-in-water emulsion. The organic solvent is acceptable to evaporate and results in the formation of polymeric nanoparticles (Panigrahi et al. 2021).

In order to create nanoparticles that would serve as carriers for the capreomycin sulfate (CMS) peptide, two distinct biodegradable polymers were used. Chitosan was chosen as a polymer that dissolves in water; in contrast, poly(lactic-co-glycolic acid) (PLGA) is soluble in organic solvents. The nanoparticles were created using a double emulsion solvent evaporation method. Numerous formulation variables for the optimization of polymeric nanoparticles were examined, including probe sonication time and intensity, homogenization speed, dispersion media *p*H, drug-polymer ratios, cross-linking agent type and concentration, and stabilizer concentration. Polymeric nanoparticles with the lowest particle diameters (343.7–1656.5 nm) and the best entrapment efficiencies (17.30–62.30%) were optimized by controlling these parameters and monitoring peptide degradation. In vitro drug release tests were performed on the created polymeric nanoparticles in phosphate buffer at *p*H 6.8. Different CMS release profiles (15.52–72.44%) were seen, and PLGA NPs were released significantly more slowly than chitosan NPs did (Alenazi et al. 2022).

10.4.4 Solvent Displacement

To create nanoparticles with increased drug loading and encapsulation effectiveness, the salting-out approach, also known as solvent displacement, is used. In this method, the polymer is dissolved into water-miscible organic solvent. Aqueous solution containing salt (e.g., magnesium sulfate) is added to the organic phase under constant stirring. This addition of salt solution leads to the precipitation of nanoparticles. Remaining salt solution or organic solvent is removed by ultrafiltration or centrifugation (Middha et al. 2019).

Yang et al. (2020) created the nanoparticles (NPs) containing the broad-spectrum antibiotic sparfloxacin (SFX) and the anti-inflammatory immunosuppressant tacrolimus (TAC) using an emulsion solvent evaporation approach; we chose poly(lactide-co-glycolide acid) (PLGA) with good biocompatibility and degradability. The grafting of the three peptides (NNQKIVNLKEKVAQLEA), which may particularly bind to the intercellular adhesion molecule-1 (ICAM-1), which is abundantly expressed on the surface of inflammatory endothelium cells, gives the NPs the potential to target inflammatory areas. The drug-loaded 3-PLGA NPs exhibit reduced hemolysis ratio, systemic toxicity, and high cytocompatibility. In addition to having strong antibacterial properties against both Gram-positive and Gram-negative bacteria, the drug-loaded 3-PLGA NPs can effectively reduce inflammation and immunological response in mice with acute lung infection. This study offers a straightforward and reliable nanoplatform to treat sepsis brought on by a lung infection, which may open the door to designing multifunctional nanomedicine for clinical application.

10.4.5 Spray Drying

In order to prepare nanoparticles for oral or inhalation drug administration, spray drying is a viable option. In this method, the polymer and drug are dissolved in a volatile organic solvent. This solution is further atomized into fine droplets by using a nozzle or spray gun. The droplets are dried rapidly in a heating chamber, which leads to the formation of solid nanoparticles (Malamatari et al. 2020).

Pedrozo et al. (2020) used the process of nano spray drying to create rutin-containing bovine serum albumin (BSA) nanoparticles. The shape, zeta potential, drug content, mean particle size, size distribution, antioxidant activity, and zeta potential of nanoparticles were all studied. Spherical particles with a mean size of 316 nm, a zeta potential of 32 mV, and an encapsulation effectiveness of about 32% were created under the ideal spray-drying conditions. Furthermore, nanoencapsulation doubled the IC50 of rutin when antioxidant activity against the ABTS+ radical was measured. The rutin-loaded BSA nanoparticles with potential antioxidant activity were produced using the nano spray-drying method, which worked well.

10.4.6 Electrospraying

With fine control over particle size, electrospraying is an electrohydrodynamic technique for creating polymeric nanoparticles. In this method, an organic solvent is used to dissolve the drug and the polymer. Feed this solution through a capillary tube and an electric field is applied to create a Taylor cone. Nanoparticles develop and are gathered on a grounded collector as the solvent is electrostatically expelled from the cone (Lee et al. 2020a).

10.5 Targeting of Polymeric Nanoparticles to the Airways

To effectively deliver medications or therapeutic agents to treat respiratory conditions like asthma, chronic obstructive pulmonary disease (COPD), and lung cancer, polymeric nanoparticles must be targeted to the airways. Numerous strategies and factors must be considered in order to achieve precise targeting to the airways (Nag et al. 2021; Kan et al. 2020; Deng et al. 2021). The following are some methods for directing polymeric nanoparticles toward the airways:

10.5.1 Particle Size and Inhalation Delivery

The nanoparticles' size is a key factor in how they settle in the respiratory tract. For inhaled administration, nanoparticles with a diameter between 1 and 5 μm are suitable. These particles have the ability to enter the lungs deeply and go to the alveoli and lower airways. The transport of nanoparticles to the airways might be affected by the inhalation device used, such as dry powder inhalers (DPIs), pressurized metered-dose inhalers (pMDIs), or nebulizers. The patient's ability to utilize the device effectively and the particle size should be considered while choosing the device.

10.5.2 Surface Modifications

The surface of the nanoparticle can be coated with ligands or antibodies that are selective for receptors or cell surface indicators in the airways. These ligands enable active targeting, guaranteeing that the nanoparticles bind to particular respiratory tract cells or tissues. The bloodstream circulation period of nanoparticles can be increased, and the pulmonary clearance rate can be decreased by coating them with polyethylene glycol (PEG). The nanoparticles' chances of entering the airways safely are increased by this "stealth" effect.

10.5.3 Mucoadhesive Properties

The mucus lining of the airways can be adhered to by adding mucoadhesive agents to the formulation of the nanoparticles. Drug absorption and retention may be improved by this extended contact with the mucosal surface.

10.5.4 Disease-Specific Targeting

Targeting disease-specific indicators or chemicals linked to inflammation, such as cytokines or enzymes, can be done with nanoparticles to treat respiratory disorders. With this strategy, the nanoparticles are delivered to the locations where the disease is currently active.

10.5.5 Compatibility with Lung Surfactant

The material that lines the alveoli, lung surfactant, should interact with nanoparticles in a positive way. Through this interaction, it is made possible for nanoparticles to get through the surfactant layer and enter the target cells.

10.6 Polymeric Nanoparticles for the Treatment of Allergic Airway Diseases

The biology of vertebrates depends primarily on the respiratory system. Due to the COVID-19 epidemic, the system's respiratory injuries have recently been continuously increasing. Both viral and noninfectious diseases can result in respiratory injury. SARS-CoV, MERS-CoV, and tuberculosis infection are examples of infectious disorders, whereas allergy diseases including allergic rhinitis and asthma are the leading causes of noninfectious diseases. These cause acute lung injury and other local lung disorders such as tracheitis, chronic obstructive pulmonary disease (COPD), and others. These respiratory illnesses and traumas can create serious conditions that necessitate hospitalization, a high mortality rate, and societal, economic, and healthcare crises (Liu et al. 2022). A simple diagrammatic representation of the treatment of allergic airway diseases is shown in Fig. 10.2.

10.6.1 Asthma

Asthma is a chronic respiratory disorder that affects nearly 339 million people currently suffering from asthma in the whole world and is estimated to rise to nearly 400 million by 2025. The symptoms of asthma include wheezing, coughing, shortness of breath, and tightness in the chest. Uncontrolled immune responses and allergic reactions are the main causes of asthma. Th1/Th2 imbalance is a crucial factor in the development of asthma. IL-25 and IL-33 are important mediators in the pathophysiology of asthma, while Th2 cell-released cytokines such as IL-4, IL-5, and IL-13 are dominating in the development. This causes structural modifications to the airways as well as eosinophilic (allergic and nonallergic) inflammation (Zuo et al. 2022).

10.6.1.1 Treatment for Asthma

Lee et al. designed double-loaded nanoparticles with dexamethasone and BM (DEX-BM nanoparticles) to synergistically exert effective antiasthma activity. To effectively scavenge H_2O_2 and release anti-inflammatory HBA that is produced by epithelium damage and inflammatory responses in the airway, BM nanoparticles are used as medication delivery systems. DEX dosage may be decreased by BM nanoparticles and DEX to mitigate its negative side effects. It was demonstrated that DEX-BM nanoparticles successfully reduce airway remodeling and inflammation in a mouse model of allergic asthma produced by OVA (Lee et al. 2020a, b).

Yang et al. used chitosan nanoparticles for the treatment of airway asthma disease. CS-BCG-PSN-OVA sustained-release vaccines significantly alleviated airway hyperresponsiveness (AHR) in asthmatic mice. To increase effectiveness even more, CS nanoparticles were encapsulated with BCG-PSN and OVA. CS is a biocompatible natural polysaccharide that aids in the safe and efficient transfer of allergens and can encourage the transmembrane absorption of peptides and proteins. After nanocrystallization, CS microspheres are created, and they can contain bases, nucleases, proteins, nucleic acids, and polysaccharide medicines in the body. The CS-BCG-PSN-OVA sustained-release vaccination was conducted that significantly reduced AHR in asthmatic mice. Additionally, it may significantly lower the amount of IL-4 and raise the level of IL-10 while also reducing the number of eosinophils and immune cells in BALF. Conversely, it might also control Tregs. The T cell proliferation rate was significantly lower in the CS-BCG-PSN-OVA group compared to the Model group, and the percentage of CD_4+CD_{25}+$Foxp_3$+ cells was higher. These findings demonstrated a superior therapeutic impact of the CS-BCG-PSN-OVA sustained-release vaccination on asthmatic mice (Yang et al. 2021).

10.6.2 Chronic Obstructive Pulmonary Disease (COPD)

COPD is a respiratory disorder that worsens over time and impairs breathing. It is a respiratory condition that has a significant death rate. Smoke from cigarettes, biomass, pollution, and tuberculosis infection are some of the causes that contribute to COPD. The two main episodes that underlie the pathogenesis of COPD are the excessive production of free radicals and the emergence of chronic inflammation. The classic signs of COPD are a persistent cough, an overabundance of phlegm, shortness of breath, and wheezing. Exacerbations of COPD, which are abrupt worsening of symptoms, can be fatal and frequently involve hospitalization (Saxena et al. 2022).

10.6.2.1 Treatment for COPD

Mohamed et al. developed poly(glycerol adipate-co-ω-pentadecalactone), PGA-co-PDL, dioleoyl trimethyl ammonium propane nanoparticles (PGA-co-PDL-DOTAP NPs) of size 244.8 ± 4.4 nm. The microRNAs (miRNAs) may be tractable targets for the treatment of chronic obstructive pulmonary disease (COPD). PGA-co-PDL-DOTAP NPs were successfully used to adsorb miR-146a in vitro, and the miR-146a still had its biological activity. The IL-1 signaling pathway was used by the miR-146a-NPs to decrease the IL-8 promoter reporter GFP. These outcomes thus show the potential of PGA-co-PDL NPs as a miR-146a delivery strategy for the treatment of COPD (Mohamed et al. 2019).

Chikuma et al. developed PLA-MnPD/DOTAP/pHDAC2 lipid nanoparticles for the treatment of COPD. Histone deacetylase 2 (HDAC2) is reduced as a result of the COPD disease's creation of ROS, which causes glucocorticoid resistance. Therefore, there is an urgent need for a novel medication that lowers ROS levels and glucocorticoid resistance at the same time. We created a delivery system for HDAC2-encoding plasmid DNA (pHDAC2) using core-shell lipid-polymer nanoparticles (LPNs) in this study. These LPNs have a poly(lactic acid) (PLA) core that contains a powerful antioxidant called Mn-porphyrin dimer (MnPD) and a cationic lipid (DOTAP) shell that binds HDAC2-encoding DNA. A considerable increase in intracellular HDAC2 expression was seen following pHDAC2 transfection and ROS removal by MnPD, indicating that MnPD's multi-antioxidative action is necessary for HDAC2 expression. Additionally, LPN therapy effectively reduced steroid resistance in COPD mice in vitro, as seen by the decreased levels of IL-8 expression. The method via which LPNs work may involve recovery from mitochondrial malfunction. Based on the synergism of ROS removal and HDAC2 expression, the PLA-MnPD/DOTAP/pHDAC2 system suggested offers a new therapeutic approach for COPD (Chikuma et al. 2020).

10.6.3 Allergic Rhinitis (AR)

Allergic rhinitis is a symptomatic condition of the nose brought on by allergen exposure through IgE-mediated hypersensitivity reactions under the chemotaxis of various inflammatory cytokines after a particular person comes into contact with antigens. According to studies, alterations in the body following exposure to allergens include an imbalance in T cell subpopulations and abnormally high levels of the inflammatory cytokines IL-6 and IL-8, which lead to an unbalanced response between Th1 and Th2 immune cells. B lymphocytes are induced to generate IgE by Th2 cytokines. IgE binds to the antigen, activating basophils and mast cells, releasing histamine from the cytoplasm, increasing vascular permeability, contracting bronchial smooth muscles, and stimulating sensory neurons, which cause symptoms including itching in the nose and sneezing. Ten percent to 30% of adults and up to 40% of children are said to have AR. There is little doubt that AR makes patients' daily lives more difficult. Anti-allergy medications that are frequently used cannot cure the condition (Paris et al. 2021). For an accurate diagnosis and a customized treatment plan, it is crucial to speak with a healthcare practitioner. Depending on the severity of your symptoms, they can advise you on the best course of action.

10.6.3.1 Treatment of Allergic Rhinitis

Abruzzo et al. prepared new mucoadhesive nasal decongestant nanoparticles obtained by direct cross-linking between the cationic polymer chitosan and the anionic drug cromolyn. This work solely used the ionic interaction between the positive charges of CH and the negative charges of Cr to prepare NPs based on CH and Cr without the need of cross-linkers. The outcomes showed that NPs based on CH and Cr can be made using a straightforward, quick, and mild procedure. The nasal mucosa and mucin glycoproteins were able to interact with NPs. This work showed that Cr released from CH_{18}/Cr_9 NPs diffused poorly across the nasal mucosa in vitro but that it remained accessible in the tissue for interactions with mast cells and the prevention of their degranulation. According to these findings, nasal administration of CH_{18}/Cr_9 NPs may be a useful strategy for the treatment of allergic rhinitis (Abruzzo et al. 2019).

Su et al. developed Gel/NPs/miR-146a polymeric nanoparticles. The effects of the gel/NPs/miR-146a binary formulation for the nasal delivery of nucleic acids in the treatment of rhinitis were then assessed. At a temperature of 34 °C, the produced binary formulation quickly created a gel in the nasal cavity with strong mucosal adherence, delivering nucleic acids to the nasal mucosa steadily and constantly. After nasal administration, gel/NPs/miR-146a was able to maintain the transport of miRNA into the mucosa. The gel/NPs/miR-146a binary formulation outperformed the monolithic formulations in terms of pharmacodynamic effects and capacity to transport nucleic acids. The gel/NPs/miR-146a binary formulation exhibits a favorable pharmacodynamic impact for the treatment of ovalbumin-induced rhinitis in rats and has a suitable nasal mucosal drug delivery ability. It was concluded that this nanoparticle is a potential nucleic acid delivery platform for the treatment of allergic rhinitis (Su et al. 2020).

Many commonly prescribed medications, such as anti-inflammatory, antioxidant, and corticosteroid medications, are dangerous, call for high doses, and have negative side effects. To treat respiratory injuries, a wide variety of nanomaterials are now being researched as pulmonary medication delivery systems. In addition to their obvious benefits for increasing pharmacokinetics, extending blood circulation time, and lowering medication toxicities, nanomaterials also provide advantages for targeted drug administration, delaying drug release, and boosting drug solubility. Polymeric nanoparticles are the most extensively employed of these due to their high degree of customizability, making them the most well-liked and efficient drug delivery materials. In addition to increasing the bioavailability of hormonal medications and reducing the number of hormonal drug administrations, nanoparticles make a significant contribution to the management of asthma by serving as a vehicle for the delivery of protein and nucleic acid medications, delaying their accelerated cellular deterioration (Zuo et al. 2022).

For transporters and medication delivery systems, polymers are widely used. Chitosan (CS), a polysaccharide, is frequently utilized to create drug delivery systems because of its ability to enhance mucus adherence and penetration. The mucus's ability to adhere provides for better local medication retention in the bronchi. Chitosan is also a biocompatible and biodegradable substance that can increase the permeability of drugs in the airway (Wang et al. 2021). The second is one of the most widely used biodegradable polymers poly(D,L-lactic-co-glycolic acid) (PLGA), which combines the benefits of both biodegradable polymers and tiny delivery vehicles. It is particularly exciting for the development of innovative drug delivery systems. Therefore, previous literature suggests a variety of polymers that could be employed for encapsulating purposes (Ding and Zhu 2018).

10.7 Conclusion

To sum up, the conversation of employing nanoparticles to address airway diseases highlights the immense possibilities of nanotechnology in transforming treatment strategies for these challenging ailments. As we've delved into throughout this section, the distinct attributes of nanoparticles like their size, surface features, and ability to release drugs in a controlled manner make them highly promising options, for delivering medications precisely to the respiratory system. Polymer-based nanoparticles have advantages compared to treatment methods. Additionally, the adaptability of these polymer materials allows for tailoring nanoparticle properties to meet the needs and disease profiles of patients. In the future, it will be crucial for scientists, engineers, and medical professionals to collaborate and conduct research in order to advance the development of polymer-based nanoparticles, for treating airway diseases. As we embark on this frontier of science, our aspiration is that these innovative approaches will ultimately enhance the quality of life for individuals grappling with allergic airway diseases by providing them with renewed hope and relief on their journey, toward better health and well-being.

References

Abruzzo A, Cerchiara T, Bigucci F, Zuccheri G, Cavallari C, Saladini B, Luppi B (2019) Cromolyn-crosslinked chitosan nanoparticles for the treatment of allergic rhinitis. Eur J Pharm Sci 131:136–145

Alavi M, Webster TJ (2021) Recent progress and challenges for polymeric microsphere compared to nanosphere drug release systems: is there a real difference? Bioorg Med Chem 33:116028

Alenazi AS, El-Bagory IM, Yassin AB, Alanazi FK, Alsarra IA, Haq N, Bayomi MA, Shakeel F (2022) Design of polymeric nanoparticles for oral delivery of capreomycin peptide using double emulsion technique: impact of stress conditions. J Drug Deliv Sci Technol 71:103326

Almoustafa HA, Alshawsh MA, Chik Z (2017) Technical aspects of preparing PEG-PLGA nanoparticles as carrier for chemotherapeutic agents by nanoprecipitation method. Int J Pharm 533(1):275–284

Babos G, Biró E, Meiczinger M, Feczkó T (2018) Dual drug delivery of sorafenib and doxorubicin from PLGA and PEG-PLGA polymeric nanoparticles. Polymers 10(8):895

Bhat G (2015) Polymeric nanofibers: recent technology advancements stimulating their growth. JTESE 5(1):1

Cheng CJ, Tietjen GT, Saucier-Sawyer JK, Saltzman WM (2015) A holistic approach to targeting disease with polymeric nanoparticles. Nat Rev Drug Discov 14(4):239–247

Chikuma K, Arima K, Asaba Y, Kubota R, Asayama S, Sato K, Kawakami H (2020) The potential of lipid-polymer nanoparticles as epigenetic and ROS control approaches for COPD. Free Radic Res 54(11–12):829–840

Deng Z, Kalin GT, Shi D, Kalinichenko VV (2021) Nanoparticle delivery systems with cell-specific targeting for pulmonary diseases. Am J Respir Cell Mol Biol 64(3):292–307

Ding D, Zhu Q (2018) Recent advances of PLGA micro/nanoparticles for the delivery of biomacromolecular therapeutics. Mater Sci Eng C 92:1041–1060

Gadad AP, Dandagi PM, Mastiholimath VS (2012) Moxifloxacin loaded polymeric nanoparticles for sustained ocular drug delivery. IJPSN 5(2):1727–1734

Jarai BM, Kolewe EL, Stillman ZS, Raman N, Fromen CA (2020) Polymeric nanoparticles. In: Nanoparticles for biomedical applications. Elsevier, pp 303–324

Kan S, Hariyadi DM, Grainge C, Knight DA, Bartlett NW, Liang M (2020) Airway epithelial-targeted nanoparticles for asthma therapy. Am J Phys Lung Cell Mol Phys 318(3):L500–L509

Kandil R, Merkel OM (2019) Recent progress of polymeric nanogels for gene delivery. Curr Opin Colloid Interface Sci 39:11–23

Lee H, Jeong SW, Jung E, Lee D (2020a) Dexamethasone-loaded H2O2-activatable anti-inflammatory nanoparticles for on-demand therapy of inflammatory respiratory diseases. Nanomedicine 30:102301

Lee SJ, Park SM, Han SJ, Kim DS (2020b) Electrolyte solution-assisted electrospray deposition for direct coating and patterning of polymeric nanoparticles on non-conductive surfaces. Chem Eng J 379:122318

Liu D, Long M, Gao L, Chen Y, Li F, Shi Y, Gu N (2022) Nanomedicines targeting respiratory injuries for pulmonary disease management. Adv Funct Mater 32(22):2112258

Mahmoud KY, Elhesaisy NA, Rashed AR, Mikhael ES, Fadl MI, Elsadek MS, Mohamed MA, Mostafa MA, Hassan MA, Halema OM, Elnemer YH, Swidan SA (2023) Exploring the potential of intranasally administered naturally occurring quercetin loaded into polymeric nanocapsules as a novel platform for the treatment of anxiety. Sci Rep 13(1):510

Malamatari M, Charisi A, Malamataris S, Kachrimanis K, Nikolakakis I (2020) Spray drying for the preparation of nanoparticle-based drug formulations as dry powders for inhalation. PRO 8(7):788
Mandal B, Bhattacharjee H, Mittal N, Sah H, Balabathula P, Thoma LA, Wood GC (2013) Core–shell-type lipid–polymer hybrid nanoparticles as a drug delivery platform. Nanomedicine 9(4):474–491
Middha E, Manghnani PN, Ng DZL, Chen H, Khan SA, Liu B (2019) Direct visualization of the ouzo zone through aggregation-induced dye emission for the synthesis of highly monodispersed polymeric nanoparticles. Mater Chem Front 3(7):1375–1384
Mohamed A, Kunda NK, Ross K, Hutcheon GA, Saleem IY (2019) Polymeric nanoparticles for the delivery of miRNA to treat Chronic Obstructive Pulmonary Disease (COPD). Eur J Pharm Biopharm 136:1–8
Nag S, Baliyan S, Pathak A, Gupta S (2021) Hybrid lipid/polymer nanoparticles for pulmonary delivery of siRNA: development and fate upon in vitro deposition on the human epithelial airway barrier. In: Handbook of lung targeted drug delivery systems. CRC Press, pp 287–302
Panigrahi D, Sahu PK, Swain S, Verma RK (2021) Quality by design prospects of pharmaceuticals application of double emulsion method for PLGA loaded nanoparticles. SN Appl Sci 3:1–21
Paris JL, de la Torre P, Flores AI (2021) New therapeutic approaches for allergy: a review of cell therapy and bio-or nano-material-based strategies. Pharmaceutics 13(12):2149
Pedrozo RC, Antônio E, Khalil NM, Mainardes RM (2020) Bovine serum albumin-based nanoparticles containing the flavonoid rutin produced by nano spray drying. Braz J Pharm Sci 56:e17692
Saxena J, Bisen M, Misra A, Srivastava VK, Kaushik S, Siddiqui AJ et al (2022) Targeting COPD with PLGA-based nanoparticles: current status and prospects. BioMed Res Int 2022:5058121
Su Y, Sun B, Gao X, Liu S, Hao R, Han B (2020) Chitosan hydrogel doped with PEG-PLA nanoparticles for the local delivery of miRNA-146a to treat allergic rhinitis. Pharmaceutics 12(10):907
Szczęch M, Szczepanowicz K (2020) Polymeric core-shell nanoparticles prepared by spontaneous emulsification solvent evaporation and functionalized by the layer-by-layer method. Nano 10(3):496
Wang D, Nasab EM, Athari SS (2021) Study effect of Baicalein encapsulated/loaded chitosan-nanoparticle on allergic asthma pathology in mouse model. Saudi J Biol Sci 28(8):4311–4317
Yang Y, Ding Y, Fan B, Wang Y, Mao Z, Wang W, Wu J (2020) Inflammation-targeting polymeric nanoparticles deliver sparfloxacin and tacrolimus for combating acute lung sepsis. J Control Release 321:463–474
Yang W, Dong Z, Li Y, Zhang Y, Fu H, Xie Y (2021) Therapeutic efficacy of chitosan nanoparticles loaded with BCG-polysaccharide nucleic acid and ovalbumin on airway inflammation in asthmatic mice. Eur J Clin Microbiol Infect Dis 40:1623–1631
Yokoyama M (2014) Polymeric micelles as drug carriers: their lights and shadows. J Drug Target 22(7):576–583
Zuo X, Guo X, Gu Y, Zheng H, Zhou Z, Wang X et al (2022) Recent advances in nanomaterials for asthma treatment. Int J Mol Sci 23(22):14427

Solid Lipid Nanoparticles for the Management of Allergic Airway Diseases

11

Girish Kumar, Reshu Virmani, Ashwani Sharma, Tarun Virmani, and Kamla Pathak

Abstract

The allergic airway diseases comprise both allergic rhinitis and asthma which are amongst the top three ailments of the twenty-first century requiring special attention for control and prevention. Various conventional drug moieties are available for treatment of allergic airway diseases, but these drug moieties show therapeutic inefficiency due to poor aqueous solubility, low bioavailability, higher adverse effects, lack of drug targeting, and drug resistance. Nanotechnology-based drug carriers like liposomes, dendrimers, solid lipid nanoparticles, nanostructured lipid carriers, polymeric micelles, polymeric nanoparticles, carbon nanotubes, metallic nanoparticles, and nanoemulsions have attracted the attention of scientists for efficacious treatment of allergic airway diseases. Solid lipid nanoparticles (SLNs) are more popular amongst all nanocarriers owing to their nanosize, higher solubility, greater bioavailability, higher entrapment efficiency, being easy to scale up, drug targeting to the lungs, and being easy to prepare. Moreover, the limitations of poor physical stability, low entrapment efficiency, and opsonization associated with liposomes, dendrimers, and polymeric nanoparticles are excluded in SLNs. Hence, this chapter was written with a view to explore the potential of SLNs in efficacious treatment of allergic airway diseases. In this chapter, various conventional drugs used for treatment of allergic airway diseases, composition of SLNs, and potential of SLNs in allergic airway diseases along with the research related to SLNs having drugs used in allergic airway diseases have been summarized.

G. Kumar · R. Virmani · A. Sharma · T. Virmani
School of Pharmaceutical Sciences, MVN University, Palwal, Haryana, India

K. Pathak (✉)
Faculty of Pharmacy, Uttar Pradesh University of Medical Sciences, Saifai, Etawah, Uttar Pradesh, India

P. Prasher et al. (eds.), *Understanding Allergic Airway Diseases*,
https://doi.org/10.1007/978-981-97-1953-2_11

Keywords

Allergic airway diseases · Asthma · Allergic rhinitis · Solid lipid nanoparticles · Corticosteroids

11.1 Introduction

Both allergic rhinitis (AR) and atopic asthma are chronic allergic airway disorders (Wikstén et al. 2018). In patients who have developed an allergy sensitivity, AR is brought on by an allergen binding to a specific IgE, which results in symptoms such rhinorrhea, obstruction, itching, sneezing, and weariness. Asthma, allergic conjunctivitis, atopic dermatitis, and chronic rhinosinusitis (CRS) are a few of the allergic and nonallergic comorbidities that are linked to AR (Hansel 1929). Asthma is described as chronic inflammation with mucus hypersecretion, variable obstruction, edema, and fatigue (Sockrider and Fussner 2020). Asthma is not a single disease entity; it covers various phenotypes, both in children and adults, with allergic and nonallergic asthma being the most often characterized phenotypes. Asthma can coarsely be categorized into Th1- and Th2-high phenotypes. The latter comprises eosinophilic and/or allergic illnesses that can range in severity from mild to extremely severe (Lemanske and Busse 2003). Adults are more likely to develop severe eosinophilic forms, such as acetylsalicylic acid (ASA)-exacerbated respiratory disease (AERD). Due to its complexity and inflammatory nature, asthma cannot be treated as a single disease. This immune illness is quite heterogeneous and dynamic, with substantial gene/environment interactions (Wikstén et al. 2018; Sharma et al. 2022).

According to the World Health Organization (WHO), allergic disorders have risen to the top three ailments requiring significant prevention and control efforts in the twenty-first century (Mazur et al. 2022). AR is a widespread illness that affects 400 million people globally since its incidence has grown over the time representing a global problem. Since the 1990s, there has been a marked rise in the prevalence of AR. Approximately 25% of children and 40% of adults worldwide are said to be affected, respectively. Eighty percent of AR symptoms appear before the age of 20 and peak between 20 and 40 before progressively fading (Nur Husna et al. 2022). More than 300 million people are suffering from asthma, and it is estimated that the number of asthmatic patients will be more than 400 million by 2025 which is very serious matter (Allan et al. 2021).

The increase in patients with allergic airway diseases is attributed to sweeping, viruses, smoking, diseases, animals, perfume odors, house dust mites, hormonal changes, rain, exercise, viral infection, etc. The control over these attributes can help in reduction of allergic airway diseases patients (Gautier and Charpin 2017).

Various therapeutic moieties are available for treatment of AR and asthma but suffer from various obstacles of lack of drug targeting, first pass metabolism, drug resistance, adverse effects, and requirement of larger dose which have prompted to develop advanced drug delivery systems for efficacious treatment of AR and asthma (Mukherjee et al. 2009; Kumar et al. 2018, 2022a). The use of nanotechnology in contemporary medicine and pharmacy has significantly increased in recent years. In addition to their obvious benefits in boosting pharmacokinetics, extending blood circulation time, and lowering medication toxicities, nanotechnology also makes it possible to deliver pharmaceuticals at targeted site, delay their release, and increase their solubility (Mathis et al. 2021; Sharma et al. 2023). Various Nanotechnology-based drug delivery systems including liposomes, solid lipid nanoparticles (SLNs), nanostructured lipid carriers (NLCs), polymeric nanoparticles (PNPs), polymeric micelles (PMs), dendrimers, niosomes, and nanoemulsions (Kumar et al. 2022b) have been employed to improve the therapeutic efficiency of drugs in AR and asthma; amongst these SLNs have attracted the attention of researchers due to possession of various characteristics such as nanosize, larger surface area, greater drug loading, reduced adverse effects, and drug targeting to the lungs (Ahmad 2022; Virmani et al. 2022a; Ahalwat et al. 2023).

11.2 Therapeutic Treatment for Allergic Airway Diseases

Various therapeutic drug moieties are available for treatment for AR and asthma. The drug moieties employed for treatment of AR include oral antihistaminics like bilastine, desloratadine, cetirizine, fexofenadine, loratadine, and levocetirizine and intranasal corticosteroids like budesonide, beclomethasone, fluticasone furoate, mometasone, triamcinolone, etc. The drug moieties for treatment of asthma include inhalable corticosteroids, muscarinic antagonists, xanthine derivatives, long-acting beta antagonists, leukotriene receptor antagonists, and antibody derivatives. Various drug moieties for treatment of AR and asthma have been summarized in Table 11.1. All conventional therapeutic moieties available for treatment of allergic airway diseases suffer from various hindrances such as lack of drug targeting, first pass metabolism, drug resistance, adverse effects, and requirement of larger dose causing therapeutic inefficiency in treatment of allergic airway diseases. Hence it is an urgent need to develop advanced drug delivery systems for efficacious delivery of available therapeutic drug moieties in treatment of allergic airway diseases.

11.3 Nanotechnology-Based Treatment for Allergic Airway Diseases

Drug delivery systems based on nanotechnology are potential alternatives for allergic airway diseases because of the advantages of drug deposition at the site of disease, controlled release of the drug, sustained release of the drug, biodegradability, improved dosing interval, reduced side effects, reduced dose, controlled size, controlled surface charge, greater surface-volume ratio, enhanced stability, and delivery of hydrophilic as well as hydrophobic drugs (Jain 2014; Virmani et al. 2023a). Due to their unique characteristics, nanocarriers serve as an effective platform for improving medication solubility, dissolution, and bioavailability (Xu et al. 2020). Nanotechnology-based drug carriers comprise liposomes, solid lipid nanoparticles

Table 11.1 List of therapeutic drugs employed for treatment of allergic airway diseases

Drug category	Drug name	Action	Adverse effects	References
AR				
Oral antihistaminics	Cetirizine	Relieves rhinorrhea, watery eyes, and sneezing	Fatigue, dizziness, pharyngitis, dryness of mouth	Meltzer and Grant (1999)
	Bilastine	Prompt to inhibition of release of histamine by mast cells	Somnolence, nausea, dry mouth, headache, dizziness, abdominal pain	Church et al. (2020)
	Loratadine	Inhibits H1 receptors present on respiratory smooth muscle cells	Sedation, dizziness, abdominal pain, GI distress, nausea, vomiting	Sidhu and Akhondi (2023)
	Fexofenadine			Meltzer et al. (2021))
	Levocetirizine			Nettis et al. (2008)
Intranasal corticosteroids	Budesonide	Inhibits airway inflammation	Blurred vision, osteoporosis, Cushing's syndrome	Wang et al. (2015)
	Fluticasone furoate	Inhibition of inflammatory cells such as mast cells, macrophages	Loss of appetite, mental depression, skin rashes, weakness	Giavina-Bianchi (2008)
	Mometasone	Activation of pathways responsible for decreasing the inflammation	Loss of appetite, candidiasis, dysphonia	Passali et al. (2016)
	Beclomethasone	Attenuation of inflammation	Body pain, congestion, cough, fever, hoarseness	Derendorf and Meltzer (2008)
Oral decongestants	Pseudoephedrine	Constriction of blood vessels in nasal passages	Nervousness, nausea, vomiting, dizziness, headache	Laccourreye et al. (2015)

(continued)

Table 11.1 (continued)

Drug category	Drug name	Action	Adverse effects	References
Intranasal cromolyns	Cromolyn sodium	Inhibition of inflammatory mediators	Nasal itching, nasal bleeding, nasal congestion, headache	Ratner et al. (2002)
Asthma				
Long-acting beta antagonists	Salmeterol xinafoate	Inhibits bronchoconstriction	Blurred vision, hives, and headache	Tashkin and Fabbri (2010)
Xanthine derivatives	Theophylline	Eases airway muscles	Abdominal pain, nausea, headache	Barnes (2010)
Leukotriene receptor antagonists	Montelukast sodium	Inhibits bronchospasm	Cough, dental pain, abdominal pain, and headache	Tamada and Ichinose (2017)
Muscarinic antagonists	Tiotropium	Dilates the bronchi	Constipation, dry mouth	Castillo et al. (2017)
Inhalable corticosteroids	Budesonide	Decreases swelling of airways	Candidiasis, dysphonia	Aalbers et al. (2016)
	Fluticasone propionate	Inhibits allergic reactions	Nausea, vomiting, and nasal dryness	Falk et al. (2016)
	Fluticasone furoate	Decreases swelling of airways	Candidiasis, dysphonia	Lilburn et al. (2019)
	Mometasone	Decreases swelling of airways	Candidiasis, dysphonia	Falk et al. (2016)
Antibody derivative	Reslizumab	Possesses anti-eosinophilic action	Pain at injection site and headache	Menzella et al. (2016)
	Mepolizumab	Inhibits airway swelling	Pain at injection site, headache, and rashes	Narendra and Hanania (2019)
	Omalizumab	Inhibits allergic reactions	Redness, pain, swelling at injection site	Okayama et al. (2020)

(SLNs), nanostructured lipid carriers (NLCs), polymeric nanoparticles (PNPs), polymeric micelles (PMs), nanoemulsions (NEs), dendrimers, niosomes, etc. (Kumar et al. 2023; Virmani et al. 2023b) out of which SLNs have provided various advantageous features over conventional drug delivery systems and earlier nanocarriers for efficient drug delivery in treatment of allergic airway diseases.

11.3.1 SLNs

Numerous colloidal carriers, primarily liposomes, niosomes, dendrimers, nanoparticles, and many more vectors, have been studied especially over the past two decades (Dua et al. 2017; Su and Kang 2020; Virmani et al. 2022b). Low physical stability, opsonization, and low entrapment efficiency (EE) of these formulations further restrict their clinical availability and economic viability (Kesharwani et al. 2015). The solid lipid-based delivery system which is solid at physiological temperatures was created in 1990 as an alternative to these delivery systems (Duong et al. 2020; Virmani et al. 2023c). Solid lipid nanoparticles (SLNs) were developed, and they show promise for long-term stability as well as medication targeting and controlled release (Dolatabadi et al. 2015; Virmani et al. 2023d). SLNs offer various notable benefits of better solubility, reduced side effects, increased bioavailability of medications, adaptability of encapsulation of both hydrophilic and hydrophobic drugs, better stability, specificity, and likelihood of large-scale production (Bayón-Cordero et al. 2019; Fathy Abd-Ellatef et al. 2020) which have been depicted in Fig. 11.1. SLNs are less toxic than other nanocarriers such as PNPs because of their biodegradability and biocompatibility (Ghasemiyeh and Mohammadi-Samani 2018). The more beneficial properties of SLNs are their nanosize (less than 400 nm), ease of functionalization, chemical and mechanical stability, and enhanced transport of lipophilic drugs (Rehman et al. 2018). By changing the lipid components, it is possible to regulate the drug release from SLNs. It is possible to alter the surface of SLNs to target particular tissues and improve stability (Uner and Yener 2007). Techniques other than solvents, like high-pressure homogenization and high-speed stirring, can be used to make SLNs (Duong et al. 2020). These benefits enable the

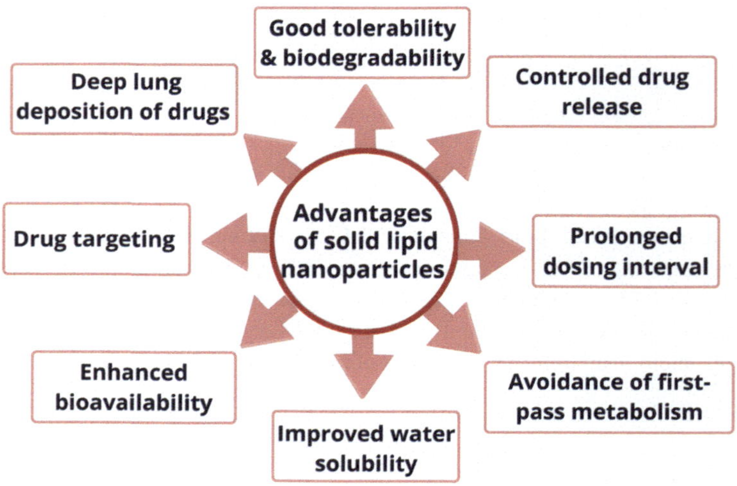

Fig. 11.1 Various benefits of SLNs as drug carrier

widespread use of SLNs in the administration of drugs via oral, parenteral, transdermal, intranasal, ophthalmic, and pulmonary routes (Scioli Montoto et al. 2020).

11.3.2 History of SLNs

Phospholipids have been intensively explored in recent years as a class of lipid molecules due to their importance in the development of lipid bilayers and their position as the primary component of cellular membrane (Müller et al. 2011). Phospholipids are a preferred class when discussing lipids and lipid-based drug delivery systems because of their wide range of qualities, including their amphiphilic nature, biocompatibility, and multifunctionality (Maheshwari et al. 2012). Lipid-based drug delivery systems comprise microemulsion, microspheres, liposomes, and SLNs (Virmani et al. 2022c). The parenteral fat emulsion which is currently often used for the parenteral delivery of poorly water-soluble medicines was the first dosage form to incorporate lipophilic medications in lipid droplets in the 1960s (Weber et al. 2014). In 1990, a patent on phospholipid-based drug carriers was filed in which drug was loaded inside the phospholipid along with covering of surface with surfactant having size between 1 and 250 μm (Tekade et al. 2017). This nanocarrier enabled the delivery of poorly aqueous soluble drugs owing to dispersibility in phospholipid and reduced particle size. After it, lipid implants came into existence for controlled delivery of drugs. But these nanocarriers suffer from the problems of production methods, poor entrapment efficiency, and lack of scale-up techniques which caused obstacles in their commercialization. In 1990, to combat the problems associated with these available nanocarriers, first-generation lipid nanocarriers, namely, solid lipid nanoparticles, were introduced (Dhakad et al. 2013). In SLN, the oil of the fat emulsion was swapped out for a solid lipid or a combination of solid lipids, making the lipid matrix of SLN solid at both room temperature and body temperature. SLNs are made up of 0.5–5% (w/w) surfactant acting as a stabilizing agent and an aqueous solution with 0.1–30% (w/w) lipid distributed in it (Dhiman et al. 2021).

11.3.3 Structural Features and Composition of SLNs

SLNs typically have smooth surfaces, spherical shapes, and mean diameters between 50 and 1000 nm (Duong et al. 2020). SLNs are spherical particles having solid lipid matrix entrapping the drug molecules along with a surfactant layer to stabilize the nanoparticles in an aqueous phase (Leong and Ge 2022). The structure of SLNs is impacted by various factors including composition of the formulation, the solubility of the compounds including the medicine, and the technique of production (Viegas et al. 2023). Three different structures of SLNs, namely, homogenous matrix model (A), drug-enriched shell model (B), and drug-enriched core model (C), have been identified which have been depicted in Fig. 11.2 (Borges et al. 2020). In homogenous matrix model, the nanoparticles are produced by hot or cold homogenization technique for highly hydrophobic drugs, while in drug-enriched

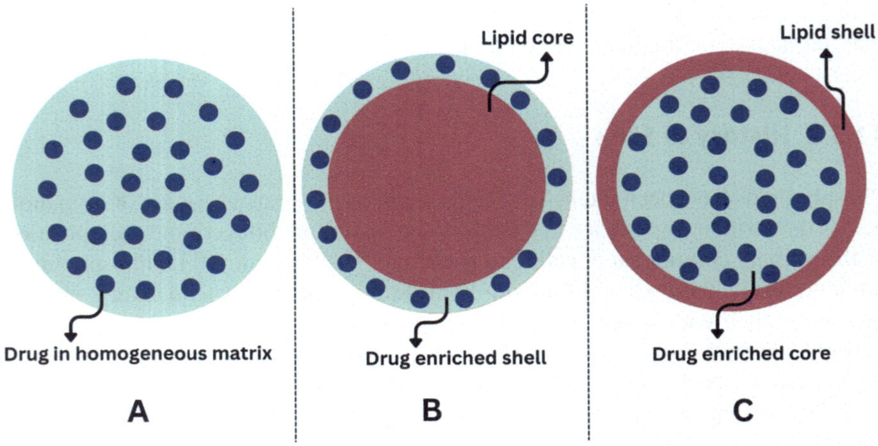

Fig. 11.2 Different structures of SLNs

Table 11.2 List of ingredients employed in fabrication of SLNs (Mishra et al. 2018)

Type of ingredients	Examples
Lipids	Beeswax, cholesterol, stearic acid, caprylic/capric triglyceride, glyceryl stearate (-mono, and -tri), cetylpalmitate, glyceryl trimyristate, glyceryl trilaurate, glyceryl tripalmitate, glyceryl behenate (Compritol), monostearate monocitrate, hardened fat (Witepsol E85, H5, and W35), behenic acid, solid paraffin
Surfactants	Soy and egg lecithin, poloxamer, polysorbate 80, poloxamine
Cosurfactants	Sodium dodecyl sulfate, sodium oleate, Tyloxapol, sodium glycocholate, taurocholate sodium salt, butanol
Cryoprotectant	Gelatin, mannose, glucose, lactose, maltose, sorbitol, polyvinyl alcohol, mannitol, polyvinyl pyrrolidone, glycine
Charge modifiers	Dipalmitoyl phosphatidyl choline, dicetylphosphate, stearylamine, dimyristoyl phosphatidylglycerol
Preservatives	Thiomersal

shell model, the nanoparticles are fabricated using hot homogenization technique only. Finally, when the drug concentration is nearly at its solubility limit in the melted lipid, the SLN drug-enriched core model produces nanoparticles (Viegas et al. 2023).

SLNs are mainly composed of solid lipids, surfactants, and solvents along with drugs. The cosurfactant, cryoprotectants, charge modifiers and preservatives may be employed on the requirement (Ebrahimi et al. 2015). Various ingredients employed in preparation of SLNs have been summarized in Table 11.2. The lipids employed in preparation of SLNs are solid at physiological temperature. Typically, when formulating SLNs, the lipids commonly thought to be safe are taken into account (Chauhan et al. 2020). Surfactants contribute to the stabilization of the SLN structure by lowering the interfacial tension between the aqueous environment and the hydrophobic surface of the lipid core (Mishra et al. 2018).

11.3.4 SLNs for Allergic Airway Diseases

Solid lipid nanoparticles (SLNs) have been the subject of substantial research for potential pulmonary medication delivery in allergic airway diseases especially in asthma for a very long time. SLNs are aqueous nanoscale suspensions made mostly of triglycerides and phospholipids which are physiological lipids (Paranjpe and Müller-Goymann 2014). Since the formulations are built around the use of physiological components, they are less toxic and, as a result, more palatable for pulmonary medication delivery in allergic airway diseases like asthma. The deep regions of the lungs contain phospholipids in abundance, which are crucial for the proper operation of the respiratory system. Alveolar surface phospholipid-based surfactant proteins are crucial for preserving ideal surface tension and minimizing friction in the lung tissue (Beck-Broichsitter et al. 2011). Nassimi et al. demonstrated the utilization of phospholipid- and triglyceride-based SLN in the ratio of 30:70 for prospective pulmonary applications. Along with assessments of cytokine activation, they also assessed the toxicity profile of these SLNs in in vitro, ex vivo, and in vivo models. When the SLN was nebulized into mice, they noticed no activation of pro-inflammatory cytokines KC and TNF-α (Nassimi et al. 2010). Moreover, the therapeutic window and retention period of naked medicines are constrained by strong pulmonary clearance mechanisms; hence the entrapment of drugs for allergic airway diseases like asthma can enhance the retention time as well as local concentration of the drugs in the lungs leading to improved therapeutic efficacy (Pilcer and Amighi 2010). SLNs provide various advantageous characteristics in delivery of anti-asthmatic drugs to the lungs which include improved tolerability of SLNs in the lungs owing to biodegradable lipids. Because of their small size, nanoparticles are easily entrapped into particles or aerosolized into droplets with appropriate aerodynamic properties allowing for adequate deep lung deposition of an active substance. Due to their small size, nanosized particles stick to the lung's mucosal surface for a longer amount of time than bigger particles do leading to drug targeting at desired site in the lungs (Iyer et al. 2015) as depicted in Fig. 11.3. Longer dose intervals and improved patient compliance are possible as a result of enhanced and sustained therapeutic effects caused by particle adhesion, accumulation, and retention in the lung as well as the prolonged release features of SLNs (Patlolla et al. 2010). Lv et al. fabricated SLNs loaded with rhynchophylline (Rhy) to enhance the therapeutic efficacy in an animal model of allergic asthma using solvent injection method. The allergic asthma model was established using ovalbumin. It was investigated that Rhy-loaded SLNs provided improved effects on preventing ovalbumin-induced inflammation of airways, airway remodeling, and oxidative stress than pure Rhy in allergic asthma model. The researchers found that Rhy-loaded SLNs reduced allergic asthma by increasing cytokine signalling 1 levels and decreasing p38 signalling pathways (Lv et al. 2021). In another study, Wang W et al demonstrated the

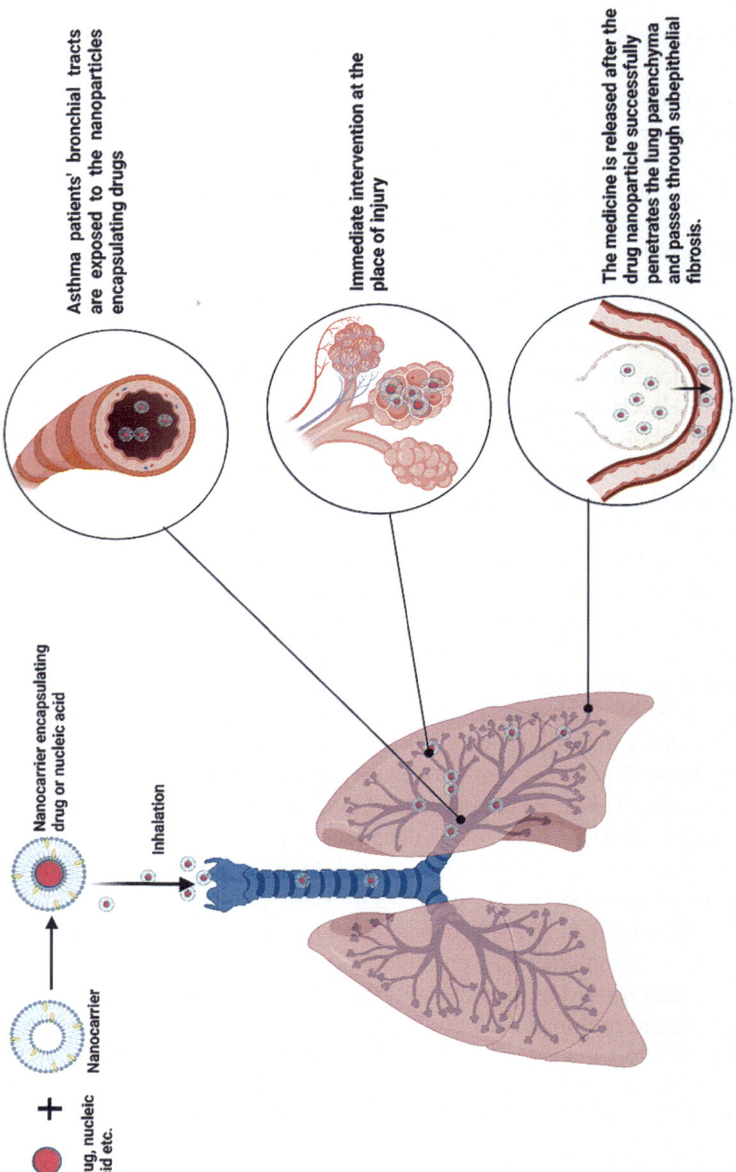

Fig. 11.3 Pictorial representation of drug targeting to the lungs employing SLNs

improved therapeutic efficacy of curcumin-loaded SLNs in ovalbumin-induced allergic rat model of asthma. The results provided that, as compared to curcumin alone, the SLNs loaded with curcumin had increased plasma concentrations. All tissue curcumin concentrations increased after the injection of the curcumin-loaded SLNs, but the liver and lung showed the greatest increases. Curcumin-loaded SLNs significantly inhibited the expression of T-helper 2-type cytokines, such as interleukin-4 and interleukin-13, in bronchoalveolar lavage fluid in the animal model of asthma when compared to the asthma group and the curcumin-treated group. These effects included the effective suppression of airway hyperresponsiveness and inflammatory cell infiltration (Wang et al. 2012).

Various drugs like antihistaminics and intranasal steroids are available for treatment of allergic rhinitis, but their therapeutic efficiency is suboptimal owing to their poor distribution within the nose and sinus (Sallam et al. 2016). Triamcinolone acetonide (TA), a synthetic corticosteroid which is used to treat rhinitis symptoms, is available as the nasal spray Nasacort® AQ. But it suffers from various drawbacks including frequent administration, bitter taste, and common side effects such nasal discomfort and epistaxis (Berger and Meltzer 2015). Along with this, defense systems such as mucociliary clearance also reduce drug residence in the nostril.

The important consideration in the management of rhinitis and nasal congestion is the longer residual period of steroids in the nasal cavity. Nanotechnology-based drug carriers mainly SLNs looked to improve the effectiveness of locally acting nasal steroids by extending the therapeutic agent's duration in the nasal cavity (Sallam et al. 2016). Furthermore, majority of the drugs employed in treatment of allergic rhinitis possess low aqueous solubility and poor bioavailability leading to therapeutic inefficiency of drug treatment in allergic rhinitis SLNs have proved their efficacy in improving the aqueous solubility and bioavailability of various drugs used in allergic airway diseases (Pohlit et al. 2017).

Madgulkar A et al. prepared mometasone furoate-loaded intranasal SLNs employing hot homogenization method which provided maximum entrapment efficiency along with reduced particle size. It was observed that glyceryl monostearate as lipid phase has maximum solubility due to which it is employed as solid lipid in preparation of SLNs loaded with mometasone furoate (Madgulkar et al. 2019). Patel R et al. fabricated cromolyn sodium-loaded SLNs with an intension of improvement in oral bioavailability using double emulsification method. It was observed that the permeability of cromolyn sodium when presented as SLNs increased by around 2.96 times, according to an ex vivo and in vivo intestinal permeation investigation. Additionally, an in vivo pharmacokinetic investigation revealed that encapsulating drug inside SLNs increased its oral bioavailability by around 2.86-fold, demonstrating that SLNs are a promising therapeutic carrier system for cromolyn sodium oral delivery (Patel et al. 2016). Various drugs used in treatment of allergic airway diseases entrapped as SLNs have been summarized in Table 11.3.

Table 11.3 A summarized list of SLNs loaded with drugs employed for allergic airway diseases

Disease	Therapeutic moiety	Method of preparation	Result outcomes	References
Asthma	Budesonide	Emulsification-solvent diffusion	Optimized formulation exhibited anticipated fine particle segment, mass median aerodynamic diameter (MMAD) along with geometric standard deviation of 49.5%, 2.06 μm, and 2.98 μm, respectively	Emami et al. (2015)
Asthma	Budesonide	Ultrasonication	Provided improved aerosolization behavior than commercial conventional formulation of budesonide showing increment in fraction of fine particles on basis of in vitro aerosolization performance testing	Esmaeili et al. (2016)
Asthma	Mometasone furoate	Solvent diffusion	Dry powder inhalations having SLNs of mometasone furoate exhibit improved efficacy than conventional dry powder inhalations of same drug	Shah et al. (2013)
Asthma	Beclomethasone	High shear homogenization	SLNs were effectively nebulized, producing aerosols with the right particle size for deep lung administration of drug as demonstrated by aerosolization and cascade impaction studies	Jaafar-Maalej et al. (2011)
Asthma	Rhynchophylline	Solvent injection	Drug-loaded SLNs reduced allergic asthma by suppressing the p38 signalling pathway and upregulating the suppressor of cytokine signalling 1 in better way	Lv et al. (2021)

(continued)

Table 11.3 (continued)

Disease	Therapeutic moiety	Method of preparation	Result outcomes	References
Asthma	Curcumin	Solvent injection	Curcumin-loaded SLNs significantly reduced the expression of T-helper 2-type cytokines, such as interleukin-4 and interleukin-13, in bronchoalveolar lavage fluid and effectively reduced airway hyperresponsiveness and inflammatory cell infiltration than asthma group and the curcumin-treated group	Wang et al. (2012)
Asthma	Montelukast sodium	Hot homogenization followed by ultrasonication	Provided enhanced particle size, entrapment efficiency along with maintenance of the sustained release of drug	Priyanka and Abdul Hasan (2012)
Allergic rhinitis	Loratadine	Emulsion congealing	Displayed increased drug entrapment, reduced particle size range, and controlled drug release Characteristics which indicate the potential of SLNs for better treatment of disease	Uner and Karaman (2013)
Allergic rhinitis	Cromolyn sodium	Double emulsification	Displayed about 2.96-fold enhancement in permeability of drug. Furthermore, 2.86-fold improved oral bioavailability of the drug was obtained as indicated by in vivo studies	Patel et al. (2016)
Asthma	Quercetin	Emulsification through phase inversion	Quercetin SLMs could be made into dry powder that is acceptable for inhalation medication delivery and was absorbed using a linear kinetic model over the Calu-3 monolayer	Scalia et al. (2013)

11.4 Conclusion and Future Perspectives

The prevalence of allergic airway diseases is constantly expanding due to which these diseases have risen to the top three ailments requiring significant prevention and control efforts in the twenty-first century. However various conventional therapeutic drugs are available to combat these diseases including antihistamines, inhaled corticosteroids, intranasal corticosteroids, leukotriene receptor antagonists, muscarinic antagonists, etc., but these drugs suffer from various hindrances of poor aqueous solubility, low bioavailability, higher adverse effects, lack of drug targeting, and drug resistance causing therapeutic inefficiency in treatment of these diseases. Nanocarrier-based formulations have shown their potential in efficient delivery of drug in treatment of allergic airway diseases; solid lipid nanoparticles (SLNs) have tremendous potential over other colloidal nanocarriers like liposomes, niosomes, dendrimers, polymeric nanoparticles, and many more because these colloidal carriers have limitations of poor physical stability, low entrapment efficiency, and opsonization. SLNs provide various benefits of reduced size, decreased adverse effects, increased bioavailability, controlled drug release, and ability of drug targeting to the lungs which are the main infected organs of allergic airway diseases. Numerous drugs employed in treatment of allergic airway diseases have been entrapped in SLNs showing potential of SLNs for better treatment of allergic airway diseases. However, the drug-entrapped SLNs have some limitations of expulsion of drug, poor drug loading capacity, unpredictable agglomeration, and increased chances of polymorphic transitions, which must be addressed for their clinical utility. Moreover, the toxicological profile of the SLNs must be established for their arrival from laboratory to consumer.

References

Aalbers R, Vogelmeier C, Kuna P (2016) Achieving asthma control with ICS/LABA: a review of strategies for asthma management and prevention. Respir Med 111:1–7. https://doi.org/10.1016/j.rmed.2015.11.002

Ahalwat S et al (2023) Mannose-functionalized isoniazid-loaded nanostructured lipid carriers for pulmonary delivery: in vitro prospects and in vivo therapeutic efficacy assessment. Pharmaceuticals 16(8):1108. https://doi.org/10.3390/ph16081108

Ahmad A (2022) Pharmacological strategies and recent advancement in nano-drug delivery for targeting asthma. Life 12(4):596. https://doi.org/10.3390/life12040596

Allan R et al (2021) Usability and robustness of the Wixela Inhub dry powder inhaler. J Aerosol Med Pulm Drug Deliv 34(2):134–145. https://doi.org/10.1089/jamp.2020.1603

Barnes PJ (2010) Theophylline. Pharmaceuticals 3(3):725–747. https://doi.org/10.3390/ph3030725

Bayón-Cordero L, Alkorta I, Arana L (2019) Application of solid lipid nanoparticles to improve the efficiency of anticancer drugs. Nanomaterials (Basel) 9(3):474. https://doi.org/10.3390/nano9030474

Beck-Broichsitter M et al (2011) Biophysical investigation of pulmonary surfactant surface properties upon contact with polymeric nanoparticles in vitro. Nanomed Nanotechnol, Biol Med 7(3):341–350. https://doi.org/10.1016/j.nano.2010.10.007

Berger WE, Meltzer EO (2015) Intranasal spray medications for maintenance therapy of allergic rhinitis. Am J Rhinol Allergy 29(4):273–282. https://doi.org/10.2500/ajra.2015.29.4215

Borges A, de Freitas V, Mateus N, Fernandes I, Oliveira J (2020) Solid lipid nanoparticles as carriers of natural phenolic compounds. Antioxidants (Basel) 9(10):998. https://doi.org/10.3390/antiox9100998

Castillo JR, Peters SP, Busse WW (2017) Asthma exacerbations: pathogenesis, prevention, and treatment. JACI: In Pract 5(4):918–927. https://doi.org/10.1016/j.jaip.2017.05.001

Chauhan I, Yasir M, Verma M, Singh AP (2020) Nanostructured lipid carriers: a groundbreaking approach for transdermal drug delivery. Adv Pharm Bull 10(2):150–165. https://doi.org/10.34172/apb.2020.021

Church MK, Tiongco-Recto M, Ridolo E, Novák Z (2020) Bilastine: a lifetime companion for the treatment of allergies. Curr Med Res Opin 36(3):445–454. https://doi.org/10.1080/03007995.2019.1681134

Derendorf H, Meltzer EO (2008) Molecular and clinical pharmacology of intranasal corticosteroids: clinical and therapeutic implications. Allergy 63(10):1292–1300. https://doi.org/10.1111/j.1398-9995.2008.01750.x

Dhakad R, Tekade R, Jain N (2013) Cancer targeting potential of folate targeted nanocarrier under comparative influence of tretinoin and dexamethasone. CDD 10(4):477–491. https://doi.org/10.2174/1567201811310040012

Dhiman N, Awasthi R, Sharma B, Kharkwal H, Kulkarni GT (2021) Lipid nanoparticles as carriers for bioactive delivery. Front Chem 9. Available https://www.frontiersin.org/articles/10.3389/fchem.2021.580118. Accessed 29 Aug 2022. [Online]

Dolatabadi JEN, Valizadeh H, Hamishehkar H (2015) Solid lipid nanoparticles as efficient drug and gene delivery systems: recent breakthroughs. Adv Pharm Bull 5(2):151–159. https://doi.org/10.15171/apb.2015.022

Dua K, Shukla SD, Tekade RK, Hansbro PM (2017) Whether a novel drug delivery system can overcome the problem of biofilms in respiratory diseases? Drug Deliv Transl Res 7(1):179–187. https://doi.org/10.1007/s13346-016-0349-0

Duong V-A, Nguyen T-T-L, Maeng H-J (2020) Preparation of solid lipid nanoparticles and nanostructured lipid carriers for drug delivery and the effects of preparation parameters of solvent injection method. Molecules 25(20):4781. https://doi.org/10.3390/molecules25204781

Ebrahimi HA, Javadzadeh Y, Hamidi M, Jalali MB (2015) Repaglinide-loaded solid lipid nanoparticles: effect of using different surfactants/stabilizers on physicochemical properties of nanoparticles. DARU J Pharm Sci 23(1):46. https://doi.org/10.1186/s40199-015-0128-3

Emami J, Mohiti H, Hamishehkar H, Varshosaz J (2015) Formulation and optimization of solid lipid nanoparticle formulation for pulmonary delivery of budesonide using Taguchi and Box-Behnken design. Res Pharm Sci 10(1):17–33

Esmaeili M, Aghajani M, Abbasalipourkabir R, Amani A (2016) Budesonide-loaded solid lipid nanoparticles for pulmonary delivery: preparation, optimization, and aerodynamic behavior. Artif Cells Nanomed Biotechnol 44(8):1964–1971. https://doi.org/10.3109/21691401.2015.1129614

Falk NP, Hughes SW, Rodgers BC (2016) Medications for chronic asthma. Am Fam Physician 94(6):454–462

Fathy Abd-Ellatef G-E et al (2020) Curcumin-loaded solid lipid nanoparticles bypass P-glycoprotein mediated doxorubicin resistance in triple negative breast cancer cells. Pharmaceutics 12(2):96. https://doi.org/10.3390/pharmaceutics12020096

Gautier C, Charpin D (2017) Environmental triggers and avoidance in the management of asthma. J Asthma Allergy 10:47–56. https://doi.org/10.2147/JAA.S121276

Ghasemiyeh P, Mohammadi-Samani S (2018) Solid lipid nanoparticles and nanostructured lipid carriers as novel drug delivery systems: applications, advantages and disadvantages. Res Pharm Sci 13(4):288–303. https://doi.org/10.4103/1735-5362.235156

Giavina-Bianchi P (2008) Fluticasone furoate nasal spray in the treatment of allergic rhinitis. TCRM 4:465–472. https://doi.org/10.2147/TCRM.S1984

Hansel FK (1929) Clinical and histopathologic studies of the nose and sinuses in allergy. J Allergy 1(1):43–70. https://doi.org/10.1016/S0021-8707(29)90083-6

Iyer R, Hsia CCW, Nguyen KT (2015) Nano-therapeutics for the lung: state-of-the-art and future perspectives. Curr Pharm Des 21(36):5233–5244

Jaafar-Maalej C, Andrieu V, Elaissari A, Fessi H (2011) Beclomethasone-loaded lipidic nanocarriers for pulmonary drug delivery: preparation, characterization and in vitro drug release. J Nanosci Nanotechnol 11(3):1841–1851. https://doi.org/10.1166/jnn.2011.3119

Jain KK (2014) Current status and future prospects of drug delivery systems. Methods Mol Biol 1141:1–56. https://doi.org/10.1007/978-1-4939-0363-4_1

Kesharwani P, Tekade RK, Jain NK (2015) Dendrimer generational nomenclature: the need to harmonize. Drug Discov Today 20(5):497–499. https://doi.org/10.1016/j.drudis.2014.12.015

Kumar G, Khar RK, Virmani T, Jogpal V, Virmani R (2018) Comparative evaluation of fast dissolving tablet of atorvastatin calcium using natural and synthetic super disintegrating agents. Res J Pharm Technol 11(11):5001–5007. https://doi.org/10.5958/0974-360X.2018.00912.5

Kumar G, Virmani T, Pathak K, Kamaly OA, Saleh A (2022a) Central composite design implemented azilsartan medoxomil loaded nanoemulsion to improve its aqueous solubility and intestinal permeability: in vitro and ex vivo evaluation. Pharmaceuticals 15(11):1343. https://doi.org/10.3390/ph15111343

Kumar G, Virmani T, Pathak K, Alhalmi A (2022b) A revolutionary blueprint for mitigation of hypertension via nanoemulsion. Biomed Res Int 2022:e4109874. https://doi.org/10.1155/2022/4109874

Kumar G, Virmani T, Sharma A, Pathak K (2023) Codelivery of phytochemicals with conventional anticancer drugs in form of nanocarriers. Pharmaceutics 15(3):889. https://doi.org/10.3390/pharmaceutics15030889

Laccourreye O, Werner A, Giroud J-P, Couloigner V, Bonfils P, Bondon-Guitton E (2015) Benefits, limits and danger of ephedrine and pseudoephedrine as nasal decongestants. Eur Ann Otorhinolaryngol Head Neck Dis 132(1):31–34. https://doi.org/10.1016/j.anorl.2014.11.001

Lemanske RF, Busse WW (2003) 6. Asthma. J Allergy Clin Immunol 111(2):S502–S519. https://doi.org/10.1067/mai.2003.94

Leong EWX, Ge R (2022) Lipid nanoparticles as delivery vehicles for inhaled therapeutics. Biomedicines 10(9):2179. https://doi.org/10.3390/biomedicines10092179

Lilburn PA, Ainge-Allen H, Thomas PS (2019) Fluticasone furoate: a once daily preparation in patients with persistent asthma. J Lung Health Dis 3(1) Available https://www.lungdiseases-journal.com/articles/fluticasone-furoate-a-once-daily-preparation-in-patients-with-persistent-asthma.html. Accessed 13 Sept 2023. [Online]

Lv C, Li H, Cui H, Bi Q, Wang M (2021) Solid lipid nanoparticle delivery of rhynchophylline enhanced the efficiency of allergic asthma treatment via the upregulation of suppressor of cytokine signaling 1 by repressing the p38 signaling pathway. Bioengineered 12(1):8635–8649. https://doi.org/10.1080/21655979.2021.1988364

Madgulkar AR, Padalkar RR, Amale SK (2019) Preformulation studies of intranasal solid lipid nanoparticles of mometasone furoate. JDDT 9(4). https://doi.org/10.22270/jddt.v9i4.3100

Maheshwari RGS et al (2012) Ethosomes and ultradeformable liposomes for transdermal delivery of clotrimazole: a comparative assessment. Saudi Pharm J 20(2):161–170. https://doi.org/10.1016/j.jsps.2011.10.001

Mathis BJ, Kusumoto M, Zaboronok A, Hiramatsu Y (2021) Packaging and delivery of asthma therapeutics. Pharmaceutics 14(1):92. https://doi.org/10.3390/pharmaceutics14010092

Mazur M, Czarnobilska M, Dyga W, Czarnobilska E (2022) Trends in the epidemiology of allergic diseases of the airways in children growing up in an urban agglomeration. J Clin Med 11(8):2188. https://doi.org/10.3390/jcm11082188

Meltzer EO, Grant JA (1999) Impact of cetirizine on the burden of allergic rhinitis. Ann Allergy Asthma Immunol 83(5):455–463. https://doi.org/10.1016/S1081-1206(10)62851-7

Meltzer EO, Rosario NA, Van Bever H, Lucio L (2021) Fexofenadine: review of safety, efficacy and unmet needs in children with allergic rhinitis. Allergy Asthma Clin Immunol 17(1):113. https://doi.org/10.1186/s13223-021-00614-6

Menzella F, Lusuardi M, Galeone C, Facciolongo N, Zucchi L (2016) The clinical profile of benralizumab in the management of severe eosinophilic asthma. Ther Adv Respir Dis 10(6):534–548. https://doi.org/10.1177/1753465816667659

Mishra V et al (2018) Solid lipid nanoparticles: emerging colloidal Nano drug delivery systems. Pharmaceutics 10(4):191. https://doi.org/10.3390/pharmaceutics10040191

Mukherjee S, Ray S, Thakur RS (2009) Solid lipid nanoparticles: a modern formulation approach in drug delivery system. Indian J Pharm Sci 71(4):349–358. https://doi.org/10.4103/0250-474X.57282

Müller RH, Shegokar R, Keck CM (2011) 20 years of lipid nanoparticles (SLN and NLC): present state of development and industrial applications. Curr Drug Discov Technol 8(3):207–227. https://doi.org/10.2174/157016311796799062

Narendra DK, Hanania NA (2019) Targeting IL-5 in COPD. Int J Chron Obstruct Pulmon Dis 14:1045–1051. https://doi.org/10.2147/COPD.S155306
Nassimi M et al (2010) A toxicological evaluation of inhaled solid lipid nanoparticles used as a potential drug delivery system for the lung. Eur J Pharm Biopharm 75(2):107–116. https://doi.org/10.1016/j.ejpb.2010.02.014
Nettis E et al (2008) Once daily levocetirizine for the treatment of allergic rhinitis and chronic idiopathic urticaria. JAA 2:17. https://doi.org/10.2147/JAA.S3022
Nur Husna SM, Tan H-TT, Md Shukri N, Mohd Ashari NS, Wong KK (2022) Allergic rhinitis: a clinical and pathophysiological overview. Front Med 9:874114. Available https://www.frontiersin.org/articles/10.3389/fmed.2022.874114. Accessed 11 Sept 2023. [Online]
Okayama Y, Matsumoto H, Odajima H, Takahagi S, Hide M, Okubo K (2020) Roles of omalizumab in various allergic diseases. Allergol Int 69(2):167–177. https://doi.org/10.1016/j.alit.2020.01.004
Paranjpe M, Müller-Goymann CC (2014) Nanoparticle-mediated pulmonary drug delivery: a review. Int J Mol Sci 15(4):5852–5873. https://doi.org/10.3390/ijms15045852
Passali D, Spinosi MC, Crisanti A, Bellussi LM (2016) Mometasone furoate nasal spray: a systematic review. Multidiscip Respir Med 11(1):18. https://doi.org/10.1186/s40248-016-0054-3
Patel RR, Chaurasia S, Khan G, Chaubey P, Kumar N, Mishra B (2016) Highly water-soluble mast cell stabiliser-encapsulated solid lipid nanoparticles with enhanced oral bioavailability. J Microencapsul 33(3):209–220. https://doi.org/10.3109/02652048.2016.1144819
Patlolla RR, Chougule M, Patel AR, Jackson T, Tata PNV, Singh M (2010) Formulation, characterization and pulmonary deposition of nebulized celecoxib encapsulated nanostructured lipid carriers. J Control Release 144(2):233–241. https://doi.org/10.1016/j.jconrel.2010.02.006
Pilcer G, Amighi K (2010) Formulation strategy and use of excipients in pulmonary drug delivery. Int J Pharm 392(1–2):1–19. https://doi.org/10.1016/j.ijpharm.2010.03.017
Pohlit H, Bellinghausen I, Frey H, Saloga J (2017) Recent advances in the use of nanoparticles for allergen-specific immunotherapy. Allergy 72(10):1461–1474. https://doi.org/10.1111/all.13199
Priyanka K, Abdul Hasan SA (2012) Preparation and evaluation of montelukast sodium loaded solid lipid nanoparticles. J Young Pharm 4(3):129–137. https://doi.org/10.4103/0975-1483.100016
Ratner PH, Ehrlich PM, Fineman SM, Meltzer EO, Skoner DP (2002) Use of intranasal cromolyn sodium for allergic rhinitis. Mayo Clin Proc 77(4):350–354. https://doi.org/10.4065/77.4.350
Rehman MU, Khan MA, Khan WS, Shafique M, Khan M (2018) Fabrication of niclosamide loaded solid lipid nanoparticles: in vitro characterization and comparative in vivo evaluation. Artif Cells Nanomed Biotechnol 46(8):1926–1934. https://doi.org/10.1080/21691401.2017.1396996
Sallam M, Helal H, Mortada S (2016) Rationally designed nanocarriers for intranasal therapy of allergic rhinitis: influence of carrier type on in vivo nasal deposition. IJN:2345. https://doi.org/10.2147/IJN.S98547
Scalia S, Haghi M, Losi V, Trotta V, Young PM, Traini D (2013) Quercetin solid lipid microparticles: a flavonoid for inhalation lung delivery. Eur J Pharm Sci 49(2):278–285. https://doi.org/10.1016/j.ejps.2013.03.009
Scioli Montoto S, Muraca G, Ruiz ME (2020) Solid lipid nanoparticles for drug delivery: pharmacological and biopharmaceutical aspects. Front Mol Biosci 7:587997. https://doi.org/10.3389/fmolb.2020.587997
Shah A, Bajaj AN, Jain DS (2013) Fabrication and in vitro evaluation of solid lipid nanoparticles of mometasone furoate for pulmonary delivery. J Nanopharm Drug Delivery 1(3):311–322. https://doi.org/10.1166/jnd.2013.1029
Sharma A et al (2022) Artificial intelligence-based data-driven strategy to accelerate research, development, and clinical trials of COVID vaccine. Biomed Res Int 2022:1. https://doi.org/10.1155/2022/7205241
Sharma J et al (2023) Statistical optimization of tetrahydrocurcumin loaded solid lipid nanoparticles using Box Behnken design in the management of streptozotocin-induced diabetes mellitus. Saudi Pharm J 31:101727. https://doi.org/10.1016/j.jsps.2023.101727

Sidhu G, Akhondi H (2023) Loratadine. In: StatPearls [Internet]. StatPearls Publishing. Available https://www.ncbi.nlm.nih.gov/books/NBK542278/. Accessed 13 Sept 2023. [Online]

Sockrider M, Fussner L (2020) What is asthma? Am J Respir Crit Care Med 202(9):P25–P26. https://doi.org/10.1164/rccm.2029P25

Su S, Kang PM (2020) Recent advances in nanocarrier-assisted therapeutics delivery systems. Pharmaceutics 12(9):837. https://doi.org/10.3390/pharmaceutics12090837

Tamada T, Ichinose M (2017) Leukotriene receptor antagonists and antiallergy drugs. Handb Exp Pharmacol 237:153–169. https://doi.org/10.1007/164_2016_72

Tashkin DP, Fabbri LM (2010) Long-acting beta-agonists in the management of chronic obstructive pulmonary disease: current and future agents. Respir Res 11(1):149. https://doi.org/10.1186/1465-9921-11-149

Tekade RK, Maheshwari R, Tekade M, Chougule MB (2017) Solid lipid nanoparticles for targeting and delivery of drugs and genes. In: Nanotechnology-based approaches for targeting and delivery of drugs and genes. Elsevier, pp 256–286. https://doi.org/10.1016/B978-0-12-809717-5.00010-5

Uner M, Karaman EF (2013) Preliminary studies on solid lipid microparticles of loratadine for the treatment of allergic reactions via the nasal route. Trop J Pharm Res 12. Available https://tspace.library.utoronto.ca/handle/1807/60945. Accessed 13 Sept 2023. [Online].

Uner M, Yener G (2007) Importance of solid lipid nanoparticles (SLN) in various administration routes and future perspectives. Int J Nanomedicine 2(3):289–300

Viegas C, Patrício AB, Prata JM, Nadhman A, Chintamaneni PK, Fonte P (2023) Solid lipid nanoparticles vs. nanostructured lipid carriers: a comparative review. Pharmaceutics 15(6):1593. https://doi.org/10.3390/pharmaceutics15061593

Virmani T, Kumar G, Virmani R, Sharma A, Pathak K (2022a) Nanocarrier-based approaches to combat chronic obstructive pulmonary disease. Nanomedicine (Lond) 17:1833. https://doi.org/10.2217/nnm-2021-0403

Virmani T, Kumar G, Pathak K (2022b) Non-aqueous nanoemulsions: an innovative lipid-based drug carrier. In: Advancements in controlled drug delivery systems. IGI Global, pp 134–158

Virmani R, Virmani T, Pathak K (2022c) Chapter 17—Nanovesicles for delivery of central nervous system drugs. In: Nayak AK, Hasnain MS, Aminabhavi TM, Torchilin VP (eds) Applications of nanovesicular drug delivery. Academic Press, pp 315–339. https://doi.org/10.1016/B978-0-323-91865-7.00004-3

Virmani T et al (2023a) Amelioration of cancer employing chitosan, its derivatives, and chitosan-based nanoparticles: recent updates. Polymers 15(13):2928. https://doi.org/10.3390/polym15132928

Virmani T, Kumar G, Virmani R, Sharma A, Pathak K (2023b) Xanthan gum-based drug delivery systems for respiratory diseases. Springer, Cham, pp 279–295. https://doi.org/10.1007/978-981-19-7656-8_16

Virmani T, Kumar G, Sharma A, Pathak K (2023c) Chapter 2—An overview of ocular drug delivery systems—conventional and novel drug delivery systems. In: Rai M, Occhiutto ML, Talegaonkar S (eds) Nanotechnology in ophthalmology. Academic Press, pp 23–48. https://doi.org/10.1016/B978-0-443-15264-1.00007-5

Virmani R, Sharma A, Sharma A, Kumar G, Virmani T, Mukherjee S (2023d) 17—Nanotechnology in pulmonary tissue engineering. In: Mondal A, Nayak AK, Chakraborty P (eds) Nanostructured Materials for Tissue Engineering. Elsevier, pp 537–556. https://doi.org/10.1016/B978-0-323-95134-0.00017-1

Wang W et al (2012) Enhanced bioavailability and efficiency of curcumin for the treatment of asthma by its formulation in solid lipid nanoparticles. Int J Nanomedicine 7:3667–3677. https://doi.org/10.2147/IJN.S30428

Wang C et al (2015) Effect of budesonide transnasal nebulization in patients with eosinophilic chronic rhinosinusitis with nasal polyps. J Allergy Clin Immunol 135(4):922–929.e6. https://doi.org/10.1016/j.jaci.2014.10.018

Weber S, Zimmer A, Pardeike J (2014) Solid lipid nanoparticles (SLN) and nanostructured lipid carriers (NLC) for pulmonary application: a review of the state of the art. Eur J Pharm Biopharm 86(1):7–22. https://doi.org/10.1016/j.ejpb.2013.08.013

Wikstén J, Toppila-Salmi S, Mäkelä M (2018) Primary prevention of airway allergy. Curr Treat Options Allergy 5(4):347–355. https://doi.org/10.1007/s40521-018-0190-4

Xu Y, Liu H, Song L (2020) Novel drug delivery systems targeting oxidative stress in chronic obstructive pulmonary disease: a review. J Nanobiotechnol 18(1):145. https://doi.org/10.1186/s12951-020-00703-5

Role of Dendrimers in Management of Allergic Airway Diseases

12

Rabab Fatima, Mousmee Sharma, and Parteek Prasher

Abstract

Dendrimers are three-dimensional polymers that are characterized by their precisely structured hyper-branches and present a cutting-edge polymeric nanoformulation with remarkable biological efficacy. Their repetitive structure and ability to encapsulate drugs adds a valuable dimension on allergic airway disorders, particularly asthma. PAMAM dendrimers, synthesized via a divergent method, effectively encapsulate anti asthmatic drugs such as beclomethasone dipropionate and methylprednisolone, ensuring targeted delivery, improved solubility, and sustained release. Particularly, surface-modified dendrimers and those loaded with diagnostic or therapeutic agents showcase outstanding biological activity against asthma. Their PEGylation further reduces toxicity and enhances water solubility. Within this chapter, we intricately explore the pivotal role and biomedical applications of dendrimers in asthma management. Rigorous clinical trials are necessary for establishing safety and efficacy, paving the way for personalized asthma treatments. Collaborative efforts among researchers, clinicians, and formulation scientists are crucial for translating dendrimer-based therapies into routine clinical practice for pulmonary diseases.

Keywords

Asthma · Dendrimer · PAMAM · Drug delivery · Toxicity

R. Fatima (✉) · P. Prasher
Department of Chemistry, University of Petroleum & Energy Studies, Dehradun, India

M. Sharma
Department of Chemistry, Uttaranchal University, Dehradun, India

P. Prasher et al. (eds.), *Understanding Allergic Airway Diseases*,
https://doi.org/10.1007/978-981-97-1953-2_12

12.1 Introduction

The introduction of nanoparticles (NP) designed for targeted drug delivery represents a novel approach in medical science. NP, with a diameter typically below 100 nanometers, are commonly composed of polymers and have shown significant utility in diverse medical applications (Wilczewska et al. 2012). Nanoparticles can be intricately and precisely loaded with therapeutic agents to facilitate effective localized drug delivery, potentially enabling sustained release when coupled with biodegradable carrier systems. These attributes render NPs capable of treating patients with minimal invasiveness while improving bioavailability by accessing targeted tissues and the bloodstream (Sahu et al. 2021). Numerous NP variants find application in pulmonary disorders, particularly in the context of chronic obstructive pulmonary disorder (COPD) and asthma. The primary category consists of polymeric NPs, which are subject to extensive research for their suitability in pulmonary diseases, stemming from their advantageous chemical and physical attributes. Polymeric NPs exhibit notable features such as high biocompatibility, sustained lung retention, and minimal presence in non-target tissues, rendering them desirable candidates for pulmonary drug delivery systems. Within the realm of polymeric nanocarriers, various subtypes such as polymeric micelles, solid lipid nanoparticles, and dendrimers have huge potential in treating respiratory ailments due to their unique features (Rytting et al. 2008).

In 1978, Fritz Vogtle pioneered the novel synthesis methods for a multitude of innovative organic structures, marking a significant advancement in the field (Tomalia 1995) which subsequently, gave rise to the emergence of supramolecular chemistry and included speleands, cavitands cyclophanes, and propeller compounds characterized by unique topology with a molecular cavity. These compounds, exhibiting repetitive motifs, functioned as precursors to the inaugural poly propylene imine dendrimers (Esfand and Tomalia 2001). Professor Tomalia in 1985 further elucidated the conceptualization, characterization, and synthesis of "starburst polymers," representing a novel category of polymeric particles whose structural components are currently recognized as dendrimers (Tomalia et al. 1985). Polyamidoamine (PAMAM) and its first through seventh synthesized generations were presented for the first time then. Dendrimers, with their distinctive three-dimensional structure, exhibit nanoscaled globular shapes, precisely defined peripheral functional groups, and internal hydrophobic or hydrophilic cavities, all accompanied by remarkably low polydispersity. This unique combination imparts versatile properties, such as modulable molecular diameters and biomimetic resemblances to proteins, making dendrimers ideal for various therapeutic applications. Their interior cavities make them effective unimolecular micelles, particularly for drug encapsulation. The minimal polydispersity further ensures consistent biodistribution, affirming their suitability as scaffolds for polymeric prodrugs.

12.1.1 Structure of Dendrimers

Dendrimers, a distinct class of nano-sized molecules, are prominently featured as a drug delivery system, standing out as one of the most promising options among all drug delivery systems due to their exceptional properties (Bae 2002). Dendrimers are highly branched macromolecules with a unique three-dimensional structure. The term "dendrimer" originates from the Greek word "dendron," which means tree, due to their tree-like appearance. They are synthetic, well-defined, and monodispersed polymers, typically created through a controlled stepwise growth process (Tomalia et al. 1985). The fundamental architecture of a dendrimer comprises three discernible components:

Core: Serving as the initiator or focal point, the central core establishes the foundation for subsequent branching within the dendrimer.

Branches: Constituting repetitive units, these branches emanate from the core in a series of generations, with each generation representing an additional layer of branches attached to the preceding one.

Periphery: The periphery consists of the outermost layer and functional end groups that can be tailored for specific applications through modification (Maiti et al. 2004) (Fig. 12.1).

Dendrimers exhibit well-defined structures characterized by precise molecular weights and sizes, a critical attribute for applications in drug delivery and nanotechnology. Introduction of minute molecules, such as ions, drugs, or biomolecules, can alter the functionality of the terminal, leading to substantial changes in the physical, chemical, biological, reactivity, and dynamic properties of dendrimer (Tomalia

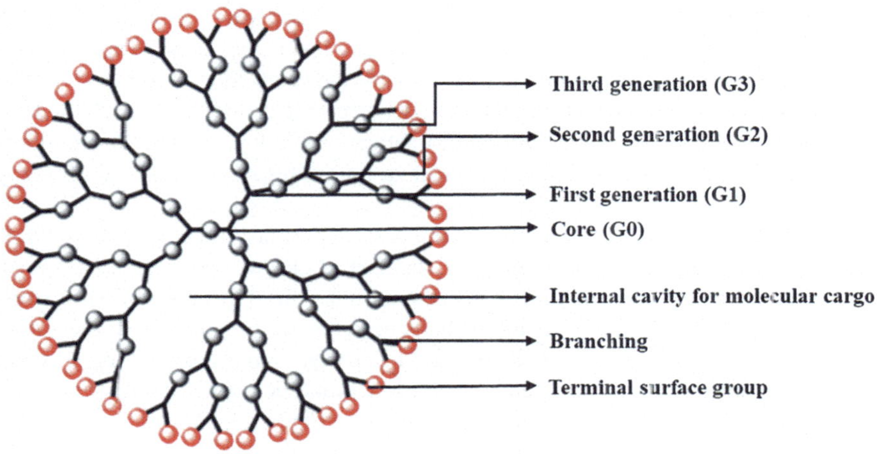

Fig. 12.1 Structure of a dendrimer representing core, first, second, and third generation as G0, G1, G2, and G3 respectively

2005). The multiple functional end groups on dendrimer result in a high density of reactive sites on its surface, providing advantageous multivalency in diverse biomedical and catalytic applications. The introduction of hydrophilic functional groups at the periphery further imparts water solubility to dendrimers, a pivotal characteristic in biomedical applications (de Araújo et al. 2018). Dendrimers, with their ability to encapsulate drug molecules within internal cavities, shield them from degradation and facilitate targeted delivery to specific locations in the body (Abbasi et al. 2014). Modification of dendrimers allows for the conjugation of imaging agents, enabling diagnostic applications such as magnetic resonance imaging (MRI) or fluorescence imaging. Some dendrimers can act as catalysts, facilitating various chemical reactions due to their multivalent nature and distinct microenvironments. Dendrimers are also used as non-viral gene delivery vectors for efficient transportation of nucleic acid into cells (Shahi et al. 2015).

12.1.2 Synthesis of Dendrimers

The first technique to create monodispersed macromolecules of dendritic structure was demonstrated in 1979 (Tomalia 2005). Their association with the realm of molecular chemistry stems from their meticulously controlled step-by-step synthesis. Simultaneously, their connection to the polymer domain arises from their repetitive structural composition, comprising monomeric units. Dendrimers can be synthesized employing two primary methodologies. The divergent approach, historically employed in the early stages of dendrimer synthesis, the process initiates from the core of the dendrimer, with arms attached incrementally through the addition of building blocks in a comprehensive and sequential manner. This core molecule engages in reactions with monomer molecules possessing one reactive and two inert groups, yielding the initial generation of dendrimer. Subsequently, the newly formed periphery of the molecule is activated for further reactions with additional monomers (Shahi et al. 2015). Conversely, in the convergent approach, synthesis commences from the periphery, originating with the molecular structure that ultimately forms the outermost arm of the final dendrimer (Liu and Fréchet 1999). This strategy entails the predetermined determination of the final generation number, mandating the prior synthesis of branches of various required sizes for each generation (Fig. 12.2).

The PAMAM dendrimer was the preliminary dendrimer to be synthesized and characterized using divergent method (Esfand et al. 1996) where generations or concentric shells of dendritic units revolving around core or primary initiator were generated (Boas Ulrik 2004). As more shells or generations were added, the architecture of the PAMAM core-shell increased linearly. Dendrimers owing to their hyper branching, symmetry, and maximized functional density at the terminals are distinct from the traditional polymers or oligomers (Fig. 12.3).

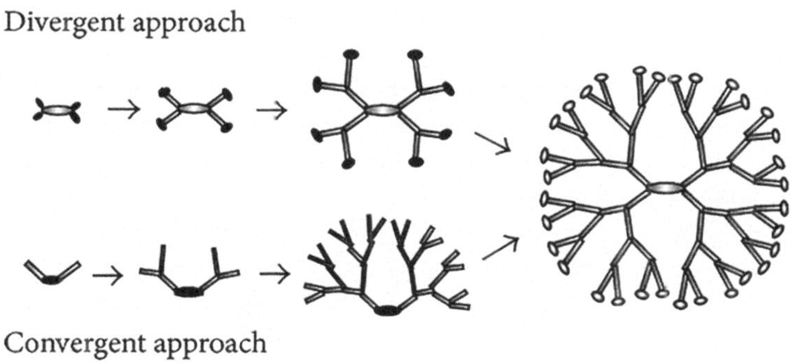

Fig. 12.2 Divergent and convergent method for synthesis of dendrimer

12.2 Lung Airway Diseases

These are the disorders that constrict the passages responsible for transporting oxygen and various gases to and from the lungs and include conditions such as bronchiectasis, emphysema, COPD, influenza, and asthma. COPD and asthma represent the foremost respiratory ailment, and their prevalence is escalating within the population due to worsening air quality and alarming pollution levels. According to a recent study conducted in 2015, COPD ranked 8th and asthma ranked 23rd in terms of global disease burden across populations. In response to this growing burden, scientists are actively engaged in the development of novel therapeutics and diagnostics to manage these diseases (Dharmage et al. 2019). Consequently, research efforts are delving into the nanoscale domain to establish efficient targeting and treatment strategies for respiratory disorders.

Lungs emerge as an appealing target for the treatment and diagnosis of respiratory diseases as they possess a substantial surface area of approximately 50 square meters for an average adult. Enhanced drug absorption within the pulmonary system is facilitated by the extensive vascularization, creating an optimal environment for therapeutic interventions (Galli et al. 2008). The lungs assume a pivotal role as a drug delivery target, given their capability to circumvent the conventional systemic delivery route, such as oral administration. This circumvention is particularly advantageous in bypassing conditions like the digestive system and mitigating the effects of first-pass metabolism. Dendrimers offer following benefits in pulmonary drug delivery:

1. Tree-like architecture improves chemical and physical properties.
2. Minimal viscosity and monodispersity.
3. Globular structure facilitates guest molecule encapsulation inside void cavity.
4. Polyvalency promotes surface group interaction with ligands.
5. Uniform drug distribution among the alveoli.
6. Improved drug solubility compared to self-aqueous solubility.
7. Sustained and controlled drug release eliminating the need for frequent dosing.

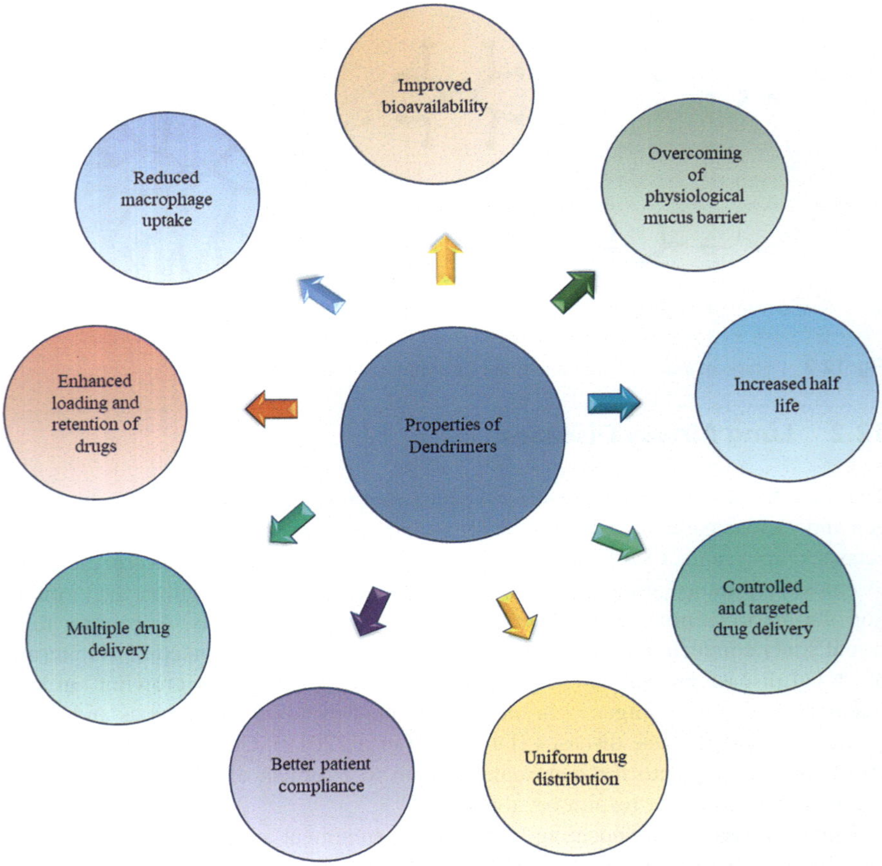

Fig. 12.3 Advantageous features and properties of dendrimers

8. Reduced side effects.
9. Improved patient compliance.

Kukowska Latallo and co-authors assessed the in vivo efficacy of PAMAM dendrimers in augmenting plasmid DNA gene transfer. The evaluation encompassed diverse administration methods to discern lung targeting dynamics. Their hypothesis posited that endobronchial administration of PAMAM predominantly directed the delivery to bronchial epithelium, while vascular administration exhibited expression predominantly within the lung parenchyma, particularly in the alveoli. These findings imply that distinct vectors are required for optimal transgene expression based on the chosen delivery route, indicating that each strategy selectively targets distinct cellular populations within the pulmonary milieu (Kukowska-Latallo et al. 1996). Negatively charged oligosaccharides known as low molecular weight heparins (LMWHs) are often used to treat pulmonary embolism and deep vein thrombosis (DVT). The need for daily subcutaneous injections has, however, restricted the use of LMWHs in outpatient settings. Pegylated dendrimeric particles were evaluated in an attempt to test the theory, and the results reveal that they can prevent deep vein thrombosis (DVT) in a mouse model by extending the duration of half-life of heparin with a low molecular weight (LMWH) and increasing the drug's pulmonary absorption (Ferruti et al. 1994; Parajapati et al. 2016). Table 12.1 summarizes the role of various types of dendrimers and effect of generation in pulmonary disorders and allergy.

Table 12.1 Role of various types of dendrimers in pulmonary disorders

Dendrimer type	Therapeutic agent	Description	Reference
PAMAM G4	siRNA	Targeting of pulmonary epithelial cells with a mass median aerodynamic diameter (MMAD) of 2.6 m and fine particle fraction of 48.9% respectively	Conti et al. (2014
PAMAM G4	siRNA	A fine 38 particle fraction of >50% results in improved cellular transport	Bielski et al. (2017)
PAMAM G4	Doxorubicin	Better transepithelial transport and pH-dependent drug release with a fine particle fraction of >80%	Zhong et al. (2017)
PAMAM G4	Beclomethasone dipropionate	Pari (air-jet) nebulizer with slow and sustained release for 8 h and a fine particle percentage of greater than 60%	Nasr et al. (2014)
Polylysine dendrimer (G5)	Doxorubicin	Enhanced circulation period achieving uniform lung drug deposition	Kaminskas et al. (2014)
Poly (propylene imine) dendrimer PAMAM	Doxorubicin	Improved alveolar cell deposition based on convex meniscus based particles	Garbuzenko et al. (2014)
PAMAM G4	Doxorubicin	Compared to doxorubicin alone, pH-dependent drug release had a better distribution pattern in lung cells and a lower rate of drug accumulation in cardiac tissues.	Zhong et al. (2016a)
PAMAM G3	Rifampicin	Sustained release with 30 and 4 times increase in MRT and plasma drug concentration as compared to intravenous injection	Rajabnezhad et al. (2016)
PAMAM G4	siRNA	Improved cellular, physicochemical, and gene silencing potential	Agnoletti et al. (2017)
PAMAM G4	Doxorubicin	pH dependent release of fine particle fraction of >62%	Zhong (2018)
PAMAM G4	Methotrexate	Showed 1.8- and 2-fold greater improvement in alveolar bioavailability of drug as compared to control	Haque et al. (2017)

12.3 Asthma

Asthma is a respiratory disease manifesting severe inflammation and constriction of the airway passages, leading to shortness of breath, wheezing, chest tightness, allergy, and coughing. The precise etiology of asthma is complex and remains incompletely understood, with indications pointing towards a multifactorial origin involving genetic, environmental, and immunological influences. Familial clustering of asthma cases suggests a genetic predisposition where specific genes linked to asthma susceptibility have been identified and their intricate interactions with environmental elements contribute to the complexity of the condition. Exposure to environmental allergens such as pollen, dust mites, and mold spores can act as triggers for asthma symptoms, particularly in individuals with allergic asthma. Early respiratory infections, especially viral infections during childhood, elevate the likelihood of asthma development. Various environmental factors such as air pollution, tobacco smoke, indoor air quality, and occupational irritants are implicated in both the genesis and exacerbation of asthma. Additionally, there also exists a correlation between obesity and an augmented risk of asthma, with obesity further exacerbating asthma symptoms (Santamaria et al. 2021).

The pathophysiology of asthma encompasses interconnected process within the respiratory system, leading to acute airway inflammation, bronchoconstriction (the constriction of airways), and heightened mucus production. The inflammatory response involves the recruitment of diverse immune cells, notably eosinophils and mast cells, which release inflammatory mediators like histamine and leukotrienes, contributing to airway inflammation. Concurrently, airway inflammation stimulates the overproduction of mucus, exacerbating airway obstruction (Mehta et al. 2019). Upon exposure to triggers such as allergens, irritants, or respiratory infections, the smooth muscles encircling the airways undergo contraction, inducing bronchoconstriction and resulting in airway narrowing. This constriction impedes the flow of air in and out of the lungs. Prolonged asthma duration can induce structural alterations in the airways, a phenomenon recognized as airway remodeling. This can result in irreversible narrowing of the airways. Asthma involves an altered immune responsiveness, with overexpressed T-helper 2 (Th2) cells responsible for airway inflammation and hypersensitivity (Graham 2016).

Asthma affects roughly 300 million people worldwide at present, and by 2025, it may affect an additional 100 million people due to an upward trend in the prevalence of the disease (Murphy et al. 2012). Despite the availability of asthma-treatment guidelines and effective therapeutic options, most asthma patients continue to experience acute or severe form of the disease symptoms, with one in two adults reporting very poorly controlled asthma (Shastri et al. 2021). Similar to adults, children account for 30–40% of all cases of severe exacerbations of asthma that are not properly controlled due to presence of inflammatory paucigranulocytic, neutrophilic, eosinophilic, and mixed granulocytic cells (Hsu et al. 2018). Due to the complex etiology of the condition, there is no specific treatment for asthma, but persistent, systematic treatments have been shown to successfully reduce symptoms, episodes, and worsen prognosis (Ratemi et al. 2016). Combined steroid treatment and bronchodilator therapy, such as SABA (short acting agonist) or LTRA (leukotriene receptor antagonist) or LABA (long acting agonist), is generally recognized as the first-line method for managing asthma (Wang et al. 2019).

One of the key treatments for mild to severe asthma is the use of corticosteroids for both initial and ongoing care. The presence of corticosteroid resistance in patients of asthma however limits the use of corticosteroids therapy, despite its acknowledged popularity and efficacy (Lommatzsch and Stoll 2016; Singh et al. 2019). While reducing negative side effects like immunosuppression, the use of NP may aid expedite distribution directly to target site, enhance the rate of alveolar drug deposition, and therapeutic effects of anti-asthmatic medications. Nanomaterials utilized in anti-asthmatic research may get past many physiological barriers such as airway mucus, to pass the endothelium and have the desired pharmacological effect (Beeraka et al. 2022). In order to reduce the development of steroid resistance or to improve the overall clinical results in asthma patients, these delivery systems can improve drug delivery, solubility, bioavailability, and pharmacokinetic characteristics using dendritic models.

Dendrimers are an important class of nanoparticles that readily promote targeted drug delivery across the lungs. Various types of dendrimers used are poly (L-lysine) (PLL), PAMAM, poly (2,2-bis(hydroxyl methyl) propionic acid), polypropyleni-mine (PPI), polyamides, polyesters (PGLSA-OH), and polyethers (Gothwal et al. 2019). Combining these polymers with PEGylation can further reduce toxicity, improve pharmacokinetic characteristics, and boost water solubility (Fig. 12.3). Dendrimers are efficient pulmonary drug delivery vehicles because they can prolong drug retention across the lungs and facilitate efficient biodistribution for controlled drug delivery during passive circulation (Zhong et al. 2016b).

Dendrimers are well investigated for the treatment of asthma. Beclomethasone diproprionate and other insoluble medicines could be transported by pEGylated PAMAM to lessen the effects of long-term high-dose hormone inhalation and further improve the bioavailability of hormones (Zhong et al. 2016b). This could increase a substance's solubility and lung accumulation capacity, which would further increase its bioavailability and reduce its dosage, frequency of administration, and toxicity. This can also be utilized to transport hydrophobic drugs such as dexamethasone directly to the lungs, reducing eosinophils and inflammatory cytokines while reducing allergic pulmonary inflammation. PAMAM dendrimers have also demonstrated to deliver biomolecules similar to siRNA when used to treat COPD or steroid resistance asthma (SSR) asthma with improved prolonged drug retention (Dua et al. 2019).

SSR is the term used to describe some asthmatics who have trouble controlling their symptoms even after utilizing the highest dose of corticosteroids and account for more than 60% of asthma-related cases (Beeraka et al. 2022). Most therapies for moderate to severe asthma include nebulized glucocorticoids, human monoclonal antibodies, and cytokine/chemokine antagonists. The great range of asthma symptoms and types further limits these strategies, despite the fact about their performance. Studies suggest that the application of nanotechnology may improve the efficacy of anti-asthma therapeutics. The inorganic NPs include iron oxide NPs, gold NPs, silica NPs, quantum dots, graphene oxide, and carbon nanotubes, whereas the organic NPs comprise liposomes, polymeric NPs, dendrimers, and micelles (Wang et al. 2019).

12.4 Significance of Dendrimer Modification

Drug molecules can either be physically enclosed inside a dendrimer core or chemically attached to the dendrimer surface. If functional groups undergo activation before coupling, a high coupling efficiency can possibly be attained in chemical conjugations. The functional groups such as primary amine (NH_2), thiol (SH), hydroxyl (OH), carboxyl (COOH), and guanidino are frequently found in medicinal compounds and polymers (Nathan et al. 1993). Active intermediates can be created from hydroxyl groups, favoring nucleophilic processes. In some circumstances, direct coupling may result in the pharmaceutical inactivation, preventing the conjugated compounds (also known as prodrugs) from exerting any therapeutic effects until the link is severed. To create linkages that can be broken down by an enzyme or hydrolyzed in vivo, a bi-functional spacer may be included to the drug delivery vehicle (Veronese and Pasut 2005). Dendrimers can be made multifunctional by simultaneously attaching moieties with a variety of functions to the surface using covalent or non-covalent bonding due to the presence of many points of contact on the dendrimer surface. The synthesis of dendrimers with catalyst-active centers on the periphery has primarily been accomplished using this method (Veronese and Morpurgo 1999).

The solubility, non-toxicity, and hypersensitivity of poly ethylene glycol (PEG) make it a popular drug delivery method. By reducing toxicity and immunogenicity, PEGylation can change dendrimers' unfavorable characteristics (Newkome et al. 1985). Dendritic micelles made of a single molecule can be produced using PEGylated dendrimers. PEGylation can boost drug loading and get around problems with drug-loaded dendrimeric systems such hemolytic toxicity, drug leaking, macrophage absorption, and a brief half-life and toxicity (Wooley et al. 1993). Dendrimers can be used for extended periods of time for medication administration since PEGylation increases their stability and effectiveness (Liu et al. 1999a). PEGylation can target active areas of action while reducing immunogenicity, antigenicity, and toxicity. It also inhibits retention and intake by the reticulo-endothelial system (Heise et al. 2001). For coupling activities, PEGylation can also offer more terminating groups.

Glycodendrimers are different dendrimer structures that have carbohydrates incorporated into them. They can be divided into three categories: carbohydrate-coated, carbohydrate-centered, and carbohydrate-based (Bhadra et al. 2005). In order to swiftly produce polyvalent glycoconjugates, dendrimers with coatings of carbohydrates are modified from pre-existing dendrimers. These kinds are primarily present in poly(propylene imine) and polyamidoamine (PAMAM) dendrimers. The small size and low polydispersity of glycodendrimers make them appealing, and the size can vary based on the dendrimer formation (Shaunak et al. 2004).

Peptide dendrimers, composed of surface peptides attached to a conventional framework, have been employed as antimicrobials, inhibitors of angiogenesis, and photoresponsive drug delivery systems. They are less harmful to human cells, more proteolysis resistant, and more water soluble (Cloninger 2002). For the delivery of photoresponsive drugs, dendrimers containing arginine and azobenzene residues on their surfaces have been developed. Up to the fourth generation, peptide dendrimers having L-lysine as the branching moiety and PEG-1000 as the core were successfully created (Kasai et al. 2002).

In order to enable water solubility and connect hydrophobic medicinal molecules, Fréchet and colleagues created polyaryl ether dendrimers (Liu et al. 1999b). They used carbonate, ester, and carbamate links to conjugate cholesterol, phenylalanine, and tryptophan (Liu et al. 2000). Crowned arborols are a new class of stable metal ion complex-forming dendritic macromolecules that Shinkai and colleagues developed (Nagasaki et al. 1994). Sialic acid coupled to dendrimers containing PAMAM completely safeguarded against an influenza pneumonitis mouse model, according to Reuter et al.'s investigation into the utilization of sialic acid processed dendrimers as an innovative therapeutic approach for managing influenza (Reuter et al. 1999).

Since dendrimers can be made of polymers like poly L lysine (PLL), poly(amidoamine) (PAMAM), polyamides, polyesters (PGLSA-OH), polyethers, polypropylenimine (PPI), poly (2,2-bis (hydroxyl methyl) propionic acid) (Gothwal et al. 2019). When paired with PEGylation, these formulations can improve pharmacokinetic profiles, reduce toxicity, and boost water solubility. When a dendrimer is PEGylated, the pattern of dendrimer degradation is altered, showing a step-by-step transition where the PEG layer is first broken down and then the dendrimer begins to shed. For example, PEGylated Poly(lysine), a highly branched structure of dendrimer, provides a variety of controlled, adaptable, peripheral functions offering a huge potential for targeted drug delivery due to ease of alterable surface chemistry. They have a greater potential than liposomes and nanoparticles due to higher cellular transit, enhanced ability to modify the local environment through functional groups, and the decreased macrophage update. Dendrimers may be considered efficient pulmonary delivery agents because they can prolong drug retention across the lungs and facilitate appropriate biodistribution for controlled drug delivery during passive circulation.

12.5 PAMAM Dendrimers in Asthma Management

Complexation of drugs with dendrimers is an emerging technique to increase the solubility of drugs, improve their stability, and achieve controlled drug release. This is primarily attributed to the remarkable monodispersity, structural homogeneity, and substantial surface area exhibited by dendrimers (Veronese and Pasut 2005).

PAMAM dendrimers have garnered attention for their potential application in asthma management, particularly concerning drug delivery and therapeutic interventions. Their precisely defined and highly branched structure facilitates the encapsulation and controlled release of therapeutic agents, including bronchodilators and anti-inflammatory drugs (Conti et al. 2014). Given the limited solubility of certain asthma drugs, impacting their bioavailability, PAMAM dendrimers offer a means to enhance solubility, thereby increasing effectiveness when administered through inhalation or alternative routes. Modifying PAMAM dendrimers further allows for tailored drug delivery to specific cells or regions within the respiratory system, such as airway epithelial cells or immune cells implicated in asthma pathogenesis. This targeted approach minimizes systemic exposure and off-target effects while elevating local drug concentrations at the desired site.

Effective asthma management often involves long-acting bronchodilators and anti-inflammatory agents. PAMAM dendrimers can be engineered to provide sustained drug release, ensuring continuous therapeutic benefits and reducing the frequency of asthma symptoms and exacerbations. Additionally, modifications can be made to PAMAM dendrimers to mitigate immunogenicity and the potential for immune system recognition, thereby enhancing their biocompatibility and safety as drug delivery carriers. These dendrimers exhibit the capacity to encapsulate multiple drugs simultaneously, facilitating combination therapy for comprehensive asthma management that addresses both bronchial constriction and inflammation in a coordinated manner (de Araújo et al. 2018). The distinctive molecular architecture of dendrimers, characterized by a substantial quantity of surface amino groups, has prompted their utilization as a substrate for binding antibodies, drugs, contrast agents, and radionucleotides across various domains of biology and medicine. Investigations involving antibody/dendrimer conjugates, conducted both in vitro and in vivo with experimental animals, have substantiated the non-toxic nature of these conjugates while demonstrating their efficacy in precisely targeting biologic agent to specific cells.

PAMAM dendrimers can also be employed in the development of diagnostic tools for asthma. They can be functionalized with specific ligands or imaging agents to detect and monitor disease progression. While PAMAM dendrimers hold promise in asthma management, it's important to note that their clinical application and safety profiles require further research and development.

Enhancing pulmonary ventilation, being immunized, helping people quit smoking, and using anti-infection and anti-inflammatory methods are some of the current therapy techniques for asthma (Murphy et al. 2012). The treatment of additional disease symptoms requires a multifaceted strategy that includes the use of oral corticosteroids like 5-lipoxygenase inhibitors, formoterol, budesonide, bronchodilators, anticholinergic medications, prolonged home oxygen therapy, as well as inhaled corticosteroids like fluticasone propionate. Additionally, combinations of theophylline with bronchodilators and corticosteroids are also considered (Suraya et al. 2021).

PAMAM dendrimers are frequently employed in asthma therapy, necessitating the modification of their cationic surface groups to neutral or anionic moieties. This modification is essential to mitigate potential toxicity and prevent excessive accumulation of dendrimers in the liver (Malik et al. 2000). Excellent complexation via PAMAM dendrimers was found in a study demonstrating the potential use of these molecules as carriers for drugs that have poor solubility, such as beclomethasone dipropionate (BDP). This complexation was found to be dependent on the generation, concentration of the dendrimers as well as the pH of the dispersion medium. Comparing the ester terminated (half-generation; G4.5) dendrimers to the amine terminated (G3, G4, and G4 (12)) dendrimers, more stable complexes with BDP were generated. Increasing the solubility of BDP required the production of dendrimers, which was also crucial. With G4 (12) dendrimers, BDP was most soluble, highlighting the fact that the hydrophobicity of the dendrimer core directly affects a drug's ability to dissolve in water (Nasr et al. 2014).

Rajabnez and colleagues conducted an investigation to assess the potential utility of spray-dried PAMAM dendrimers combined with rifampicin as an efficient delivery system for the targeted delivery of rifampicin to the pulmonary region. Rifampicin was co-spray dried with polyamidoamine dendrimers (PAMAM) of the G1, G2, and G3 generations to create inhalable microspherical particles for pulmonary administration. By using scanning electron microscopy (SEM), laser diffraction, and a helium densitometer, the particle size distribution, shape, and density of the developed mixtures were assessed. By applying an Andersen cascade impactor, it was determined how well various compositions aerosolized. The efficient aerosol formulations with favorable fine particle fraction (FPF) and emission fraction (EF) indicate that the powders are safe for inhalation. The in situ pulmonary absorption technique was used to look at the absorption patterns of rifampicin after pulmonary administration of various formulations. After intrapulmonary doses for 60 h, the pharmacokinetic parameters of various rifampin formulations were investigated. It was clear that the formulations' tmax values had fallen while their Cmax values had risen, and due to this phenomenon that PAMAM dendritic generations had gone from G1 to G3. The pharmacokinetics characteristics of rifampicin and eventually drug bioavailability were found to be significantly impacted by the smaller generation PAMAM microspheres (Rajabnezhad et al. 2016). The results demonstrated that generation-3 PAMAM dendrimers possess the capacity to effectively target rifampicin to lung tissues.

Methylprednisolone (MP) is a potential corticosteroid that is used to treat asthmatic lung irritation via inhalation. In a pulmonary inflammatory murine model based on an 11-fold increase in eosinophil lung accumulation after five daily inhalation exposures of sensitized mice to the experimental allergen, ovalbumin, MP-polyamidoamine (PAMAM) dendrimer conjugate's capacity to enhance airway delivery was assessed. Twelve MP molecules were effectively attached to PAMAM G4-OH dendrimers, and these molecules were then solubilized in lysine carrier (Ahmad 2022). While free MP reduced baseline phagocytic cell recoveries by airway lavage and tissue collagenase dispersion, it did not cause further lung inflammation over five daily trans-nasal treatments with the carrier alone, free MP, or MP-dendrimer at 5 mg kg^{-1}. Ovalbumin-associated airway and tissue eosinophil recoveries were reduced by 71% and 47%, respectively, by MP treatments alone. With reductions of 87% and 67%, respectively, equivalent daily MP dose with MP-dendrimer conjugate further decreased these values. These results show that MP conjugation with a dendrimer improves MP's capacity to reduce allergen-induced inflammation, possibly by lengthening drug residence time in the lung. This is corroborated by the fact that over a 3-day period, only 24% of a single dose of dendrimer administered to the peripheral lung is lost. Therefore, drug conjugation to a dendrimer may offer a better way to sustain drug in the lung for treating inflammatory illnesses like asthma (Inapagolla et al. 2010).

G3, G4, and G4 (12) dendrimers were used to test polyamidoamine (PAMAM) dendrimers as nanocarriers for the pulmonary delivery of the model anti-asthma medication beclometasone dipropionate (BDP). Three nebulizers were used to characterize BDP-loaded dendrimers for drug solubility, in vitro drug release, and aerosolization properties: Pari LC Sprint (air-jet), Aeroneb Pro (actively vibrating-mesh), and Omron MicroAir (passively vibrating-mesh). By boosting dendrimer synthesis and employing fluids with a higher pH, BDP solubilization using dendrimers was boosted. Studies of the in vitro release of BDP complexed with dendrimers revealed a sustained release, with less than 35% of the medication released after 8 h for all dendrimer formulations. Nebulization experiments showed that nebulizer performance, as opposed to dendrimer formation, was what determined aerosol performance. When compared to the Omron (passive mesh) nebulizer, the Pari (air-jet) and Aeroneb Pro (active mesh) nebulizers had nebulization output values in the range of 90–92% and 85–89%, respectively. Comparing the jet nebulizer to the two mesh devices, the size of the droplets it produced was a little bit smaller, and its aerosol polydispersity was lower. This study shows that BDP-dendrimers have the ability to be inhaled into the lungs when used with vibrating mesh and air jet nebulizers. Additionally, nebulizer design, not dendrimer synthesis, has an impact on aerosol properties (Nasr et al. 2014).

To address cytotoxicity concerns associated with cationic dendrimers, strategies such as incorporating lauroyl chains and polyethylene glycol (PEG) have been employed. These modifications effectively shield the excess positive charge on the dendrimer's surface, rendering them more biocompatible and safer for use (Dias et al. 2020). These nanoparticles serve as carriers for therapeutic agents, enhancing drug delivery to respiratory tissues and promoting drug absorption by lung cells. This approach significantly improves the safety and efficacy of dendrimer-based drug delivery to the lung (Jevprasesphant et al. 2003).

Modified PAMAM dendrimers have proven to be versatile agents for a range of biomolecules, including proteins and nucleic acids (Paleos et al. 2007). Amino-terminated PAMAM dendrimers possess surface groups that undergo protonation under physiological conditions (Albertazzi et al. 2010). These positively charged surface groups can interact with nucleic acids, facilitating the formation of nanoscale complexes. This interaction presents a promising avenue for improving the targeted delivery of small interfering RNA (siRNA) to airway epithelial cell cytosols, potentially enhancing the efficacy of gene therapy for respiratory diseases, such as asthma.

PAMAM dendrimers exhibit notable solubility and readily modifiable surface functional terminal groups, rendering them valuable tools for solubilizing hydrophobic drugs utilized in asthma treatment, including rifampicin, dexamethasone, beclomethasone dipropionate, and methylprednisolone (Kenyon et al. 2013). Upon conjugation to PAMAM dendrimers, these therapeutics significantly enhance alveolar accumulation and improved bioavailability, presenting a promising strategy for optimizing drug delivery. The conjugation of functionalized molecules with PAMAM dendrimers also tends to shield or diminish the cationic charge on the PAMAM surface, preventing interactions with the anionic cell membrane surface, thereby mitigating PAMAM toxicity and augmenting its biocompatibility.

12.5.1 Dendrimer Drug Conjugation

Dendrimers are highly appealing for drug delivery due to their three-dimensional structural feature and abundant surface terminal functional groups, enabling a substantial drug-loading capacity. This capacity arises from the ability to load drug molecules both within the dendrimer's interior and onto its surface groups. Dendrimers function as drug carriers through two primary methods: non-covalent encapsulation and covalent conjugation. In the non-covalent approach, drugs are encapsulated within the dendritic structure through physical entrapment, utilizing the steric bulk of the dendrimer's exterior or interactions between the dendrimer and the drug. Dendrimers serve as unimolecular micelles and "dendritic boxes," maintaining their structure due to covalently connected hydrophobic segments. The covalent approach involves attaching drug molecules to the dendrimer's periphery, forming complexes through electrostatic interactions or covalent bonds. Dendrimers can be conjugated with various biologically active molecules, allowing tunable drug loading by adjusting the dendrimer's generation. Controlled drug release is achievable by incorporating degradable linkages such as amide or ester, between the drug and dendrimer.

The strong covalent characteristics further enable effective management of drug loading invading particles' entry and exit through broad surface changes. These qualities result in a diverse range of defined shapes and size, corresponding to cargo spaces. Discrimination between free and complexed guest molecules, particularly pharmaceuticals, relies on limited energy barriers associated with their ease of entry and exit from dendrimer cavities (Esfand and Tomalia 2001). The formation of a complex or partial entrapment of a drug guest within the dendrimer host is contingent on factors such as the drug molecule's size, hydrophilic or lipophilic nature, and compatibility with the dendrimer cavity. For optimal interaction between a hydrophobic drug' and corresponding spaces inside dendrimer, one anticipates the drug associating with the interior. To minimize contact with aqueous and polar domains in physiological setting, the hydrophobic part of the guest molecule should separate from the external interface of dendrimer complex. The hydrophobic and hydrophilic characteristics of spatial binding sites, along with other supramolecular features like non-covalent binding and acid-base reactions, are anticipated to significantly impact guest-host interactions (Parajapati et al. 2016). Analysis of a typical symmetrically branching dendritic lattice reveals minute yet significant factors that regulate the dendrimer's inner space and influence interactions between guests and host. Notably, these spatial binding-sites' hydrophobic and hydrophilic characteristics as well as other supramolecular (such as non-covalent binding and acid-base reaction) characteristics are anticipated to have a significant impact on these guest-host interactions (D'Emanuele and Attwood 2005). It is clear from analysis of a typical symmetrically branching dendritic lattice that there are other minute but significant factors that could regulate a dendrimer's inner space and affect interactions between guests and hosts. In this context, Kumar et al. devised a formulation involving rifampicin-loaded mannosylated dendrimers designed to achieve selective drug delivery and enhance uptake by alveolar macrophages (Kumar et al. 2006). This approach holds promise for improving the delivery of rifampicin to its intended target within the asthmatic lung.

12.5.2 Effect of Generation on Dendrimer Drug Conjugation

External surface conjugation could enhance biomedically significant functionalities on the surface of the dendrimers, such as photon absorption paramagnetic imaging targeting, chelation fluorescence. On the other hand, these functionalities could be selectively internalized using flexible and open scaffolding (generation = 0–3) to semi-rigid container like structures (generation = 4–6) (Esfand et al. 1996). The gating qualities for entry and exit from can be further controlled by a wide variety of surface functions (Fréchet 1994; Duncan and Izzo 2005). Moreover, many novel choices for drug presentation and/or confinement, targeting populations, or imaging moieties are provided by these distinctive architectural elements.

In one study, Roflumilast (RofA), a selective phosphodiesterase 4 (PDE4) inhibitor was conjugated onto PAMAM dendrimer, enhancing its potential for asthma/COPD management. Systematic assessments included encapsulation efficiency, loading capacity, in vitro release kinetics, solubilization behavior, and stability of synthesized conjugates, in the study. Notably, entrapment of RofA increased proportionally with the molecular weight and size of PAMAM dendrimers, demonstrating a direct correlation and was also approximately two and four times proportional with fourth generation dendrimer when compared to 2.0 and 3.0G dendrimers. The encapsulation efficiency and loading capacity varied in complexes from 72.64% to approximately 20.16%, and from 63.60% to 19.88% respectively. In vitro release studies further revealed sustained release from RofA-PAMAM complexes, with reduced release observed in full-generation compared to half-generation complexes due to strengthened bond formation. The inhibitory activity against protein denaturation supported their anti-inflammatory potential. RofA solubility in PAMAM dendrimer solutions correlated positively with increasing pH, with PAMAM dendrimers (4.0G) exhibiting optimal water solubility. Weak anti-bacterial effects further suggested these complexes as candidates for adjuvant antibiotic treatment for protection against microbial infection. Comparative analysis demonstrated the efficacy and simplicity of the proposed synthesis method for enhancing roflumilast solubility (Chawla et al. 2022).

Numerous common pulmonary diseases, such as lung cancer, asthma, and cystic fibrosis, are treatable with small interfering RNA (siRNA)-based therapeutics. However, the absence of effective carriers for siRNA delivery to lung epithelial cells and the difficulties in creating formulations for oral inhalation (OI) for the local administration of siRNA and their carriers to the lungs have slowed progress in this area. The enhanced green fluorescent protein (eGFP) gene can be silenced on A549 lung alveolar epithelial cells by using generation four, amine-terminated poly(amidoamine) (PAMAM) dendrimer (G4NH2) siRNA complexes (dendriplexes). It was revealed that the N/P ratio did not significantly affect the size or capacity of dendriplexes to silence genes. Up to 40% efficiency in silencing has been reported. Pressurized metered-dose inhalers (pMDIs) based on hydrofluoroalkane (HFA) were used to create stable dispersions of dendriplexes that were both encapsulated in mannitol and a biodegradable and water-soluble co-oligomer. Their aerosol properties, with respirable fractions of up to 77%, were exceedingly favorable and conducive to deep lung deposition. Importantly, it was demonstrated that siRNA that was created as dendriplexes in pMDIs maintained its integrity both throughout and after the particle manufacturing processes as well as after prolonged exposures to HFA. This study is important because it is the first to describe the creation of inhalable siRNA with aerosol characteristics appropriate for deep lung deposition utilizing pMDIs, the least priced and most popular portable inhalers. This discovery is significant because it demonstrates for the first time how effectively siRNAG4NH2 dendriplexes can target lung alveolar epithelial A549 cells and mute genes even after siRNA has been exposed to the external environment (Conti et al. 2014).

12.6 Toxicity and Biocompatibility

The cytotoxicity, biocompatibility, and biodistribution of dendrimers must be evaluated in vitro and in vivo in order to determine their suitability as drug carriers. The impact of dendrimers on cell survival, form, hemolysis, amount of protein, and various other biological characteristics warrants further study in dendrimers as drug delivery systems. How much the potential safety/efficacy of any new technology may be justified relies on a number of variables, including pharmacokinetics, biodistribution, fate, and the sort of disease that needs to be treated. In terms of cells, organs, or even the individual's body as a multi-organ system, cytotoxicity provides a measure of the general, undesirable harm it may do. The pharmacokinetics and biodistribution of polymers have a significant impact on their toxicity in vivo; as a result, biodistribution studies are a crucial component of the initial polymer screening process (Mishra et al. 2009). In example, if a polymer is carrying a pharmacological payload that is surface-dispersed, characteristics like the molecular weight, charge, and hydrophobicity of the polymer as well as other subtle elements of the behavior of physico-chemical solutions can have a significant impact on biodistribution. According to the definition of biocompatibility, it is the capacity of a material to elicit the proper host reaction during a certain application (Yang and Kao 2012). It emphasizes the necessity of weighing a material's suitability in relation to both its potential for harm to the body (toxicity) and the potential for harm or benefit to the environment that depends upon the effectiveness of the material (Table 12.2).

Table 12.2 Cytotoxicity of various surface modified dendrimers in vitro

Study/experiment	Cell lines	Incubation time	Polymer types	Concentrations	Assay method	Key findings	Reference
In vitro cytotoxicity	B16F10, CCRF, HepG2	72 h	Dendrimers (NH2, PAMAM, DAB, DAE) and reference polymers (PLL, PEI, dextran)	0–5 mg/ml	MTT assay	NH2-terminated dendrimers showed concentration and generation-dependent cytotoxicity. Anionic dendrimers not cytotoxic, but cationic ones, especially PAMAM and DAB generations 4, displayed marked cytotoxicity. Morphological changes observed in cells exposed to certain dendrimers. PEO modification decreased toxicity	Roberts et al. (1996)
PAMAM dendrimers cytotoxicity	Clone 9	72 h	PAMAM generations 1–4	50–300 μg/ml	MTT assay	Cationic PAMAM generations 1–4 displayed concentration-dependent cytotoxicity, similar to poly(L-lysine). PAMAM dendrimers slightly less toxic than DAB dendrimers with equivalent surface functionality	Chen et al. (2004)
Effect of surface modification	Hep G2,293 HEK cells	3 h	Cationic dendrimers with various surfaces (amine, guanidine, carboxylate, sulfonate, phosphonate, PEGylated)	NA	MTT assay	Cationic dendrimers were more cytotoxic than anionic or PEGylated dendrimers. Quaternised PAMAM–OH derivatives showed lower cytotoxicity than PAMAM–NH2. Surface modification using PEO reduced toxicity	Prieto et al. (2006)
PEGylation extent and cytotoxicity	NT2 cells	NA	PEI with different degrees of PEGylation	Up to 2 mM	MTT assay	Lower degrees of PEGylation did not significantly improve cell viability. Higher degrees of PEGylation reduced PEI-induced cell death	Tang et al. (2003)

Study/experiment	Cell lines	Incubation time	Polymer types	Concentrations	Assay method	Key findings	Reference
PAMAM–NH2 and PAMAM–COOH cytotoxicity	Caco-2 cells	90–210 min	PAMAM generations 0.5–4.5, PAMAM–OH (G2–G4)	NA	LDH assay	PAMAM–NH2 and PAMAM–COOH dendrimers showed generation-, incubation-time-, and concentration-dependent cytotoxicity. G0.5 dendrimers were non-toxic at certain conditions	El-Sayed et al. (2003)
Mannosylated dendrimers cytotoxicity	Vero cells	NA	Mannosylated dendrimers	100 μg/ml	NA	Mannosylated dendrimers at 100 μg/ml showed negligible cytotoxicity. RIF-loaded mannosylated dendrimer had lower cytotoxicity than drug solution	Kumar et al. (2006)
Drug conjugated dendrimer	MT2 cells	NA	Lamivudine-loaded mannosylated PPI dendrimers	Up to 2.5 nM/ml	MTT assay	No toxicity observed up to 0.625 nM/ml of free lamivudine. Blank PPI and mannosylated PPI dendrimers were subtoxic at certain concentrations	Dutta and Jain (2007)

NA not available

A dendrimer's structure and surface properties play a major role in determining its toxicity. It is widely known that cationic agents interact with negatively charged biomembranes inducing cell lysis and destabilization. Dendrimers with cationic charge have an increased tendency to lyse cell membrane (Jain et al. 2010). As such, dendrimers have greater binding affinity for heavy metal ions nucleic acids, amphiphilic lipids, proteins bile acids, and, vitamins and, indicating that the administration of dendrimers may cause these important molecules and ions to lose function. The interaction of these compounds with dendrimers may impact oxygen transport, peroxide metabolism, and cell membrane integrity, resulting in leakage of vital chemicals and intracellular ions. Additionally, daily treatment of such drugs may disrupt protein synthesis and cell cycle processes, leading to acute hypoalbuminemia.

Zebrafish model was used to investigate PAMAM-G3.5 and PAMAM-G4 dendrimers (Martinez et al. 2017; Heiden et al. 2019). PAMAM-G4 with functional amine at terminal groups demonstrated dosage, exposure period, and lethal and sublethal parameters dependence. On the other hand, PAMAM-G3.5-COOH was not poisonous or lethal at any concentration. Zebrafish models were established by Pryor and colleagues to examine the toxicity of these dendrimers from various generations (G3, G4, G5, and G6) with anionic (succinic acid), cationic (-NH2), and neutral (amidoethanol) terminal groups (Pryor et al. 2014). Anionic or neutral dendrimers did not exhibit appreciable death at the tested doses. In contrast, positively charged terminal groups caused pericardial edema and mortality (sublethal effects), as well as important cardiac effects. The authors claimed that the toxicities of compounds in zebrafish models were correlated with their terminal group, generation number, and size (Jones et al. 2012). It was established that for cationic dendrimers, lesser production and size were associated with greater toxicity. Higher-generation dendrimers are more hazardous in cell culture, in contradiction to these studies. This most likely occurred because the negatively charged cell membrane was destabilized by the binding of the cationic moieties leading to cell lysis (Bodewein et al. 2016). Moreover, PAMAM dendrimers displayed generation-dependent toxicity and a decline in cell viability in Caco-2 models. Higher generations, such as PAMAM-G3 and PAMAM-G4, displayed worse cell viability than G0, G1, or G2. G3 dendrimers, which are less cytotoxic and hemolytic than other polycations, did not, however, display substantial cytotoxicity up to 1 mg/ml or hemolytic effects up to 10 mg/ml in a similar investigation with L929 murine fibroblasts (Malik et al. 2000; El-Sayed et al. 2003). Additionally, the toxicity and immune cell activation of the dendrimer surface was evaluated, suggesting the potential induction of inflammatory responses. According to Wang et al., several investigations have showed that PEGylation of PAMAM reduces its toxicity (Wang et al. 2010). When PEGylated, these formulations can reduce toxicity, improve pharmacokinetic profiles, and boost water solubility. Notably, when a dendrimer is PEGylated, the pattern of dendrimer degradation is altered, showing a step-by-step transition where the PEG layer is first broken down and then the dendrimer begins to shed. A straightforward technique, red blood cell (RBC) lysis, is frequently used to explore the interaction between polymers and blood cell membranes. It provides a numerical assessment of the release of hemoglobin (Hb). Because PAMAM dendrimers have NH_2 groups on their perimeter, a coating of carbohydrates around the dendrimeric framework has demonstrated a reduction in the hemolytic toxicity through neutralizing the cationic charges (Table 12.3). The various factors affecting dendrimer compatibility are generation number, charge of the functional group, and degree of substitution at surface group. The biocompatibility of such structures necessitated more intelligent alterations, despite the fact that dendrimers were effective for drug loading. A few smart techniques to improve its biocompatibility comprise (i) surface group derivatization, where terminal positively charged molecules can be transformed through PEGs as well as fatty acids; (ii) surface group replacement where terminal positively charged groups may be substituted by hydroxyl groups; (iii) surface group concealment where the final positively charged groups shall be masked; and (iv) creating a neutral or anionic dendrimer to address safety concerns associated with dendrimer based nanotherapeutics.

12.7 Future Prospects and Challenges

The effective delivery of therapeutics to the lungs through aerosol relies on the synergy of four key elements: formulation, the metering system aerosol device design, and the patient's responsiveness. Achieving optimal pulmonary performance requires a comprehensive approach, treating both the formulation and aerosol device as integral components of the medicinal product. Despite considerable efforts by formulation scientists and chemists in exploring respirable dendrimers for enhanced outcomes in asthma and related pulmonary disorders, this field remains in its early stages of exploration. Numerous research articles emphasize the importance of factors such as particle size, aerodynamic performance, and pulmonary pharmacokinetics in dendrimer-based formulations. However, beyond these considerations, it is crucial to address various issues and challenges during the formulation of dendrimers for asthma therapy. Ultimately, the continued advancement of inhaled dendrimers especially when treating allergic airway disease such as asthma hinges on progress in three major critical areas: (i) the design and modification of dendrimers architecture; (ii) a nuanced understanding of responses observed in human clinical trials; and (iii) a thorough exploration of pulmonary toxicity.

Dendrimers are being advocated for their utilization as nanocarriers in lung cancer drug delivery, owing to their valuable attributes since their discovery. However, despite their numerous advantageous qualities, dendrimers exhibit certain drawbacks, with cytotoxicity being a significant concern. Specifically, cationic dendrimers, while effective against cancerous cells, can also manifest toxicity towards healthy normal cells, particularly at elevated concentrations. While dendrimers hold tremendous promise, their synthesis can be complex and expensive. Additionally, their potential toxicity and immunogenicity need to be thoroughly studied for biomedical applications. Nonetheless, research into dendrimers continues, and they remain an exciting area of study in nanotechnology, pulmonary drug delivery, and other fields that benefit from precise and controlled molecular structures.

Table 12.3 Impact of surface modification of dendrimers on hemolysis

Study	Dendrimer type	Active moiety	Coating/ modification	Hemolysis percentage	Comments	Reference
Haemolysis reduction with carbohydrate coating	Poly(L-lysine) dendrimers	Primaquine phosphate	Carbohydrate (galactose) coating	35.7% (4.0G) to 49.2% (5.0G) for uncoated dendrimers	Carbohydrate coating drastically reduced haemolysis, with 10–7.1% residual haemolysis for coated 4.0G and 5.0G, respectively	Mishra et al. (2009)
Galactose conjugation for reduced toxicity	Poly(L-lysine) dendrimers	Chloroquine phosphate	Galactose conjugation	Less toxic than uncoated dendrimers of other compounds	Galactose conjugation reduced toxicity, providing a safe, effective, and stable sustained release system for chloroquine phosphate	Agrawal et al. (2007)
Mannosylation for Haemolytic toxicity reduction	Poly(L-lysine) dendrimers	Rifampicin	Mannosylation	15.6 ± 3.7% (uncoated 5.0G), less than 2.2 ± 0.8% (Mannosylated)	Mannosylation significantly decreased haemolysis, attributed to the inhibition of interaction with quaternary ammonium ions. Mannosylated dendrimer suppressed the haemolytic toxicity of rifampicin to 6.5 ± 1.5%	Kumar et al. (2006)
PEGylation for reduced hemolysis	PAMAM (4.0G)	5-fluorouracil	PEG	<5 for coated and 15.3–17.3 for uncoated dendrimers	PEGylation improved their ability to load drugs loading capacity, reduced drug release rate and hemolytic toxicity. By conducting in vitro tests and hemolysis experiments on albino rats, the devices were found to be adequate for sustained delivery of 5 fluorouracil	Bhadra et al. (2003)

12.8 Conclusion

Dendrimers have emerged as promising candidates for managing allergic asthmatic disorders, presenting a multifaceted approach to address the intricate pathophysiology of this condition. Scientific advancements in nanotechnology have further positioned dendrimers as highly versatile entities within the realm of allergic asthma. From a scientific perspective, dendrimers serve as efficacious carriers for targeted and controlled drug delivery to pathological alveolar tissues. Their precisely defined structure further provides meticulous control over drug release kinetics, ensuring optimal therapeutic concentrations at the targeted site while minimizing systemic side effects. This approach proves particularly advantageous in allergic asthma, where localized inflammation and bronchoconstriction necessitate a focused therapeutic intervention. Additionally, dendrimers can be tailored to encapsulate various traditional anti-asthmatic drugs, including β2 adrenergic agonists and corticosteroids, as well as novel biologics. This enhances the potential of dendrimers to address the heterogeneity of allergic asthma, accommodating individualized treatment strategies based on the specific needs and characteristics of patients. The integration of dendrimers into the management of allergic asthma underscores the significance of collaborative efforts among researchers, clinicians, and formulation scientists. The translation of dendrimer-based therapies from bench to bedside necessitates a seamless interplay of expertise, encompassing an understanding of intricate immunological mechanisms underlying allergic asthma to optimizing the pharmacokinetic profiles of dendrimer-drug conjugates. Substantial emphasis should be placed on rigorous clinical trials to ensure the safety and efficacy of dendrimer-based interventions. Long-term studies, comprehensive patient monitoring, and robust data analysis are imperative to establish the viability of dendrimers in routine clinical practice. Ultimately, the role of dendrimers in managing allergic asthmatic disorders is poised to revolutionize therapeutic strategies. Leveraging their unique properties, dendrimers provide a targeted and versatile platform for pulmonary drug delivery, aligning with the complexities of allergic asthma. The successful integration of dendrimers into clinical practice demands a concerted effort from the scientific and medical communities, signifying a collaborative stride towards more effective and personalized treatments for asthma.

References

Abbasi E et al (2014) Dendrimers: synthesis, applications, and properties. Nanoscale Res Lett 9(1):1–10. https://doi.org/10.1186/1556-276X-9-247

Agnoletti M et al (2017) Inhalable siRNA-loaded nano-embedded microparticles engineered using microfluidics and spray drying. Eur J Pharm Biopharm 120:9–21. https://doi.org/10.1016/j.ejpb.2017.08.001

Agrawal P, Gupta U, Jain NK (2007) Glycoconjugated peptide dendrimers-based nanoparticulate system for the delivery of chloroquine phosphate. Biomaterials 28(22):3349–3359. https://doi.org/10.1016/j.biomaterials.2007.04.004

Ahmad A (2022) Pharmacological strategies and recent advancement in nano-drug delivery for targeting asthma. Life 12(4):596. https://doi.org/10.3390/life12040596

Albertazzi L, Storti B, Marchetti L, Beltram F (2010) Delivery and subcellular targeting of dendrimer-based fluorescent pH sensors in living cells. J Am Chem Soc 132(51):18158–18167. https://doi.org/10.1021/ja105689u

Bae YH (2002) Smart polymers in drug delivery. Pharm News 9(6):417–424. https://doi.org/10.1081/e-ebpp-120050557

Beeraka NM et al (2022) Immune repertoire and advancements in nanotherapeutics for the impediment of severe steroid resistant asthma (SSR). Int J Nanomedicine 17:2121–2138. https://doi.org/10.2147/IJN.S364693

Bhadra D, Bhadra S, Jain S, Jain NK (2003) A PEGylated dendritic nanoparticulate carrier of fluorouracil. Int J Pharm 257(1–2):111–124. https://doi.org/10.1016/S0378-5173(03)00132-7

Bhadra D, Yadav AK, Bhadra S, Jain NK (2005) Glycodendrimeric nanoparticulate carriers of primaquine phosphate for liver targeting. Int J Pharm 295(1–2):221–233. https://doi.org/10.1016/j.ijpharm.2005.01.026

Bielski E et al (2017) TPP-dendrimer nanocarriers for siRNA delivery to the pulmonary epithelium and their dry powder and metered-dose inhaler formulations. Int J Pharm 527(1–2):171–183. https://doi.org/10.1016/j.ijpharm.2017.05.046

Bodewein L, Schmelter F, Di Fiore S, Hollert H, Fischer R, Fenske M (2016) Differences in toxicity of anionic and cationic PAMAM and PPI dendrimers in zebrafish embryos and cancer cell lines. Toxicol Appl Pharmacol 305:83–92. https://doi.org/10.1016/j.taap.2016.06.008

Chawla M, Kaushik RD, Malik MK, Pundir V, Singh J, Rehmaan H (2022) Development and optimization of RofA-PAMAM dendrimer complex materials for sustained drug delivery. Mater Today Commun 33:104881. https://doi.org/10.1016/j.mtcomm.2022.104881

Chen H, Neerman MF, Parrish AR, Simanek EE (2004) Cytotoxicity, hemolysis, and acute in vivo toxicity of dendrimers based on melamine, candidate vehicles for drug delivery. J Am Chem Soc 126(32):10044–10048. https://doi.org/10.1021/ja048548j

Cloninger MJ (2002) Biological applications of dendrimers. Curr Opin Chem Biol 6(6):742–748. https://doi.org/10.1016/S1367-5931(02)00400-3

Conti DS, Brewer D, Grashik J, Avasarala S, Da Rocha SRP (2014) Poly(amidoamine) dendrimer nanocarriers and their aerosol formulations for siRNA delivery to the lung epithelium. Mol Pharm 11(6):1808–1822. https://doi.org/10.1021/mp4006358

D'Emanuele A, Attwood D (2005) Dendrimer-drug interactions. Adv Drug Deliv Rev 57(15):2147–2162. https://doi.org/10.1016/j.addr.2005.09.012

de Araújo RV, da Silva Santos S, Ferreira EI, Giarolla J (2018) New advances in general biomedical applications of PAMAM dendrimers. Molecules 23(11):1–27. https://doi.org/10.3390/molecules23112849

Dharmage SC, Perret JL, Custovic A (2019) Epidemiology of asthma in children and adults. Front Pediatr 7. https://doi.org/10.3389/fped.2019.00246

Dias AP et al (2020) Dendrimers in the context of nanomedicine. Int J Pharm 573:118814. https://doi.org/10.1016/j.ijpharm.2019.118814

Dua K et al (2019) The potential of siRNA based drug delivery in respiratory disorders: recent advances and progress. Drug Dev Res 80(6):714–730. https://doi.org/10.1002/ddr.21571

Duncan R, Izzo L (2005) Dendrimer biocompatibility and toxicity. Adv Drug Deliv Rev 57(15):2215–2237. https://doi.org/10.1016/j.addr.2005.09.019

Dutta T, Jain NK (2007) Targeting potential and anti-HIV activity of lamivudine loaded mannosylated poly (propyleneimine) dendrimer. Biochim Biophys Acta—Gen Subj 1770(4):681–686. https://doi.org/10.1016/j.bbagen.2006.12.007

El-Sayed M, Rhodes CA, Ginski M, Ghandehari H (2003) Transport mechanism(s) of poly(amidoamine) dendrimers across Caco-2 cell monolayers. Int J Pharm 265(1–2):151–157. https://doi.org/10.1016/S0378-5173(03)00391-0

Esfand R, Tomalia DA (2001) Poly(amidoamine) (PAMAM) dendrimers: from biomimicry to drug delivery and biomedical applications. Drug Discov Today 6(8):427–436. https://doi.org/10.1016/S1359-6446(01)01757-3

Esfand R, Beezer AE, Mitchell JC, Twyman LJ (1996) Synthesis, complexation and pharmaceutical applications of tetra-directional cascade dendrimers. Pharm Sci 2(3):157–159

Ferruti P et al (1994) Recent results on functional polymers and macromonomers of interest as biomaterials or for biomaterial modification. Biomaterials 15(15):1235–1241. https://doi.org/10.1016/0142-9612(94)90275-5

Fréchet JMJ (1994) Functional polymers and dendrimers: reactivity, molecular architecture, and interfacial energy. Science 263(5154):1710–1715. https://doi.org/10.1126/science.8134834

Galli SJ, Tsai M, Piliponsky AM (2008) The development of allergic inflammation. Nature 454(7203):445–454. https://doi.org/10.1038/nature07204

Garbuzenko OB, Mainelis G, Taratula O, Minko T (2014) Inhalation treatment of lung cancer: the influence of composition, size and shape of nanocarriers on their lung accumulation and retention. Cancer Biol Med 11(1):44–55. https://doi.org/10.7497/j.issn.2095-3941.2014.01.004

Gothwal A, Malik S, Gupta U, Jain NK (2019) Toxicity and biocompatibility aspects of dendrimers. Elsevier Inc.

Graham GG (2016) Compendium of inflammatory diseases. Springer Basel, Basel

Haque S et al (2017) Effect of increased surface hydrophobicity via drug conjugation on the clearance of inhaled PEGylated polylysine dendrimers. Eur J Pharm Biopharm 119:408–418. https://doi.org/10.1016/j.ejpb.2017.07.005

Heiden TCK, Dengler E, Kao WJ, Heideman W, Peterson RE (2019) Developmental toxicity of low generation PAMAM dendrimers in zebrafish. Toxicol Appl Pharmacol 225(1):70–79. https://doi.org/10.1016/j.taap.2007.07.009.Developmental

Heise A, Trollsås M, Magbitang T, Hedrick JL, Frank CW, Miller RD (2001) Star polymers with alternating arms from miktofunctional μ-initiators using consecutive atom transfer radical polymerization and ring-opening polymerization. Macromolecules 34(9):2798–2804. https://doi.org/10.1021/ma0012954

Hsu J, Chen J, Mirabelli MC (2018) Asthma morbidity, comorbidities, and modifiable factors among older adults. J Allergy Clin Immunol Pract 6(1):236–243.e7. https://doi.org/10.1016/j.jaip.2017.06.007

Inapagolla R et al (2010) In vivo efficacy of dendrimer-methylprednisolone conjugate formulation for the treatment of lung inflammation. Int J Pharm 399(1–2):140–147. https://doi.org/10.1016/j.ijpharm.2010.07.030

Jain K, Kesharwani P, Gupta U, Jain NK (2010) Dendrimer toxicity: let's meet the challenge. Int J Pharm 394(1–2):122–142. https://doi.org/10.1016/j.ijpharm.2010.04.027

Jevprasesphant R, Penny J, Jalal R, Attwood D, McKeown NB, D'Emanuele A (2003) The influence of surface modification on the cytotoxicity of PAMAM dendrimers. Int J Pharm 252(1–2):263–266. https://doi.org/10.1016/S0378-5173(02)00623-3

Jones CF et al (2012) Cationic PAMAM dendrimers aggressively initiate blood clot formation. ACS Nano 6(11):9900–9910. https://doi.org/10.1021/nn303472r

Kaminskas LM et al (2014) Pulmonary administration of a doxorubicin-conjugated dendrimer enhances drug exposure to lung metastases and improves cancer therapy. J Control Release 183(1):18–26. https://doi.org/10.1016/j.jconrel.2014.03.012

Kasai S, Nagasawa H, Shimamura M, Uto Y, Hori H (2002) Design and synthesis of antiangiogenic/heparin-binding arginine dendrimer mimicking the surface of endostatin. Bioorg Med Chem Lett 12(6):951–954. https://doi.org/10.1016/S0960-894X(02)00066-5

Kenyon NJ et al (2013) Self-assembling nanoparticles containing dexamethasone as a novel therapy in allergic airways inflammation. PLoS One 8(10):e77730. https://doi.org/10.1371/journal.pone.0077730

Kukowska-Latallo JF, Bielinska AU, Johnson J, Spindle R, Tomalia DA, Baker JR (1996) Efficient transfer of genetic material into mammalian cells using starburst polyamidoamine dendrimers. Proc Natl Acad Sci USA 93(10):4897–4902. https://doi.org/10.1073/pnas.93.10.4897

Kumar PV, Asthana A, Dutta T, Jain NK (2006) Intracellular macrophage uptake of rifampicin loaded mannosylated dendrimers. J Drug Target 14(8):546–556. https://doi.org/10.1080/10611860600825159

Liu M, Fréchet JMJ (1999) Designing dendrimers for drug delivery. Pharm Sci Technol Today 2(10):393–401. https://doi.org/10.1016/S1461-5347(99)00203-5

Liu H, Jiang A, Guo J, Uhrich KE (1999a) Unimolecular micelles: synthesis and characterization of amphiphilic polymer systems. J Polym Sci Part A Polym Chem 37(6):703–711. https://doi.org/10.1002/(SICI)1099-0518(19990315)37:6<703::AID-POLA5>3.0.CO;2-O

Liu M, Kono K, Fre JMJ (1999b) Conjugates as potential drug carriers, pp 3492–3503

Liu M, Kono K, Fréchet JMJ (2000) Water-soluble dendritic unimolecular micelles: their potential as drug delivery agents. J Control Release 65(1–2):121–131. https://doi.org/10.1016/S0168-3659(99)00245-X

Lommatzsch M, Stoll P (2016) Novel strategies for the treatment of asthma. Allergo J Int 25(1):11–17. https://doi.org/10.1007/s40629-016-0093-5

Maiti PK, Çağin T, Wang G, Goddard WA (2004) Structure of PAMAM dendrimers: generations 1 through 11. Macromolecules 37(16):6236–6254. https://doi.org/10.1021/ma035629b

Malik N et al (2000) Dendrimers: relationship between structure and biocompatibility in vitro, and preliminary studies on the biodistribution of 125I-labelled polyamidoamine dendrimers in vivo. J Control Release 65(1–2):133–148. https://doi.org/10.1016/S0168-3659(99)00246-1

Martinez CS et al (2017) Relation between biophysical properties of nanostructures and their toxicity on zebrafish. Biophys Rev 9(5):775–791. https://doi.org/10.1007/s12551-017-0294-2

Mehta M et al (2019) Interactions with the macrophages: an emerging targeted approach using novel drug delivery systems in respiratory diseases. Chem Biol Interact 304:10–19. https://doi.org/10.1016/j.cbi.2019.02.021

Mishra V, Gupta U, Jain NK (2009) Surface-engineered dendrimers: a solution for toxicity issues. J Biomater Sci Polym Ed 20(2):141–166. https://doi.org/10.1163/156856208X386246

Murphy KR, Meltzer EO, Blaiss MS, Nathan RA, Stoloff SW, Doherty DE (2012) Asthma management and control in the United States: results of the 2009 asthma insight and management survey. Allergy Asthma Proc 33(1):54–64. https://doi.org/10.2500/aap.2011.32.3518

Nagasaki T, Kimura O, Ukon M, Arimori S, Hamachi I, Shinkai S (1994) Synthesis, metal-binding properties and polypeptide solubilization of 'crowned' arborols. J Chem Soc Perkin Trans 1(1):75–81. https://doi.org/10.1039/p19940000075

Nasr M, Najlah M, D'Emanuele A, Elhissi A (2014) PAMAM dendrimers as aerosol drug nanocarriers for pulmonary delivery via nebulization. Int J Pharm 461(1–2):242–250. https://doi.org/10.1016/j.ijpharm.2013.11.023

Nathan A, Zalipsky S, Ertel SI, Agathos SN, Yarmush ML, Kohn J (1993) Copolymers of lysine and polyethylene glycol: a new family of functionalized drug carriers. Bioconjug Chem 4(1):54–62. https://doi.org/10.1021/bc00019a008

Newkome GR, Yao ZQ, Baker GR, Gupta VK (1985) Cascade molecules: a new approach to micelles.1aA [27]-arborol. J Org Chem 50(11):2003–2004. https://doi.org/10.1021/jo00211a052

Paleos CM, Tsiourvas D, Sideratou Z (2007) Molecular engineering of dendritic polymers and their application as drug and gene delivery systems. Mol Pharm 4(2):169–188. https://doi.org/10.1021/mp060076n

Parajapati SK, Maurya SD, Das MK, Tilak VK, Verma KK, Dhakar RC (2016) Potential application of dendrimers in drug delivery: a concise review and update. J Drug Deliv Ther 6(2):71–88. https://doi.org/10.22270/jddt.v6i2.1195

Prieto MJ et al (2006) Nanomolar cationic dendrimeric sulfadiazine as potential antitoxoplasmic agent. Int J Pharm 326(1–2):160–168. https://doi.org/10.1016/j.ijpharm.2006.05.068

Pryor JB, Harper BJ, Harper SL (2014) Comparative toxicological assessment of PAMAM and thiophosphoryl dendrimers using embryonic zebrafish. Int J Nanomedicine 9(1):1947–1956. https://doi.org/10.2147/IJN.S60220

Rajabnezhad S et al (2016) Pulmonary delivery of rifampicin microspheres using lower generation polyamidoamine dendrimers as a carrier. Powder Technol 291:366–374. https://doi.org/10.1016/j.powtec.2015.12.037

Ratemi E, Sultana Shaik A, Al Faraj A, Halwani R (2016) Alternative approaches for the treatment of airway diseases: focus on nanoparticle medicine. Clin Exp Allergy 46(8):1033–1042. https://doi.org/10.1111/cea.12771

Reuter JD et al (1999) Inhibition of viral adhesion and infection by sialic-acid-conjugated dendritic polymers. Bioconjug Chem 10(2):271–278. https://doi.org/10.1021/bc980099n

Roberts JC, Bhalgat MK, Zera RT (1996) Preliminary biological evaluation of polyamidoamine (PAMAM) Starburst™ dendrimers. J Biomed Mater Res 30(1):53–65. https://doi.org/10.1002/(SICI)1097-4636(199601)30:1<53::AID-JBM8>3.0.CO;2-Q

Rytting E, Nguyen J, Wang X, Kissel T (2008) Biodegradable polymeric nanocarriers for pulmonary drug delivery. Expert Opin Drug Deliv 5(6):629–639. https://doi.org/10.1517/17425247.5.6.629

Sahu T, Ratre YK, Chauhan S, Bhaskar LVKS, Nair MP, Verma HK (2021) Nanotechnology based drug delivery system: current strategies and emerging therapeutic potential for medical science. J Drug Deliv Sci Technol 63:102487. https://doi.org/10.1016/j.jddst.2021.102487

Santamaria F, Borrelli M, Baraldi E (2021) GINA 2021: the missing pieces in the childhood asthma puzzle. Lancet Respir Med 9(10):e98. https://doi.org/10.1016/S2213-2600(21)00275-7

Shahi SR, Kulkarni MS, Karva GS, Giram PS, Gugulkar RR (2015) Review article: DENDRIMERS. Int J Pharm Sci Res 33:187–198

Shastri MD et al (2021) Emerging concepts and directed therapeutics for the management of asthma: regulating the regulators. Inflammopharmacology 29(1):15–33. https://doi.org/10.1007/s10787-020-00770-y

Shaunak S et al (2004) Polyvalent dendrimer glucosamine conjugates prevent scar tissue formation. Nat Biotechnol 22(8):977–984. https://doi.org/10.1038/nbt995

Singh AP, Biswas A, Shukla A, Maiti P (2019) Targeted therapy in chronic diseases using nanomaterial-based drug delivery vehicles. Signal Transduct Target Ther 4(1):1–21. https://doi.org/10.1038/s41392-019-0068-3

Suraya R, Nagano T, Katsurada M, Sekiya R, Kobayashi K, Nishimura Y (2021) Molecular mechanism of asthma and its novel molecular target therapeutic agent. Respir Investig 59(3):291–301. https://doi.org/10.1016/j.resinv.2020.12.007

Tang GP et al (2003) Polyethylene glycol modified polyethylenimine for improved CNS gene transfer: effects of PEGylation extent. Biomaterials 24(13):2351–2362. https://doi.org/10.1016/S0142-9612(03)00029-2

Tomalia DA (1995) Dendrimer molecules. Sci Am 272(5):42–48. https://doi.org/10.1038/scientificamerican0595-62

Tomalia DA (2005) Birth of a new macromolecular architecture: dendrimers as quantized building blocks for nanoscale synthetic polymer chemistry. Prog Polym Sci 30(3–4):294–324. https://doi.org/10.1016/j.progpolymsci.2005.01.007

Tomalia DA et al (1985) A new class of polymers: starburst-dendritic. Polym J 17(1):117–132. [Online]. Available http://www.japsonline.com/abstract.php?article_id=1465

Ulrik B, Heegaard PMH (2004) Dendrimers in drug research. Chem Soc Rev 33(1):43–63

Veronese FM, Morpurgo M (1999) Bioconjugation in pharmaceutical chemistry. Farmaco 54(8):497–516. https://doi.org/10.1016/S0014-827X(99)00066-X

Veronese FM, Pasut G (2005) PEGylation, successful approach to drug delivery. Drug Discov Today 10(21):1451–1458

Wang W, Xiong W, Zhu Y, Xu H, Yang X (2010) Protective effect of PEGylation against poly(amidoamine) dendrimer-induced hemolysis of human red blood cells. J Biomed Mater Res Part B Appl Biomater 93(1):59–64. https://doi.org/10.1002/jbm.b.31558

Wang L, Feng M, Li Q, Qiu C, Chen R (2019) Advances in nanotechnology and asthma. Ann Transl Med 7(8):180–180. https://doi.org/10.21037/atm.2019.04.62

Wilczewska AZ, Niemirowicz K, Markiewicz KH, Car H (2012) Nanoparticles as drug delivery systems. Pharmacol Rep 64(5):1020–1037. https://doi.org/10.1016/S1734-1140(12)70901-5

Wooley KL, Hawker CJ, Pochan JM, Fréchet JMJ (1993) Physical properties of dendritic macromolecules: a study of glass transition temperature. Macromolecules 26(7):1514–1519. https://doi.org/10.1021/ma00059a006

Yang H, Kao WJ (2012) Dendrimers for pharmaceutical and biomedical applications. J Biomater Sci 2013:37–41

Zhong Q (2018) Co-spray dried mannitol/poly(amidoamine)-doxorubicin dry-powder inhaler formulations for lung adenocarcinoma: morphology, in vitro evaluation, and aerodynamic performance. AAPS PharmSciTech 19(2):531–540. https://doi.org/10.1208/s12249-017-0859-1

Zhong Q, Bielski ER, Rodrigues LS, Brown MR, Reineke JJ, da Rocha SRP (2016a) Conjugation to poly(amidoamine) dendrimers and pulmonary delivery reduce cardiac accumulation and enhance antitumor activity of doxorubicin in lung metastasis. Mol Pharm 13(7):2363–2375. https://doi.org/10.1021/acs.molpharmaceut.6b00126

Zhong Q, Merkel OM, Reineke JJ, Da Rocha SRP (2016b) Effect of the route of administration and PEGylation of poly(amidoamine) dendrimers on their systemic and lung cellular biodistribution. Mol Pharm 13(6):1866–1878. https://doi.org/10.1021/acs.molpharmaceut.6b00036

Zhong Q, Humia BV, Punjabi AR, Padilha FF, da Rocha SRP (2017) The interaction of dendrimer-doxorubicin conjugates with a model pulmonary epithelium and their cosolvent-free, pseudo-solution formulations in pressurized metered-dose inhalers. Eur J Pharm Sci 109:86–95. https://doi.org/10.1016/j.ejps.2017.07.030

Nanofibers for the Management of Allergic Airway Disease

13

Nitin Sharma, Akshita Arora, Amit Chaudhary, Anjana Sharma, and Namita Aggarwal

Abstract

A subset of airway illnesses known as allergic airway disease or allergic respiratory disease is characterized by heightened immunological reactions to particular allergens. The process of allergy starts with the entry of allergen into the airways, which triggers the immunogenic reaction though several processes and may cause several pathological morbidities including inflammation. Treating allergic responses in airways is the biggest challenge as the formulation is exposed to several physiological challenges such as mucociliary clearance and continuous contraction and dilatation of respiratory muscles. In this challenging context, nanofibers show multiple potentials not only as a prominent carrier system for therapeutic substances but also as an efficient protective system in devices. This chapter highlights the several applications of nanofibers in the management of allergic diseases associated with the respiratory tract.

N. Sharma (✉)
Department of Pharmaceutics, Amity Institute of Pharmacy, Amity University, Noida, India

A. Arora
Department of Pharmaceutics, ISF College of Pharmacy, Moga, Punjab, India

A. Chaudhary · A. Sharma · N. Aggarwal
Chitkara University School of Pharmacy, Chitkara University, Baddi, Himachal Pradesh, India

A. Sharma
Chitkara University School of Pharmacy, Chitkara University, Baddi, Himachal Pradesh, India

Department of Pharmacy, Lloyd Institute of Management and Technology, Greater Noida, India

P. Prasher et al. (eds.), *Understanding Allergic Airway Diseases*, https://doi.org/10.1007/978-981-97-1953-2_13

Keywords

Nanofibers · Respiratory disease · Airway disease

13.1 Introduction

13.1.1 Allergic Airway Disease

Allergic airway disease, often referred to as allergic respiratory disease, is a subset of airway diseases characterized by exaggerated immune responses to specific allergens. The immune system of individuals with allergic airway disease reacts abnormally to substances like pollen, dust mites, pet dander, and mold spores, triggering inflammation and symptoms in the respiratory tract (D'Amato et al. 2015). This immune response leads to inflammation and hypersensitivity of the airways, resulting in various symptoms such as sneezing, coughing, wheezing, shortness of breath, and nasal congestion. Common allergic airway diseases include allergic rhinitis (hay fever), allergic asthma, and other related conditions where exposure to allergens triggers a cascade of immune reactions that affect the upper and lower respiratory tracts (Pawankar et al. 2012).

The general mechanism of allergic airways started when inhaled allergens reach the respiratory pathways are captured and processed by dendritic cells (DCs), which are sensitized by thymic stromal lymphopoietin (TSLP) produced by the airway's epithelial cells. These allergens additionally prompt mast cells to release CCL17 and CCL22 which interact with CCR4 (CC chemokine receptor 4), initiating the chemotactic movement of TH-2 cells. TH-2 cells play vital roles in coordinating the inflammatory response triggered by allergens through the secretion of IL-4 and IL-13. These interleukins stimulate the generation of IgE antibodies by B cells to produce IL-5 (essential for eosinophilic inflammation) and IL-9 (a factor that spurs mast cell proliferation) (Fig. 13.1) (Reddel et al. 2022).

The primary cause of allergic airway disease is an exaggerated immune response to specific allergens in susceptible individuals. When someone with a predisposition to allergies comes into contact with these allergens, their immune system overreacts, leading to inflammation and various symptoms in the airways. The exact cause of this heightened immune response is complex and involves a combination of genetic, environmental, and immunological factors (Fokkens et al. 2020; Geldof et al. 2019; Kidwell et al. 2000). Some key contributors to the development of allergic airway disease include:

13.1.1.1 Genetics

Family history plays a significant role in predisposing individuals to allergic conditions. If a person has close relatives with allergies, they are more likely to develop allergic airway disease themselves.

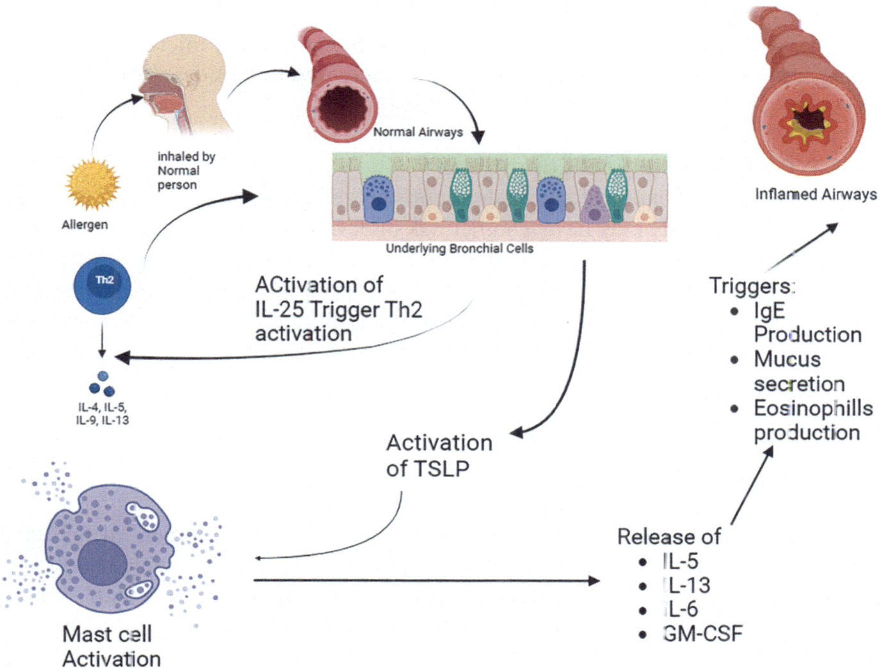

Fig. 13.1 General mechanism of allergic airways

13.1.1.2 Environmental Allergens

Exposure to common environmental allergens such as pollen, dust mites, pet dander, mold spores, and certain foods can trigger allergic reactions in susceptible individuals. These allergens can be inhaled, ingested, or come into contact with the skin.

13.1.1.3 Immune System Sensitization

In allergic individuals, the immune system mistakenly identifies harmless substances (allergens) as threats. This triggers the production of specific antibodies (IgE antibodies), which attach to immune cells and mast cells, priming them to release inflammatory mediators upon allergen exposure.

13.1.1.4 Air Pollution and Irritants

Environmental factors like air pollution and exposure to tobacco smoke can worsen allergic airway disease symptoms and increase the risk of developing these conditions

13.1.1.5 Early Childhood Exposures

Exposures during early childhood, including exposure to allergens, viral infections, and diet, can influence the development of allergies and allergic airway disease.

13.1.1.6 Hygiene Hypothesis

Some theories suggest that reduced exposure to infections and microbial diversity during early childhood might contribute to an increased risk of developing allergic conditions. This is known as the "hygiene hypothesis."

13.1.1.7 Other Medical Conditions

Individuals with a history of atopic dermatitis (eczema) or allergic sensitization in other parts of the body may have a higher risk of developing allergic airway disease.

While these factors contribute to the development of allergic airway disease, it's important to note that each individual's response is unique. The interplay between genetic susceptibility and environmental triggers ultimately leads to the immune system's aberrant response, resulting in the characteristic symptoms of allergic airway disease.

13.1.2 Nanofibers

The rapid development of nanotechnology has led to the development of multiple techniques for the production of diverse nanoscale composites, with nanofibers which are receiving particular attention. Polymeric or composite fibers with a diameter of the nanoscale are known as nanofibers. They have a wide range of porous structures and a large surface. Nanomaterials with at least one dimension of 100 nm or fewer and a length-to-diameter ratio of 100 times are considered nanofibers. However, in general, all fibers with a diameter of less than one micron are regarded as nanofibers. Nanomaterials with large surface areas, high surface-to-volume ratios, and adaptable mechanical and morphological properties are produced as a result of a special combination of construction materials and fabrication techniques. These materials can be used for a variety of applications in the fields of water and air filtration, automotive and transportation, textiles, medical and biosensor appliances, electronics, and energy storage. The market for nanofibers has been expanding steadily in recent years and is anticipated to develop significantly in the years to come. The market is currently worth $785 million, and between 2022 and 2027, it is anticipated to increase at a compound yearly growth rate of more than 15%. Despite the market's dominance by filtration devices, long-term market growth will mostly be driven by the needs of the medical and pharmaceutical sectors. Chemical plasticity can be paired with materials that have adjustable surface topology, porosity, and pore size to create materials that can carry pharmaceuticals and biomedicines, as well as be utilized to develop biomedical devices for tissue engineering and dressing wounds (Anup et al. 2021; Stojanov and Berlec 2020).

13.1.3 Application of Nanofibers in Allergic Airway Disease

Nanofibers hold promise in the management of allergic airway diseases due to their unique properties and potential applications in air filtration, drug delivery, and tissue engineering. Here is the discussion how nanofibers could contribute to managing allergic airway diseases:

13.1.3.1 Air Filtration

The major environmental issue in recent years has been air pollution, notably the global haze issue that is primarily caused by fine particulate matter (PM), which is growing as a result of industrialization and urbanization. Airborne fine particle pollution (PM2.5 and PM1) have been linked to a number of health issues, including cancer, fibrosis, and chronic lung disease, according to numerous exposure studies. In addition, airborne particle pollution contains a range of germs, pollen, fungi, spores, and allergens that can cause allergic reactions and respiratory illnesses. Specifically, PM 0.3, which are solid particles or droplets in the air that have a diameter of less than or equal to 0.3 m, can disperse widely, transport a range of bacteria and viruses, enter the bronchial system, and even enter the blood (Wang and Otani 2013; Huang et al. 2014).

Filtration is the better way to protect individual persons from PM pollutants. Nano fibers can be used to create highly efficient air filters that capture allergens such as pollen, dust mites, pet dander, and other particles which are causing the allergic airways disease. These filters can be integrated into ventilation systems, air purifiers, and face masks to reduce allergen exposure and improve indoor air quality. Such fibrous filters, however, typically display insufficient removal effectiveness for tiny particles. The comparatively wide pores, uneven fiber diameter, and poor mechanical performance are the main causes of the poor filtration performance. Manufacturing filtration materials with high filtration performance is therefore crucial. As a result, more and more research is being done on the creation of electrospun air filters. Numerous polymer materials have been effectively converted into nanofibrous membranes and used in the field of air filtration, including polyimide, polyurethane, polyacrylonitrile, polyamide-66, polyamide-56, polysulfone, etc. (Stanford et al. 2019; Zhang et al. 2020).

13.1.3.2 Drug Delivery

Nanofibers are effective drug delivery systems because of their large surface area, high loading capacity, simplicity of functionalization, and potential to simultaneously include a variety of therapeutic compounds. Furthermore, the various instrumental configurations and programmable experimental parameters allow for the design of tunable morphologies for the optimum drug loading and release. Drug distribution via oral, topical, transdermal, and local administration routes can be facilitated by adjusting the mechanical properties of nanofibers. It is undeniable that the increased bioavailability, specific drug release, longer drug release profile, minimal toxicity, and lower dosing frequency offered by new nanomaterials such as nanofibers has improved patient compliance and (Li et al. 2022) Nanofibers can be administered in oral DDSs as nanofiber scaffolds, oral rapid delivery systems using electrosponge nanofibers, multilayered nanofiber loaded mashes, and surface-modified cross-linked electrospun nanofibers (Kapahi et al. 2015). Application of electrospun polymeric nanofibers has shown to be an intriguing method for drug delivery systems. The fibers' high surface-to-volume ratio can enhance a number of activities, including mass transfer, drug loading, and cell attachment and proliferation (Torres-Martínez et al. 2018). Drug delivery methods including nanoscale formulations (such liposomes), polymer micelles, some complexes, and nanofibers have all been created or are actively being researched. Due to their potential to enhance therapeutic benefits and lessen the toxicity of traditional dose forms, they have drawn significant attention in recent years (Huang et al. 2003).

The incidence of side effects associated with treatment with conventional dosage forms has been reduced, and controlled release systems that release the active substance at the desired rate for the desired time have been developed to increase patient compliance with the treatment by reducing the frequency of drug use. Controlled release systems allow the active ingredient to be directed to a specific location of the body and the release timing and rate to be modified in accordance with the intended purpose (Uhrich et al. 1999). A prolonged release of bioactive molecules is made possible by controlled release drug delivery systems, which also boost bioavailability over an extended period of time. The development of controlled release systems requires the use of biocompatible and biodegradable components like polycaprolactone (PCL) nanofibers and alginate hydrogel (Rasouli et al. 2019). Pharmaceutical dosage forms that are capable of regulated, delayed, and fast release can be created using nanofibers. The advantages of conventional solid dosage forms, such as simplicity of processing, stability, and convenience of transportation and storage, are also carried by nanofibers, which boost clinical efficacy by increasing biopharmaceutical characteristics. For the creation of controlled release nanofibers, the compatibility of the active component, polymer, and solvent is significant. The electrospinning technique is the most popular method for producing nanofibers among the many nanofiber preparation techniques because it enables the manufacture of nanofibers using a variety of materials, including ceramics and natural or synthetic polymers, all while being quick, easy, and inexpensive to produce. Therapeutic compounds are frequently integrated to the nanofibers for controlled release in electrospun nanofiber applications in regenerative medicine and wound dressings (Lv 2015). It is well known that electrospun implants deliver a localized, long-term, regulated release of the active ingredient and are superior to the products on the market when implanted into the tissue. Nanotubes are resilient and continue to exist in tissues even after being administered by aerosol inhalation, making them potentially more effective than other nanoformulations (Ranganath and Wang 2008).

Varenicline (one of the most popular medications in the family of smoking cessation aids) loaded nanomicelles-in-coaxial nanofibers were created by Yahia et al. The Pluronic F127 nanomicelles served as the basis for the nanofibers' cores, while cellulose acetate (CA) and polycaprolactone (PCL) polymeric solutions served as the material for the fibers' outer sheaths. The resulting NFs transdermal patches had an ultra-porous structure, exceptional mechanical characteristics, and a strong attraction for water (Yahia et al. 2022). In another study, ovalbumin loaded carbon nanotube possess significant increment in ovalbumin-induced T-lymphocyte proliferation and further increment in lung Th2 cytokines and chemokines as compare to the ovalbumin alone. These researchers' more recent research suggested that nanotubes can further aggravate murine airway inflammation caused by allergies via increased Th2 immunity stimulation and increased oxidative stress may occur in part due to the improper activation of antigen-presenting cells, such as dendritic cells (Inoue et al. 2009, 2010).

13.1.3.3 Tissue Engineering

Nano fibers can be used in tissue engineering to create scaffolds for regenerating damaged respiratory tissues. This could potentially aid in repairing airway linings damaged by inflammation in conditions like allergic bronchitis. However, using scaffold at respiratory region is challenging due to several factors such as regularly expansion and contraction of the lung tissue due to the inhalation process, where the cells are continuously under stress condition. The nanofibers-based scaffolds have shown the potential applications in several respiratory diseases including chronic obstructive pulmonary disease, and asthma using variety of polymers such as chitosan and polycaprolacton. Rezaei et al. reported the chitosan and polycaprolactone based nanofibers scaffolds for lung tissue engineering. They conclude that the prepared scaffold can be used in lungs effectively in case of even severe conditions (Rezaei et al. 2021). Young et al. developed poly-l-lactic acid based scaffold for the lugs which could be relevant in vitro model of small airway smooth muscle. The scaffolds were developed by electrospun method to investigate cells-extracellular matrix interactions in the small airways (Young et al. 2017).

13.1.3.4 Immunotherapy Delivery

Modern, effective methods for the prevention and treatment of a number of respiratory disorders have been made available by the advent of immunotherapies. Cell immunotherapy involves giving patients live cells to trigger their immune systems to fight against the disease. The mucosal barrier of the respiratory tract performs as an immune organ and is capable of identifying and eliminating any harmful exposures. Pathogen recognition receptors are expressed by epithelial cells and a variety of immune cells in the respiratory system, including alveolar macrophages, to support strong immunological responses, such as during an infection, which cause immune cell recruitment and inflammation (Daly et al. 2021). The local immune system works in conjunction with the physical mucus layer covering the mucosal epithelium, mucosal immune cells, lymphoid organs, and underlying blood and lymphatic vasculature to operate as a barrier against infections. The above-mentioned barriers, however, pose a challenge for immunotherapy, which must first travel through mucus and epithelium before reaching its localized therapeutic target. It is widely recognized that the induction of mucosal immunity through vaccination is a difficult task and that particular delivery methods and adjuvants are frequently needed for maximum effectiveness (McCright and Maisel 2020). Therefore, an effective delivery system is essential to ensure the delivery of immunogen as desired site by avoiding the physiological berries. Nano fibers can play a role in the controlled delivery of allergen extracts or immunotherapeutic agents, facilitating desensitization and immune modulation to treat allergic airway diseases.

Mercel et al. developed peptide amphiphile nanofibers for the targeting of proteins (angiotensin-converting enzyme and receptor for advanced glycation end products) upregulated during smoke inhalation injury. The targeting efficiency was determined in rat model after smoke inhalation injury. The authors conclude that the nanofibers based formulation was found effective in smoke inhalation injury (Mercel et al. 2021). In a similar kind of work, Marulanda et al. developed self-assembled peptide amphiphile nanofibers for the enhancement of therapeutic value in pulmonary hypertension and reduction in systemic toxicity. The lungs targeting was actively controlled through angiotensin I-converting enzyme and the receptor for advanced glycation end-products. The authors confirmed the improved efficacy of developed formulation in a chronic hypoxia-induced pulmonary hypertension mouse model (Marulanda et al. 2021).

13.1.3.5 Respiratory Sensing Diagnostic Equipment

Nano fiber-based sensing diagnostic could offer enhanced protection against allergens and airborne irritants, especially in environments with high allergen exposure which may cause abnormal respiratory patterns. Respiratory rate monitoring is just one of the many uses for which airflow sensors have been created. To assess respiratory rate, tidal volume, breathing frequency, and minute ventilation, airflow sensors are frequently used in real-time respiration monitoring. Minute ventilation is the total amount of air exhaled from the lungs in a minute and it is inversely related to blood carbon dioxide levels. The entire amount of air moved during resting inhalation and exhalation is represented by the tidal volume. Breathing problems are indicated by abnormal tidal volume and minute ventilation ranges. Airflow sensors must have quick response times and great sensitivity in order to accurately record breathing patterns for the early detection and diagnosis of respiratory illnesses, which may help doctors be better prepared.

Moshizi et al. also developed novel airflow sensor based on platinum (Pt) thin films for respiratory monitoring. Polydimethylsiloxane, which gives the sensors their exceptional bendability and ultrahigh sensitivity, supported the conductive Pt thin sheets (Moshizi et al. 2022). Li et al. proposed automatic auxiliary diagnosis system makes use of a flexible, self-powered sensor built on an electret, signal processing, and machine learning techniques. With high accuracy rates of 99.43%, sensitivity of 98.30%, and specificity of 99.02%, the suggested system may detect and prevent a variety of respiratory disorders. It is based on a strain-enhanced laminated electret with an edge-to-edge hollow hemisphere array construction (Li et al. 2023).

13.2 Conclusion

Allergic diseases associated with the airway are very common due to the direct exposure of allergen to the organ/site. This is difficult to manage with the conventional mode of the treatment as the cells are already stress conditions with multiple of pathological and physiological factors. Nanofibers have proven their applicability is this gray area by providing multiple applications includes drug delivery, diagnosis, immunotherapy, and tissue engineering. However further in-depth knowledge of their clinical acceptability should be documented by trials.

References

Anup N, Chavan T, Chavan S, Polaka S, Kalyane D, Abed SN, Venugopala KN, Kalia K, Tekade RK (2021) Reinforced electrospun nanofiber composites for drug delivery applications. J Biomed Mater Res A 109:2036–2064. https://doi.org/10.1002/jbm.a.37187

D'Amato G, Holgate ST, Pawankar R et al (2015) Meteorological conditions, climate change, new emerging factors, and asthma and related allergic disorders. A statement of the World Allergy Organization. World Allergy Organ J 8:1–52. https://doi.org/10.1186/s40413-015-0073-0

Daly S, O'Sullivan A, Mac Loughlin R (2021) Cellular immunotherapy and the lung. Vaccines (Basel) 9(9):1018. https://doi.org/10.3390/vaccines9091018

Fokkens WJ, Lund VJ, Hopkins C et al (2020) European position paper on rhinosinusitis and nasal polyps 2020. Rhinology 58(Suppl S29):1–464

Geldof T, Rawal S, Dyck WV, Huys I (2019) Comparative and combined effectiveness of innovative therapies in cancer: a literature review. J Comp Eff Res 8:205–216. https://doi.org/10.2217/cer-2018-0131

Huang ZM, Zhang YZ, Kotaki M, Ramakrishna S (2003) A review on polymer nanofibers by electrospinning and their applications in nanocomposites. Compos Sci Technol 63:2223–2253. https://doi.org/10.1016/S0266-3538(03)00178-7

Huang RJ, Zhang Y, Bozzetti C et al (2014) High secondary aerosol contribution to particulate pollution during haze events in China. Nature 514:218–222. https://doi.org/10.1038/nature13774

Inoue K, Koike E, Yanagisawa R, Hirano S, Nishikawa M, Takano H (2009) Effects of multi-walled carbon nanotubes on a murine allergic airway inflammation model. Toxicol Appl Pharmacol 237:306–316. https://doi.org/10.1016/j.taap.2009.04.003

Inoue K, Yanagisawa R, Koike E, Nishikawa M, Takano H (2010) Repeated pulmonary exposure to single-walled carbon nanotubes exacerbates allergic inflammation of the airway: possible role of oxidative stress. Free Radic Biol Med 48:924–934. https://doi.org/10.1016/j.freeradbiomed.2010.01.013

Kapahi H, Mansoor Khan N, Bhardwaj A, Mishra N (2015) Implication of nanofibers in oral drug delivery. Curr Pharm Des 21:2021–2036

Kidwell CS, Starkman S, Eckstein M, Weems K, Saver JL (2000) Identifying stroke in the field: prospective validation of the Los Angeles Prehospital Stroke Screen (LAPSS). Stroke 31:71–76. https://doi.org/10.1161/01.STR.31.1.71

Li J, Liu Y, Abdelhakim HE (2022) Drug delivery applications of coaxial electrospun nanofibres in cancer therapy. Molecules 27:1803. https://doi.org/10.3390/molecules27061803

Li X, Qi B, Wan X, Zhang J, Yang W, Xiao Y, Mao F et al (2023) Electret-based flexible pressure sensor for respiratory diseases auxiliary diagnosis system using machine learning technique. Nano Energy 114:108652. https://doi.org/10.1016/j.nanoen.2023.108652

Lv Y (2015) Editorial (Thematic issue: nanofiber-based drug design, delivery and application). Curr Pharm Des 21:1918–1919

Marulanda K, Mercel A, Gillis DC, Sun K, Gambarian M, Roark J, Weiss J et al (2021) Intravenous delivery of lung-targeted nanofibers for pulmonary hypertension in mice. Adv Healthc Mater 10(13):2100302. https://doi.org/10.1002/adhm.202100302

McCright JC, Maisel K (2020) Engineering drug delivery systems to overcome mucosal barriers for immunotherapy and vaccination. Tissue Barriers 8:1695476. https://doi.org/10.1080/21688370.2019.1695476

Mercel AI, Marulanda K, Gillis DC et al (2021) Development of novel nanofibers targeted to smoke-injured lungs. Biomaterials 274:120862. https://doi.org/10.1016/j.biomaterials.2021.120862

Moshizi SA, Abedi A, Pastras CJ, Peng S, Wu S, Sanaeepur M, Asadnia M (2022) Carbon nanofiber-reinforced Pt thin film-based airflow sensor for respiratory monitoring. Sens Actuators A Phys 347:113969. https://doi.org/10.1016/j.sna.2022.113969

Pawankar R, Bunnag C, Khaltaev N, Bousquet J (2012) Allergic rhinitis and its impact on asthma in Asia Pacific and the ARIA update 2008. World Allergy Organ J 5:S212–S217. https://doi.org/10.1186/1939-4551-5-S3-S212

Ranganath SH, Wang CH (2008) Biodegradable microfiber implants delivering paclitaxel for post-surgical chemotherapy against malignant glioma. Biomaterials 29:2996–3003. https://doi.org/10.1016/j.biomaterials.2008.04.002

Rasouli R, Barhoum A, Bechelany M, Dufresne A (2019) Nanofibers for biomedical and healthcare applications. Macromol Biosci 19:1800256. https://doi.org/10.1002/mabi.201800256

Reddel HK, Bacharier LB, Bateman ED et al (2022) Global initiative for asthma strategy 2021: executive summary and rationale for key changes. Am J Respir Crit Care Med 205:17–35. https://doi.org/10.1164/rccm.202109-2205PP

Rezaei FS, Khorshidian A, Beram FM, Derakhshani A, Esmaeili J, Barati A (2021) 3D printed chitosan/polycaprolactone scaffold for lung tissue engineering: hope to be useful for COVID-19 studies. RSC Adv 11:19508–19520. https://doi.org/10.1039/D1RA03410C

Stanford MG, Li JT, Chen Y, McHugh EA, Liopo A, Xiao H, Tour JM (2019) Self-sterilizing laser-induced graphene bacterial air filter. ACS Nano 13:11912–11920. https://doi.org/10.1021/acsnano.9b05983

Stojanov S, Berlec A (2020) Electrospun nanofibers as carriers of microorganisms, stem cells, proteins, and nucleic acids in therapeutic and other applications. Front Bioeng Biotechnol 8:130. https://doi.org/10.3389/fbioe.2020.00130

Torres-Martínez EJ, Cornejo Bravo JM, Serrano Medina A, Pérez González GL, Villarreal Gómez LJ (2018) A summary of electrospun nanofibers as drug delivery system: drugs loaded and biopolymers used as matrices. Curr Drug Deliv 15:1360–1374. https://doi.org/10.2174/1567201815666180723114326

Uhrich KE, Cannizzaro SM, Langer RS, Shakesheff KM (1999) Polymeric systems for controlled drug release. Chem Rev 99:3181–3198

Wang CS, Otani Y (2013) Removal of nanoparticles from gas streams by fibrous filters: a review. Ind Eng Chem Res 52:5–17. https://doi.org/10.1021/ie300574m

Yahia S, Dena ASA, El Nashar RM, El-Sherbiny IM (2022) Nanomicelles-in-coaxial nanofibers with exit channels as a transdermal delivery platform for smoking cessation. J Mater Chem B 10:4984–4998. https://doi.org/10.1039/D2TB00818A

Young BM, Shankar K, Allen BP, Pouliot RA, Schneck MB, Mikhaiel NS, Heise RL (2017) Electrospun decellularized lung matrix scaffold for airway smooth muscle culture. ACS Biomater Sci Eng 3:3480–3492. https://doi.org/10.1021/acsbiomaterials.7b00384

Zhang L, Li L, Wang L, Nie J, Ma G (2020) Multilayer electrospun nanofibrous membranes with antibacterial property for air filtration. Appl Surf Sci 515:145962

Quantum Dots for the Management of Allergic Airway Diseases

14

Rohit Y. Sathe

Abstract

Quantum dots are one of the most substantial nanomaterials that have emerged as a powerful tool engaging the recent materials science advancements in nanotechnology. They are essentially composed of approximately 10–100 atoms of semiconductor materials. Due to their highly specific emission spectra, photostability, unique optoelectronic properties, and ease of active-surface engineering, they are seen as promising tools in the medical field for the diagnosis and treatment of various diseases. Quantum dots, being size-controlled nanoparticles, can be administered easily to the airways and are approached for a number of allergic airway diseases. They are predominantly employed in molecular imaging, regenerative medicines, and targeted drug delivery. It is possible to use quantum dots to reduce the time required to identify disease by shifting the early diagnosis from the tissue level to the cellular level. Many recent examples suggest that a variety of bioactive compounds can be functionalized over quantum dots to specifically target allergic airway diseases and also provide a sustained and controlled release. There are a few known issues pertaining to the biocompatibility of functionalized quantum dots and the bioavailability of drugs functionalized over quantum dots, which have attracted significant research attention in the past decade. This chapter discusses the structure, synthesis, mechanism, properties, biomedical applications, and recent advances in the quantum dots-based management of allergic airway diseases. The observed immunological responses associated with quantum dots-based disease management are noted. Based on the pros and cons involved in the current methodologies, future prospects in the

R. Y. Sathe (✉)
Department of Biological Sciences and Biotechnology, Institute of Chemical Technology, Mumbai, Maharashtra, India
e-mail: ry.sathe@rs.ictmumbai.edu.in

P. Prasher et al. (eds.), *Understanding Allergic Airway Diseases*,
https://doi.org/10.1007/978-981-97-1953-2_14

management of airway diseases using quantum dots are also mentioned in this chapter.

Keywords
Quantum dots · Allergic airway diseases · Semiconductor crystals · Optical properties · Electronic properties · Surface engineering · Nanotechnology · Emission spectra

14.1 Introduction

14.1.1 Structure and Synthesis of Quantum Dots

Quantum dots are extremely small particles (crystals) composed of semiconductor elements. These particles exhibit very unique quantum mechanical properties. They are typically nanoscale in size, with dimensions on the order of a few nanometers (2–10 nm). The electronic and physical behavior of quantum dots is governed by the principles of quantum physics, particularly the confinement of electrons and holes (absence of an electron) within a small volume (Jacak et al. 2013; Alhassid 2000; Lim et al. 2015). The most common type of quantum dots is composed of semiconducting materials, such as elements from groups II–VI (e.g., cadmium selenide, cadmium sulfide) or III–V (e.g., indium arsenide, gallium arsenide) of the periodic table (Kambhampati 2011; Semonin et al. 2012; Reimann and Manninen 2002). The structure of quantum dots can be described as follows:

1. *Core*: The core is the central part of the quantum dot and is typically a semiconductor nanocrystal. It consists of a crystalline lattice of atoms arranged in a specific pattern, where the atoms are confined in all three dimensions. The core size is typically in the range of a few nanometers.
2. *Type-I quantum dots (shell)*: Some quantum dots have an additional layer around the core, known as the shell or capping layer. The shell can be made of a different semiconductor material than the core, and its purpose is to modify the quantum dot's properties further. The shell can help improve the quantum dots' photoluminescence efficiency, stability, and biocompatibility. The structure of a quantum dot is shown in Fig. 14.1.
3. *Type-II quantum dots*: In some cases, the shell material has a lower electron affinity than the core material, leading to the formation of type-II quantum dots. In type-II quantum dots, the electron and hole wave functions are spatially separated between the core and shell regions. This unique configuration can result in longer carrier lifetimes and reduced non-radiative recombination, leading to interesting optoelectronic properties.
4. *Ligands*: To stabilize the quantum dots and prevent them from aggregating, they are often coated with organic molecules called ligands. These ligands form a protective layer around the quantum dot surface, providing stability and enabling

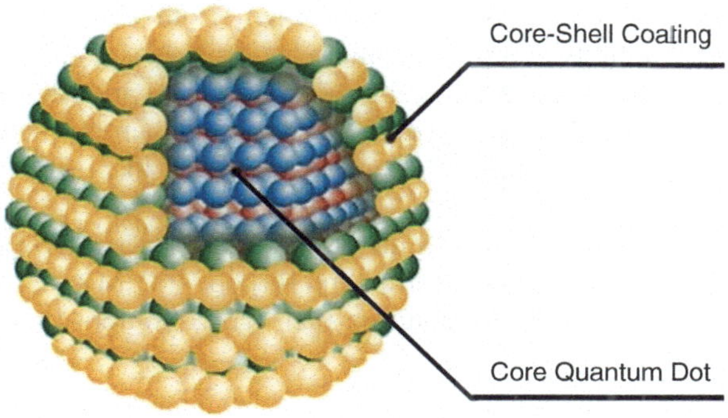

Fig. 14.1 The architecture of a type-I quantum dot (Mohamed et al. 2021). Adopted and reprinted with permission from De Gruyters

their solubility in various solvents. Ligands can also be functionalized with specific biomolecules or targeting agents for biomedical applications.

They can be synthesized using various methods, but one common approach involves chemical synthesis. The steps involved in the typical synthesis of quantum dots are as follows:

1. *Selection of Semiconductor Material*: As noted, quantum dots can be made from a variety of semiconductor materials, including cadmium selenide (CdSe), lead sulfide (PbS), and indium phosphide (InP). The choice of material determines the quantum dots' properties, such as their size and the wavelengths of light they can emit.
2. *Precursor Chemicals*: The semiconductor material is obtained in the form of precursor chemicals. For example, CdSe quantum dots can be synthesized from cadmium acetate and selenium powder.
3. *Reaction in a Solvent*: The precursors are dissolved in a solvent, often a high-boiling-point organic solvent like trioctylphosphine oxide (TOPO) or octadecene (ODE). This solution is then heated to create a reaction mixture.
4. *Injection of Precursors*: A separate solution, often containing a stabilizing agent and a reducing agent, is injected into the reaction mixture. The reducing agent helps convert the precursors into nanocrystals by providing electrons needed for the semiconductor material to grow.
5. *Nucleation and Growth*: The injection of the reducing agent leads to the nucleation of nanocrystals. At this stage, tiny crystalline seeds begin to form. These seeds then grow into larger quantum dots by absorbing more precursor material from the reaction solution. The growth time and temperature can be controlled to determine the size of the quantum dots.
6. *Surface Passivation*: Quantum dots typically have a core-shell structure. After the core quantum dots have formed, a shell is grown around them. The shell is often made of different semiconductor materials and is used to passivate the

surface of the quantum dots, preventing them from clumping together and improving their stability and photoluminescent properties.
7. *Purification and Isolation*: The resulting quantum dots are often coated with a layer of organic ligands to stabilize them. They are then purified and isolated from the reaction mixture through processes like precipitation and centrifugation.
8. *Characterization*: The synthesized quantum dots are characterized using techniques such as transmission electron microscopy (TEM), X-ray diffraction (XRD), and spectroscopy to confirm their size, composition, and optical properties.

The combination of the core's size, composition, and optional shell and ligands results in a wide variety of quantum dots with unique optoelectronic properties. By engineering the structure, researchers can tune the emission wavelength, quantum yield, stability, and other properties of quantum dots for specific applications in fields such as optoelectronics, imaging, and biomedicine.

14.1.2 Mechanism of Quantum Dots

The fundamental principle of the mechanism of quantum dots can be acknowledged by considering the concept of "quantum confinement." Two principal factors cause the properties of quantum dots to differ significantly from other materials: increased relative surface area and quantum effects. At the bulk level, the electronic properties of (any) semiconductor materials are considered to be "continuous." These properties can be defined by energy bands. However, when a semiconductor material is reduced to a nanosize (exciton Bohr radius), the energy levels become discrete (each energy level can only possess one of a distinct number of levels, and the value cannot be in between these numbers) due to the confinement of charge carriers (essentially, electrons and holes) within a small volume (Reimann and Manninen 2002; García de Arquer et al. 2021). The confinement of these charge carriers in quantum dots leads to several important effects:

1. *Quantum confinement of carriers*: Due to the small size of quantum dots, electrons, and holes, they are spatially restricted to a tiny volume. This confinement curbs their motion. The confinement maximizes the interaction between charge carriers, which leads to unique electronic and optical phenomena (Liu et al. 2021; Rasal et al. 2021).
2. *Size dependency of energy levels*: Since the energy levels of electrons and holes in quantum dots can only occupy specific energy levels, the energy spacing between these levels increases as the dimensions of the quantum dot decrease. As a result of this, the electronic and optical properties of quantum dots, e.g., bandgap and emission wavelength, can be tuned by controlling the quantum dot size (Bacon et al. 2014; Xu et al. 2022).

3. *Size-dependent fluorescence*: The emission properties of quantum dots are size-dependent. This is known as the quantum size effect or size-dependent fluorescence. When quantum dots are excited with light or an electrical current, they emit light with wavelengths determined by their size. Quantum dots with larger dimensions emit light at longer wavelengths (e.g., red), while the ones with smaller dimensions emit light with shorter wavelengths (e.g., blue). This tunability makes quantum dots attractive for applications in display technologies, lighting, and biological imaging.
4. *Enhanced optical properties*: Quantum dots efficiently convert absorbed light into emitted light (high photoluminescence quantum yields) (Wang et al. 2020; Shu et al. 2020).

The optoelectronic properties of quantum dots play a crucial role in their functionality and are highly dependent on the energy difference between the conductance band and the valence band, often referred to as the "bandgap." The bandgap is, in turn, dependent upon the shape, size, volume, composition of quantum dots, and thermodynamic parameters to which the quantum dots are exposed (Tong et al. 2018; Wu et al. 2015; Chistyakov et al. 2017). These characteristics make quantum dots promising for a wide range of applications in electronics, photonics, and other fields (e.g., displays, lighting, solar cells, medical, as well as medicinal).

Quantum confinement and fluorescence phenomena can be better understood by referring to Fig. 14.2.

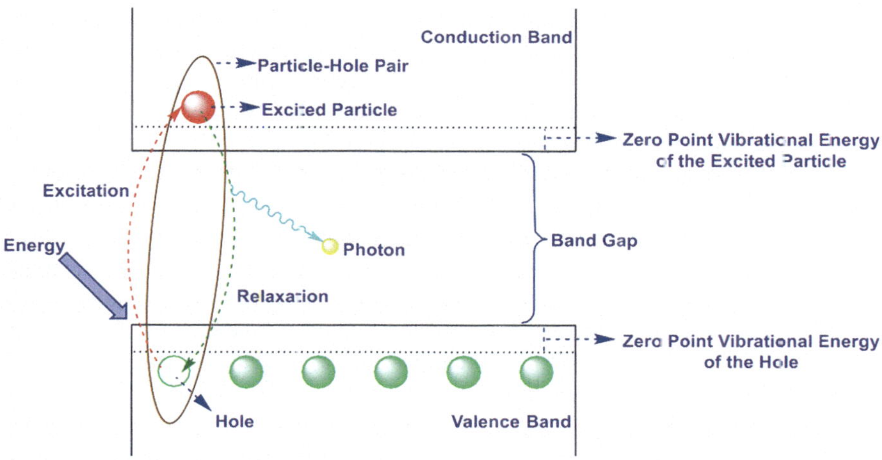

Fig. 14.2 Mechanism of the release of a photon in quantum dots through excitation and relaxation (de-excitation) of a valance electron, i.e., fluorescence due to particle-hole pair between the valance and conduction bands

14.1.3 Properties of Quantum Dots

Quantum dots exhibit a range of optoelectronic properties that arise from their unique quantum confinement effects. These properties make quantum dots attractive for various applications in optoelectronic devices. The key optoelectronic properties of quantum dots are mentioned below:

1. *Size-tunable absorption and emission*: One of the most significant optoelectronic properties of quantum dots is their size-dependent absorption and emission spectra. The energy bandgap of a quantum dot, which determines the wavelengths of light it can absorb and emit, can be finely tuned by controlling the size of the dot. Larger quantum dots have smaller bandgaps, absorbing and emitting longer-wavelength light (e.g., red), while smaller dots have larger bandgaps, absorbing and emitting shorter-wavelength light (e.g., blue). This tunability allows quantum dots to cover a wide spectral range and makes them useful for applications such as light-emitting diodes (LEDs), displays, and biological imaging (Reimann and Manninen 2002; Shu et al. 2020; Ratemi et al. 2016; Ren et al. 2022).
2. *High photoluminescence quantum yield*: Quantum dots exhibit high photoluminescence quantum yield (PLQY), which refers to the efficiency with which they convert absorbed photons into emitted photons. This high PLQY is a result of reduced non-radiative recombination processes due to the unique energy levels and electronic structure of quantum dots. High PLQY enables quantum dots to emit bright and intense light, making them valuable for applications such as displays, lighting, and imaging (Yang et al. 2021).
3. *Broad absorption spectrum*: Quantum dots can have broad absorption spectra that extend over a wide range of wavelengths. This broad absorption allows them to efficiently absorb light from different parts of the electromagnetic spectrum. By choosing the appropriate size and composition, quantum dots can be engineered to absorb light from ultraviolet (UV) to near-infrared regions (NIR). This property is useful for applications such as photodetectors, solar cells, and sensing devices that require broad spectral coverage (Duan et al. 2021).
4. *Fast and efficient carrier dynamics*: Quantum dots exhibit fast carrier dynamics, including fast carrier capture and relaxation processes. The carriers (electrons and holes) in quantum dots are confined within a small volume, leading to enhanced carrier-carrier and carrier-phonon interactions. These interactions enable efficient carrier capture, relaxation, and radiative recombination processes, resulting in rapid response times and high emission rates. Fast carrier dynamics are advantageous for applications such as optical switches, lasers, and photovoltaics.
5. *Size-dependent nonlinear optical properties*: Quantum dots can exhibit size-dependent nonlinear optical properties, such as nonlinear absorption and nonlinear refraction. Nonlinear optical phenomena arise due to the intense interaction between light and matter at the nanoscale. By controlling the size and composition of quantum dots, their nonlinear optical properties can be tailored. These

properties find applications in areas such as ultrafast optical switches, optical data processing, and nonlinear imaging.

The optoelectronic properties of quantum dots, including size-tunable absorption and emission, high photoluminescence quantum yield, broad absorption spectrum, fast carrier dynamics, and size-dependent nonlinear optical properties, make them versatile materials for optoelectronic devices. Their unique characteristics enable advancements in areas such as displays, lighting, solar cells, photodetectors, lasers, and optical data processing (Chistyakov et al. 2017; Duan et al. 2021; Lin et al. 2021).

14.2 Biomedical Applications of Quantum Dots for Management of Airway Diseases

Quantum dots have shown promise in various biomedical applications, including disease diagnosis (Jamieson et al. 2007). While quantum dots have not been extensively studied yet specifically for the management of airway diseases, their unique properties make them potentially useful in this context (Gidwani et al. 2021; Cotta 2020). Quantum dots can be utilized for the management of airway diseases as follows:

14.2.1 Biomarker Detection and Therapy Monitoring

Quantum dots can be functionalized with specific antibodies or biomolecules to detect and quantify biomarkers associated with asthma. For asthma diagnosis, relevant biomarkers may include specific cytokines, chemokines, or immunoglobulins that indicate airway inflammation or allergic responses. By conjugating quantum dots with these biomarkers, they can be used in diagnostic assays to detect and measure their concentrations in patient samples, such as blood or sputum (Mousavi et al. 2022; Wei et al. 2022). The high photoluminescence quantum yield of quantum dots can enhance the sensitivity of these assays, allowing for accurate and precise biomarker detection (Gidwani et al. 2021).

14.2.1.1 Inflammatory Marker Visualization

Quantum dots can be designed to bind to specific biomarkers associated with inflammation in respiratory diseases. For example, they could target cytokines or chemokines that are elevated during inflammatory responses. By visualizing the spatial distribution and concentration of these biomarkers using quantum dots, researchers can gain a better understanding of the extent and nature of inflammation in the respiratory system.

14.2.1.2 Monitoring Treatment Response

Quantum dots can be used to track the effects of treatments in real time. By labeling cells or biomarkers with quantum dots, researchers can observe how different therapies affect disease-related changes over time. This can help assess treatment efficacy and guide therapeutic strategies.

14.2.2 Imaging and Visualization

Quantum dots can be employed for imaging and visualization of airway disease-related features. For instance, they can be used in fluorescence microscopy or imaging techniques to visualize airway inflammation, cellular responses, or structural changes associated with asthma (Matea et al. 2017; Zrazhevskiy et al. 2010). Quantum dots with different emission wavelengths can be used simultaneously to label different cell types or markers, enabling multiplexed imaging. This can provide valuable insights into the pathophysiology of airway diseases and aid in diagnosis (Wagner et al. 2019; Biswas et al. 2021).

14.2.2.1 Targeted Cellular Imaging

Quantum dots can be functionalized with specific targeting molecules, such as antibodies or peptides, that recognize surface markers on cells involved in respiratory diseases (e.g., immune cells, inflamed airway epithelial cells). These functionalized quantum dots can then be introduced into the body, where they will bind specifically to the targeted cells. This allows researchers and clinicians to visualize and track the distribution of these cells in the respiratory system, providing insights into the disease process and its spatial distribution. One such example of targeted imaging can be understood from Fig. 14.3, where Gao et al. have studied the tumor growth in a live mouse model using quantum dot probes conjugated to a prostate-specific membrane antigen (PSMA) monoclonal antibody (Gao et al. 2004).

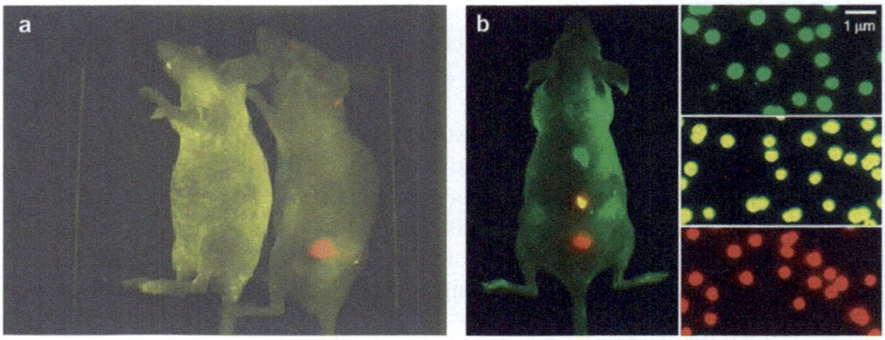

Fig. 14.3 Targeted cellular imaging with quantum dots. Here, (**a**) molecular targeting of a tumor using quantum dot-antibody conjugate and (**b**) in vivo image of microbeads encoded with multicolor quantum dots injected into a live mouse (Gao et al. 2004). (Adopted and reprinted with permission from Springer Nature)

14.2.2.2 Multiplexed Imaging

Quantum dots of different emission wavelengths can be used simultaneously to label different types of cells or biomarkers. This multiplexed imaging approach enables the visualization of multiple components within the same sample. In respiratory diseases, it could be used to simultaneously visualize different cell types, inflammatory markers, and structural features associated with the disease, providing a comprehensive view of the pathology.

14.2.2.3 Dynamic Imaging

Quantum dots with fast fluorescence decay times can be used for dynamic imaging of cellular processes, such as immune cell migration and interactions. This can provide insights into how cells move and communicate within the respiratory system during disease progression or immune responses.

14.2.2.4 Nanoparticle-Based Contrast Agents

Quantum dots can also be used as contrast agents in imaging modalities such as fluorescence imaging, confocal microscopy, and even potentially in emerging techniques like quantum dot-based magnetic resonance imaging (MRI). These contrast agents can enhance the visualization of specific areas within the respiratory system.

Various combinations of formulations of quantum dots to be utilized in the bioimaging of lung tissue structures or related lung diseases in the respiratory system and lung cell signaling are available in the current literature. Therefore, the use of such nanotechnological advances helps overcome traditional chest radiology by providing a noninvasive imaging and radiation-free alternative (Ren et al. 2022). The hemodynamical conditions of the lungs can be observed under pathologic or normal conditions through light microscopic imaging of blood vessels. Saitoh et al (Saitoh et al. 2012). injected glutathione-decorated quantum dots into the right ventricle of the heart and observed precise peripheral pulmonary alveoli blood flow at different time courses. They performed in vivo cryotechnique (IVCT) in abnormal and normal lung conditions. This facilitated the investigation of mice lung microvascular hemodynamics and the altered structures (Saitoh et al. 2016). A three-dimensional single-particle tracking technique (SPT) and quantum dots-labeled avian influenza H9N2 virus were used by Wang et al. to prove that sialic acid receptors are directly proportional to the number of influenza viruses in human bronchial epithelial cells. Their findings prove helpful to monitor the avian influenza viral infection by monitoring sialic acid receptors (Wang et al. 2014). The H5N1 pseudotype of avian influenza is in vivo labeled with quantum dots by Pan et al (Pan et al. 2014). They observed that the quantum dots-labeled H5N1p exhibits bright and sustained fluorescence in mice pulmonary tissues. This eases out the noninvasive study of viral infections in real time. Gorshkov et al. synthesized angiotensin-converting enzyme 2 (ACE2)-specific recombinant spike receptor conjugated with quantum dots to track the SARS-CoV-2. This probe was found to immediately bind the target, followed by endocytosis (Gorshkov et al. 2020).

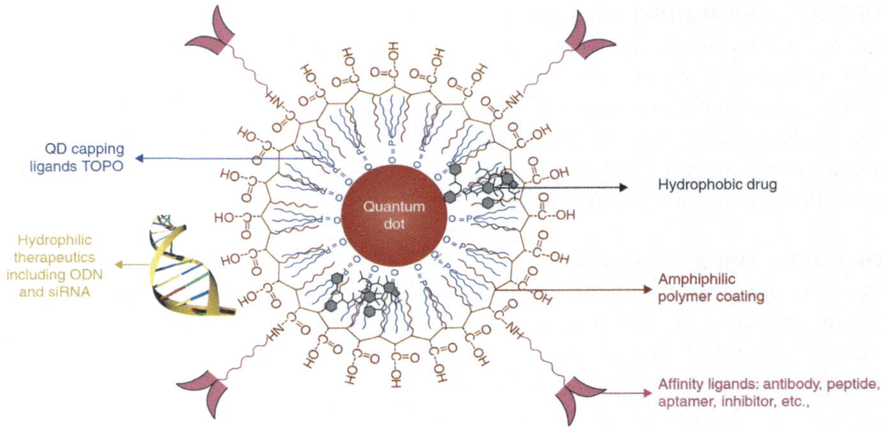

Fig. 14.4 Schematic representation of a quantum dot coated with amphiphilic layer (Qi and Gao 2008). Hydrophobic drug molecules can be sandwiched between an amphiphilic polymer coat and the hydrophobic core. Hydrophilic molecules and drug compounds can be functionalized on the outer surface. (Adopted and reprinted with permission from *Taylor & Francis Online*)

14.2.3 Drug Delivery

Quantum dots can serve as carriers for targeted drug delivery in the treatment of airway diseases. By functionalizing quantum dots with specific ligands or antibodies, they can be directed to target specific cells or tissues in the respiratory system, such as airway smooth muscle cells or inflamed airway epithelium (Qi and Gao 2008). This targeted delivery can enhance the efficacy of asthma medications and minimize off-target effects. Quantum dots can also be used for monitoring therapeutic responses by labeling drugs or therapeutic agents and tracking their distribution within the respiratory system (Biswas et al. 2021; Kang et al. 2022; Probst et al. 2013). Drug delivery using quantum dots can be understood with the illustration provided in Fig. 14.4. Quantum dots can be used for drug delivery in the following ways:

14.2.3.1 Targeted Drug Delivery

Quantum dots can be functionalized with specific ligands, antibodies, or peptides that recognize receptors on the surface of cells involved in respiratory diseases. For instance, in asthma, immune cells or inflamed airway epithelial cells could be targeted. These functionalized quantum dots can carry drugs and deliver them directly to the target cells, enhancing the drug's effectiveness while minimizing off-target effects.

14.2.3.2 Enhanced Permeation

Quantum dots can be used to enhance the permeation of drugs through biological barriers, such as the airway epithelium in respiratory diseases. Their nanoscale size allows them to penetrate tissues more efficiently and access regions that might be difficult to reach with conventional drug formulations.

14.2.3.3 Sustained Release

Quantum dots can be engineered to encapsulate drugs and release them in a controlled and sustained manner. This sustained release can prolong the drug's therapeutic effect, reducing the frequency of administration and maintaining steady drug concentrations in the target site.

14.2.3.4 Real-Time Monitoring

Quantum dots can serve a dual purpose as drug carriers and imaging agents. By incorporating imaging molecules or by using the inherent fluorescence of quantum dots, researchers can track the distribution of both the quantum dots and the loaded drug in real time. This monitoring can provide insights into drug release kinetics and accumulation in target tissues.

14.2.3.5 Combination Therapy

Quantum dots can be used to deliver multiple therapeutic agents simultaneously. This could be especially beneficial in respiratory diseases where multiple factors contribute to the disease's progression. For instance, quantum dots could carry both anti-inflammatory drugs and bronchodilators for a synergistic therapeutic effect.

14.2.3.6 Personalized Medicine

Quantum dots could be used in conjunction with diagnostic techniques to enable personalized drug delivery. By targeting specific disease markers, quantum dots could deliver drugs to patients with a particular molecular profile, maximizing treatment efficacy.

For example, Zheng et al. have revealed the cell entry pathway of the respiratory syncytial virus by tracking the quantum dot-labeled single virus (Zheng et al. 2017). Their findings are crucial for the development of new antiviral vaccines and drugs. The trajectory analysis and time-lapse fluorescence imaging of the quantum dot-labeled respiratory syncytial virus were performed. They proved that the entry of respiratory syncytial virus into HEp-2 cells takes place via a typical endocytosis trafficking.

14.2.4 Point-of-Care Diagnostics

Quantum dots offer the potential for developing portable and rapid diagnostic devices for airway diseases. Their size-tunable emission allows for the development of miniaturized sensors that can detect allergic airway disease-related biomarkers in breath samples or saliva. These point-of-care devices could provide quick and non-invasive diagnostic information, enabling early detection, monitoring, and management of airway diseases (Singh et al. 2020; Sahoo et al. 2022).

Fig. 14.5 The ellume·lab is a quantum dots-based miniaturized digital platform at the point of care. This device provides very accurate and rapid results for the detection of various infectious diseases (Mahe 2019)

14.2.4.1 Miniaturized Detection Platforms

Quantum dot-based point-of-care devices can be designed as miniaturized platforms, which can include microfluidic chips, paper-based assays, or portable handheld devices. These platforms are designed to handle small sample volumes and require minimal sample preparation, making them suitable for rapid diagnostics at the point of care. Figure 14.5 shows an example of a commercially available digital point-of-care diagnostic device called ellume·lab (Mahe 2019) (www.ellumehealth.com) that can be used to detect various infectious diseases, including airway diseases.

14.2.4.2 Fluorescence Detection

Quantum dots emit bright and stable fluorescence when excited by a specific wavelength of light. Point-of-care devices equipped with light sources and detectors can measure the fluorescence emitted by the functionalized quantum dots in response to the target biomarkers. The intensity of fluorescence can be correlated with the concentration of the biomarkers, allowing for quantitative measurement.

14.2.4.3 Smartphone-Based Detection

Advances in smartphone technology have enabled the development of point-of-care diagnostic devices that can be connected to smartphones for data analysis and communication. Quantum dot-based assays could be integrated with smartphone cameras to quantify fluorescence and interpret results, making diagnostics accessible even in resource-limited settings.

14.2.4.4 Rapid Results

One of the main advantages of quantum dot-based point-of-care diagnostics is their potential for rapid results. The specific binding of quantum dots to target biomarkers enables sensitive detection within a short time frame, allowing healthcare providers to make quick decisions about patient management.

In conclusion, the potential applications of quantum dots in the field of respiratory diseases hold significant promise but are still in the early stages of research and development. Quantum dots, with their unique optical and nanoscale properties, offer innovative opportunities for diagnosis, imaging, drug delivery, and potentially even treatment of respiratory conditions. They could revolutionize how we understand, monitor, and manage diseases like asthma and other respiratory disorders. While the concept is exciting, it's crucial to emphasize that practical implementation requires overcoming various challenges. These include addressing concerns about biocompatibility, safety, specificity of targeting, regulatory approvals, and the complexity of translating laboratory findings into effective clinical solutions.

As research in the field of quantum dots and respiratory diseases progresses, collaboration between experts in nanotechnology, medicine, and clinical practice will play a pivotal role. The future holds potential breakthroughs that could transform the landscape of respiratory disease management, offering improved diagnostics, more effective treatments, and personalized approaches that enhance patients' quality of life.

It's important to stay informed about the latest developments in this rapidly evolving field through updated scientific literature, ongoing clinical trials, and the insights of medical professionals and researchers. This dynamic intersection of quantum technology and respiratory medicine holds the promise of advancing our understanding of diseases and improving patient outcomes in the years to come.

References

Alhassid Y (2000) The statistical theory of quantum dots. Rev Mod Phys 72(4):895

Bacon M, Bradley SJ, Nann T (2014) Graphene quantum dots. Part Part Syst Charact 31(4):415–428

Biswas MC, Islam MT, Nandy PK, Hossain MM (2021) Graphene Quantum Dots (GQDs) for bioimaging and drug delivery applications: a review. ACS Mater Lett 3(6):889–911

Chistyakov AA, Zvaigzne MA, Nikitenko VR, Tameev AR, Martynov IL, Prezhdo OV (2017) Optoelectronic properties of semiconductor quantum dot solids for photovoltaic applications. J Phys Chem Lett 8(17):4129–4139

Cotta MA (2020) Quantum dots and their applications: what lies ahead? ACS Appl Nano Mater 3(6):4920–4924

Duan L, Hu L, Guan X, Lin CH, Chu D, Huang S, Liu X, Yuan J, Wu T (2021) Quantum dots for photovoltaics: a tale of two materials. Adv Energy Mater 11(20):2100354

Gao X, Cui Y, Levenson RM, Chung LW, Nie S (2004) In vivo cancer targeting and imaging with semiconductor quantum dots. Nature Biotechnol 22(8):969–976

García de Arquer FP, Talapin DV, Klimov VI, Arakawa Y, Bayer M, Sargent EH (2021) Semiconductor quantum dots: technological progress and future challenges. Sci. *373*(6555):8541

Gidwani B, Sahu V, Shukla SS, Pandey R, Joshi V, Jain VK, Vyas A (2021) Quantum dots: prospectives, toxicity, advances and applications. J Drug Deliv Sci Technol 61:102308

Gorshkov K, Susumu K, Chen J, Xu M, Pradhan M, Zhu W, Hu X, Breger JC, Wolak M, Oh E (2020) Quantum dot-conjugated SARS-CoV-2 spike pseudo-virions enable tracking of angiotensin converting enzyme 2 binding and endocytosis. ACS Nano 14(9):12234–12247
Jacak L, Hawrylak P, Wojs A (2013) Quantum dots. Springer Science & Business Media
Jamieson T, Bakhshi R, Petrova D, Pocock R, Imani M, Seifalian AM (2007) Biological applications of quantum dots. Biomaterials 28(31):4717–4732
Kambhampati P (2011) Unraveling the structure and dynamics of excitons in semiconductor quantum dots. Acc Chem Res 44(1):1–13
Kang SH, Shin YS, Lee D-H, Park IS, Kim SK, Ryu D, Park Y, Byun S-H, Choi J-H, Hong SJ (2022) Interactions of nanoparticles with macrophages and feasibility of drug delivery for asthma. Int J Mol Sci *23*(3):1622
Lim SY, Shen W, Gao Z (2015) Carbon quantum dots and their applications. Chem Soc Rev 44(1):362–381
Lin J, Lu Y, Li X, Huang F, Yang C, Liu M, Jiang N, Chen D (2021) Perovskite quantum dots glasses based backlit displays. ACS Energy Lett 6(2):519–528
Liu M, Yazdani N, Yarema M, Jansen M, Wood V, Sargent EH (2021) Colloidal quantum dot electronics. Nat Electron 4(8):548–558
Mahe L (2019) Quantum dots: a diagnostic tool. Team Consulting. https://www.team-consulting.com/insights/quantum-dots-new-dimension-diagnostics/. Accessed 1 Sept 2023
Matea CT, Mocan T, Tabaran F, Pop T, Mosteanu O, Puia C, Iancu C, Mocan L (2017) Quantum dots in imaging, drug delivery and sensor applications. Int J Nanomedicine Volume 12:5421–5431
Mohamed WAA, El-Gawad HA, Mekkey S, Galal H, Handal H, Mousa H, Labib A (2021) Quantum dots synthetization and future Prospect applications. Nanotechnol Rev 10(1):1926–1940. https://doi.org/10.1515/ntrev-2021-0118. Accessed 19 Jul 2023
Mousavi SM, Hashemi SA, Yari Kalashgrani M, Omidifar N, Lai CW, Vijayakameswara Rao N, Gholami A, Chiang W-H (2022) The pivotal role of quantum dots-based biomarkers integrated with ultra-sensitive probes for multiplex detection of human viral infections. Pharmaceuticals 15(7):880
Pan H, Zhang P, Gao D, Zhang Y, Li P, Liu L, Wang C, Wang H, Ma Y, Cai L (2014) Noninvasive visualization of respiratory viral infection using bioorthogonal conjugated near-infrared-emitting quantum dots. ACS Nano 8(6):5468–5477
Probst CE, Zrazhevskiy P, Bagalkot V, Gao X (2013) Quantum dots as a platform for nanoparticle drug delivery vehicle design. Adv Drug Deliv Rev 65(5):703–718
Qi L, Gao X (2008) Emerging application of quantum dots for drug delivery and therapy. Expert Opin Drug Deliv 5(3):263–267
Rasal AS, Yadav S, Yadav A, Kashale AA, Manjunatha ST, Altaee A, Chang J-Y (2021) Carbon quantum dots for energy applications: a review. ACS Appl Nano Mater 4(7):6515–6541
Ratemi E, Sultana Shaik A, Al Faraj A, Halwani R (2016) Alternative approaches for the treatment of airway diseases: focus on nanoparticle medicine. Clin Exp Allergy 46(8):1033–1042
Reimann SM, Manninen M (2002) Electronic structure of quantum dots. Rev Mod Phys 74(4):1283
Ren L, Wang L, Rehberg M, Stoeger T, Zhang J, Chen S (2022) Applications and immunological effects of quantum dots on respiratory system. Front Immunol 12:795232
Sahoo S, Nayak A, Gadnayak A, Sahoo M, Dave S, Mohanty P, Mohanty JN, Das J (2022) Quantum dots enabled point-of-care diagnostics: a new dimension to the nanodiagnosis. In: Advanced nanomaterials for point of care diagnosis and therapy. Elsevier, pp 43–52
Saitoh Y, Terada N, Saitoh S, Ohno N, Jin T, Ohno S (2012) Histochemical analyses and quantum dot imaging of microvascular blood flow with pulmonary edema in living mouse lungs by "in vivo cryotechnique". Histochem Cell Biol 137:137–151
Saitoh Y, Terada N, Ohno N, Ohno S (2016) Histochemical analyses and quantum dot imaging of microvascular blood flow with pulmonary edema. In: In vivo cryotechnique in biomedical research and application for bioimaging of living animal organs. Springer, Cham, pp 77–81
Semonin OE, Luther JM, Beard MC (2012) Quantum dots for next-generation photovoltaics. Mater Today 15(11):508–515

Shu Y, Lin X, Qin H, Hu Z, Jin Y, Peng X (2020) Quantum dots for display applications. Angew Chem 132(50):22496–22507

Singh S, Dhawan A, Karhana S, Bhat M, Dinda AK (2020) Quantum dots: an emerging tool for point-of-care testing. Micromachines 11(12):1058

Tong X, Kong X-T, Wang C, Zhou Y, Navarro-Pardo F, Barba D, Ma D, Sun S, Govorov AO, Zhao H et al (2018) Optoelectronic properties in near-infrared colloidal heterostructured pyramidal "Giant" core/shell quantum dots. Adv Sci 5(8):1800656

Wagner AM, Knipe JM, Orive G, Peppas NA (2019) Quantum dots in biomedical applications. Acta Biomater 94:44–63

Wang ZG, Liu SL, Zhang ZL, Tian ZQ, Tang HW, Pang DW (2014) Exploring sialic acid receptors-related infection behavior of avian influenza virus in human bronchial epithelial cells by single-particle tracking. Small 10(13):2712–2720

Wang L, Li W, Yin L, Liu Y, Guo H, Lai J, Han Y, Li G, Li M, Zhang J (2020) Full-color fluorescent carbon quantum dots. Sci Adv 6(40):6772

Wei J, Zhang X, Mugo SM, Zhang Q (2022) A portable sweat sensor based on carbon quantum dots for multiplex detection of cardiovascular health biomarkers. Anal Chem 94(37):12772–12780

Wu J, Chen S, Seeds A, Liu H (2015) Quantum dot optoelectronic devices: lasers, photodetectors and solar cells. J Phys D Appl Phys 48(36):363001

Xu Q, Niu Y, Li J, Yang Z, Gao J, Ding L, Ni H, Zhu P, Liu Y, Tang Y (2022) Recent progress of quantum dots for energy storage applications. Carbon Neutrality 1(1):13

Yang H, Li R, Zhang Y, Yu M, Wang Z, Liu X, You W, Tu D, Sun Z, Zhang R et al (2021) Colloidal alloyed quantum dots with enhanced photoluminescence quantum yield in the NIR-II window. J Am Chem Soc 143(6):2601–2607

Zheng LL, Li CM, Zhen SJ, Li YF, Huang CZ (2017) A dynamic cell entry pathway of respiratory syncytial virus revealed by tracking the quantum dot-labeled single virus. Nanoscale 9(23):7880–7887

Zrazhevskiy P, Sena M, Gao X (2010) Designing multifunctional quantum dots for bioimaging, detection, and drug delivery. Chem Soc Rev 39(11):4326–4354

Future Directions in Allergic Airways Diseases

15

Tanisqa Mall, Mousmee Sharma, and Parteek Prasher

Abstract

This review explores the prospective avenues in the domain of allergic airways diseases, specifically focusing on asthma and allergic rhinitis. It discusses the forthcoming advancements in precision medicine, personalized therapies, and immunotherapy. The integration of biomarkers, genetic profiling, and environmental data is underscored for refining disease categorization and enhancing treatment efficacy. Additionally, it delves into the transformative influence of emerging technologies such as digital health, artificial intelligence, and big data analytics on disease management and patient care. Moreover, it examines the potential of nanomedicine for targeted drug delivery and the exploration of innovative therapeutic modalities such as gene editing and microbiome modulation. Understanding and embracing these future directions are vital for progressing towards improved outcomes and enhanced quality of life for patients globally.

Keywords

Precision medicine · Personalized therapy · Immunotherapy · Biomarker · Genetic Profiling · Environmental factors · Digital health · Artificial intelligence · Big data analytics

---

T. Mall (✉) · P. Prasher
Department of Chemistry, University of Petroleum & Energy Studies, Dehradun, India

M. Sharma
Department of Chemistry, Uttaranchal University, Dehradun, India

P. Prasher et al. (eds.), *Understanding Allergic Airway Diseases*,
https://doi.org/10.1007/978-981-97-1953-2_15

15.1 Introduction

Allergy is characterized as an immune-mediated hypersensitive state induced by exposure to an allergen, defined as a biological or chemical agent provoking an allergic reaction. The predominant allergic response involves the IgE-mediated activation of effector cells, notably mast cells and basophils. This activation precipitates an inflammatory cascade, culminating in discernible clinical manifestations. Prevalent and extensively researched allergic airway diseases include allergic rhinoconjunctivitis (commonly referred to as hay fever) and allergic asthma. Allergic rhinoconjunctivitis primarily impacts the upper airways, encompassing the nasal passages and conjunctiva, resulting in symptoms such as sneezing, rhinorrhea, nasal congestion, pruritus, lacrimation, and conjunctival redness. Conversely, allergic asthma predominantly affects the lower airways, specifically the bronchial tubes, leading to symptoms such as wheezing, dyspnea, coughing, and chest tightness. Asthma exacerbations span from mild to severe, posing significant public health concerns due to potential life-threatening consequences. Both allergic rhinoconjunctivitis and allergic asthma share common underlying mechanisms characterized by an exaggerated immune response to typically innocuous environmental substances referred to as aeroallergens. These aeroallergens include pollens from trees, grasses, and weeds, dust mites, pet dander, mold spores, and specific occupational allergens. The incidence of allergic airway diseases has exhibited an upward trend, particularly in urban settings, imposing substantial burdens on healthcare systems, productivity, and the overall well-being of affected individuals.

Air pollution, global warming, and climate change are primary factors contributing significantly to the escalating prevalence of allergic airway diseases, such as asthma and allergic rhinitis. These issues are poised to become central challenges for health systems in the twenty-first century. Asthma, identified as an early-onset non-communicable environmental disease, has evolved into a global epidemic, imposing substantial psycho-social and economic burdens. Globally, there is a rising epidemic of airway allergies such as asthma and allergic rhinitis, along with associated comorbidities, leading to substantial psycho-social, economic, and health burdens on societies and nations. The principal factor contributing significantly to the increased prevalence of allergic airway diseases is air pollution (Lee et al. 2021). Approximately 23% of all global deaths are attributable to environmental pollution, equating to approximately 12.6 million deaths annually. Notably, 3.8 million of these deaths occur in India and other countries in South East Asia (Bangalore Thimmadasiah and Joshi 2020). Air pollution, responsible for nearly eight million global deaths, affected over one billion individuals in 2015, a figure projected to rise to four billion by 2050. Beyond triggering asthma episodes, air pollution induces alterations in genetic patterns that initiate the disease process. The paradigm in managing allergic airway diseases has shifted from symptom control to disease prevention over time. Achieving this necessitates positive advancements in clean air policies, adherence to standardized norms, systematic progress tracking, monitoring and evaluation, and collaborative efforts through partnerships and conventions with local and global authorities (Paramesh 2018).

Our strategic focus should prioritize sustainable, secure human settlements, cities, sustainable energy practices, industrialization, and research endeavors. Through the collective and effective efforts of all stakeholders, the reduction of air pollution and prevention of global warming by 2030 can be achieved, aligning with the objectives endorsed by 194 countries, as established by the World Health Organization in May 2015.

15.2 Allergic Rhinitis

15.2.1 Current Treatment and Strategies

15.2.1.1 Pharmacotherapy

Treatment options for allergic rhinitis (AR) encompass allergen avoidance, nasal saline rinses, pharmacotherapy, and allergen immunotherapy (AIT). First-line treatments for mild seasonal and perennial AR include second-generation oral antihistamines (AHs) and intranasal antihistamines (INAH). Intranasal corticosteroids (INCS) are recommended for moderate to severe AR, exhibiting superior efficacy compared to oral AHs or leukotriene receptor antagonists (LTRAs). Daily use of INCS is most effective, and combination therapy with intranasal azelastine and fluticasone is approved for moderate to severe AR. Montelukast, an LTRA, is approved for mild asthma associated with SAR and PAR. Saline irrigation is beneficial as an adjunct treatment for AR. Intranasal ipratropium bromide is approved for rhinorrhea in various forms of AR, while intranasal decongestant use is cautioned beyond 3–5 days due to concerns about rebound nasal congestion. However, it can be safely used in conjunction with INCS for adult and adolescent patients with persistent nasal congestion. Studies support the efficacy and safety of combining intranasal decongestants with INCS (Brożek et al. 2010).

15.2.1.1.1 Anti-H1 Antihistamines

More than 45 H1 antihistamines constitute the largest class of medications for allergic diseases, acting by blocking H1 receptors and downregulating allergic inflammation. While first-generation H1 antihistamines have limited use due to sedative effects, second-generation ones are highly selective, minimizing adverse events. They are more effective than chromones and montelukast but less so than intranasal corticosteroids (INCSs) (Grubbe et al. 2009; Simons and Simons 2011). Newer agents, like rupatadine, show multiple receptor antagonism, targeting platelet-activating factor (PAF) receptors and histamine H1 receptors hence providing additional anti-inflammatory effects. Bilastine, approved in Europe, has a favorable safety profile. Advances include rapid-onset nasal and ophthalmic H1 antihistamines for allergic rhinitis (AR) and conjunctivitis, with nasal formulations showing efficacy comparable to or exceeding oral counterparts, particularly for nasal congestion relief. Ongoing research explores H1 antihistamine/glucocorticoid nasal formulations (Alevizos et al. 2013).

15.2.1.1.2 Intranasal Corticosteroids

Nasal steroids are highly effective in treating mild-to-acute rhinitis in both children as well as adults, offering advantages such as cost-effectiveness, reduced risk of side effects, and higher adherence. The newer formulations, like fluticasone furoate or mometasone, exhibit reduced systemic bioavailability and moderate adverse effects. Patients with both allergic rhinitis (AR) and asthma should be monitored for systemic side effects when using both inhaled and intranasal steroids. Meta-analyses confirm the significant efficacy of mometasone furoate, fluticasone, and other nasal steroids in relieving nasal congestion, rhinorrhea, sneezing, and nasal itching. Some, like ciclesonide, administered and activated on-site, show targeted activation with minimal systemic adverse effects. Nasal steroids may also have a positive impact on asthma symptoms, as reducing upper airway inflammation may improve asthma outcomes. Pharmacoeconomic data are limited, but some trials suggest cost-effectiveness compared to antihistamines like terfenadine or levocabastine and disodium cromoglycate. Overall, nasal steroids demonstrate significant therapeutic efficacy, safety, and potential cost-effectiveness in the management of allergic rhinitis (Mygind et al. 2001; Nielsen et al. 2001).

15.2.1.1.3 Intramuscular and Oral Corticosteroids

Intramuscular corticosteroid injections should be avoided because of the potential for severe adverse events, including systemic side effects and the risk of subcutaneous and muscular necrosis. Consideration for a brief course of oral corticosteroids is contingent upon patients with allergic rhinitis (AR) exhibiting mild-to-acute nasal symptoms that remain refractory to alternative treatments. It is imperative to exercise caution in recommending oral corticosteroids, particularly in the case of children, pregnant individuals, and patients with established contraindications (Bousquet et al. 2006).

15.2.1.1.4 Anticholinergics, Decongestants, and Mast Cell Stabilizers

Stimulation of nasal glands through cholinergic pathways results in watery secretion. Intranasal anticholinergics effectively mitigate rhinorrhea with minimal side effects, but their impact is confined to this specific symptom, leaving other nasal manifestations unaffected. Ipratropium bromide has demonstrated efficacy comparable to beclomethasone dipropionate in controlling rhinorrhea in children with perennial allergic rhinitis (AR). Nasal decongestants, such as xylometazoline, ephedrine, and pseudoephedrine, exhibit strong vasoconstrictive properties, providing instant relief for nasal congestion. In a placebo-controlled study, a single dose of pseudoephedrine, but not phenylephrine, significantly improved nasal congestion measures. However, oral administration of decongestants is associated with various side effects and contraindications, including headache, tachycardia, insomnia, and hypertension. Nasal decongestant sprays are limited to short-term use to avoid rhinitis medicamentosa and rebound congestion upon prolonged usage (Horak et al. 2009; Milgrom et al. 1999).

The stabilization of mast cells can ameliorate nasal indications by diminishing the synthesis and release of mediators like leukotrienes, tryptase, and histamine, after IgE antibody binding to the FcRI receptor. Intranasal chromones have the ability to constrain both late and initial phases of allergic reactions, thereby reducing nasal symptoms. However, due to their requirement for frequent applications (three or four times daily), their efficacy in alleviating AR symptoms is comparatively weaker when contrasted with other medications (Horak 2010; Varghese et al. 2010).

15.2.1.1.5 Leukotriene Inhibitors

Leukotriene inhibitors, exemplified by montelukast, act by competitively blocking cysteinyl leukotrienes' binding to end-organ receptors, targeting potent inflammatory mediators. Montelukast, the sole FDA-approved leukotriene inhibitor for allergic rhinitis (AR) in both adults and pediatrics, exhibits an optimal safety profile, even during pregnancy (FDA risk category B). Clinical studies and data related to it have shown its efficacy in reducing all AR symptoms, including nasal congestion, although its effectiveness is deemed inferior to antihistamines and intranasal corticosteroids (INCS). Therefore, combining oral H1 antihistamines or intranasal corticosteroids (INCS) is viewed as a more effective approach than using a single therapy for the treatment of allergic rhinitis (AR). The added advantage of montelukast in relieving bronchial symptoms in individuals with both asthma and AR makes it a sensible option (Li et al. 2009).

15.2.1.2 Allergen Immunotherapy

15.2.1.2.1 Peptide and Intralymphatic Immunotherapy

Intralymphatic immunotherapy (ILIT) involves administering allergen extract directly into the inguinal or cervical lymph nodes. It has shown efficacy for outdoor and indoor allergens. ILIT is considered safe and cost-effective, but its adoption may be limited due to the need for specialized staff and equipment (Konradsen et al. 2020). Another approach, peptide immunotherapy (PIT), injects short synthetic peptides to elicit an immune response to specific allergens. Studies on grass and birch allergen peptides suggest long-term symptom improvement even after treatment cessation. PIT offers a shorter treatment schedule, potentially improving patient compliance. Further research is needed to explore PIT's efficacy across various allergens (Mösges et al. 2018).

15.2.1.2.2 Subcutaneous Immunotherapy and Sublingual Immunotherapy

Recent advancements in SCIT research involve exploring dosing schedules and allergoids. A retrospective claims-based analysis revealed discontinuation rates of 41–77% for SCIT and 45–93% for SLIT, with cost and inconvenience as common reasons. Developing shorter SCIT protocols, such as using depigmented-polymerized allergen extracts (DPAEs) for ultra rush up dosing, has shown promise, significantly reduced symptoms and improving quality of life. SLIT, widely accepted by allergists, is considered more cost-effective than SCIT, and combination therapy with multiple allergens is being investigated. Studies on dual grass and ragweed SLIT-tablet administration and single vs. multiallergen SLIT in polysensitized patients suggest potential benefits, but further research is needed, especially for multiallergen SLIT, which is not yet FDA-approved (Mösges et al. 2019).

15.2.2 Future Directions

15.2.2.1 Diagnosis

The current diagnostic methods for allergic airway diseases involve skin testing by allergists and blood tests for allergen-specific IgE, with the ImmunoCAP Specific IgE test replacing the outdated RAST test. However, there is a challenge in interpreting these results effectively, particularly among general practitioners and pediatricians. The future of diagnostic testing appears to be transitioning toward microarrayed recombinant allergens, offering a multiplexed platform capable of simultaneously assessing specific IgE responses to numerous allergens. A noteworthy advancement in diagnostics is the emergence of molecular allergy diagnosis, specifically component-resolved diagnostics (CRD). This innovative approach involves characterizing a patient's IgE specificity using individual allergen molecules, providing a more precise understanding of their allergic sensitivities. While CRD is still in its early stages, it holds promise in offering greater discrimination between clinical reactivity and asymptomatic sensitization (Shreffler 2011).

The potential advantages of CRD include its ability to assay specific IgE to a large number of allergens with just a small amount of serum. Studies comparing CRD to traditional methods have shown promising results, indicating similar dynamic ranges, sensitivities, and specificities. However, the lack of provocation testing in some studies underscores the need for larger, well-designed investigations to thoroughly evaluate the diagnostic and prognostic capabilities of CRD in comparison to existing diagnostic modalities. In summary, the diagnostic landscape for allergic airway diseases is evolving towards more advanced and precise methodologies, with CRD and microarrayed recombinant allergens offering potential breakthroughs. Continued research and comprehensive studies will be crucial in determining the clinical utility and effectiveness of these emerging diagnostic approaches in the future (Sastre 2010).

15.2.2.2 The Anticipated Impact of Healthcare Reform on the Healthcare System

The restructuring of the American healthcare system is anticipated to significantly impact access to allergists and the management of Allergic Rhinitis (AR). The diverse sources of payment, encompassing governmental, employer-based, insurance, and individual contributions, will play a pivotal role in shaping the trajectory of AR healthcare in the United States. Recent research conducted by Truven Health Analytics, Harvard, and the University of Michigan underscores that escalating copayments correlate with increased work absenteeism and diminished productivity, potentially stemming from individuals opting to forego seeking care.

Presently, medical decision-making for AR is predominantly vested in nonspecialist practitioners, with a potential shift towards heightened involvement in primary care or even a scenario where physician engagement becomes unnecessary, particularly with the transition of numerous AR medications from prescription to over-the-counter (OTC) status. The methodologies employed by third-party payers to determine healthcare professionals' reimbursement, including fee-for-service, managed care, capitation systems, Diagnosis-Related Groups, Resource-Based Relative Value Scale, and Ambulatory Payment Classifications, will wield considerable influence on the evolving landscape of AR management in the United States (Gibson et al. 2012).

15.2.2.3 Combined Therapies

Approval of a novel combination therapy with fluticasone propionate and azelastine hydrochloride nasal spray, granted by FDA and EMA, demonstrated superior efficacy in treating seasonal allergic rhinitis compared to individual components or placebo. Meta-analysis across three trials further supported the combination's effectiveness, showcasing a quicker onset of action and improved total nasal symptom scores. Another study explored a combination of oxymetazoline and mometasone furoate nasal spray, revealing reduced nasal congestion and improved quality of life without significant adverse effects (Linton et al. 2021).

15.2.2.4 Anti-IgE

Omalizumab has proven efficacy in addressing both seasonal and persistent allergic rhinitis (AR) through multiple clinical trials. These studies consistently demonstrated enhancements in AR symptoms, improvements in the quality of life, and a decrease in the need for rescue medications among patients treated with omalizumab. A meta-analysis of randomized clinical trials further supported these findings, indicating a significant reduction in daily nasal symptom severity scores and the use of rescue medication. This suggests that omalizumab holds promise as a therapeutic option for inadequately controlled AR, particularly in individuals with moderate to severe AR, confirmed allergen-specific antibodies, unresponsive to standard treatments, and those with concurrent AR and asthma. While the European Academy of Allergy and Clinical Immunology recognizes omalizumab as a potential treatment for pediatric rhinitis and moderate-to-severe asthma, its application for AR lacks FDA approval, and its high cost limits its prolonged use for persistent AR (Casale et al. 2006).

15.2.2.5 Alternative Pathways for Delivering Specific Immunotherapy

In addition to traditional administration routes, novel approaches are being explored to enhance adherence and tolerability of Specific Immunotherapy (SIT). These approaches leverage the skin and lymph nodes' characteristics, emphasizing high dendritic cell density and low mast cell and basophil numbers to optimize antigen presentation and minimize side effects. Intralymphatic administration, tested in patients with grass pollen allergy, proved safe and effective, achieving symptom improvement comparable to conventional Subcutaneous Immunotherapy (SCIT) after only three injections (Senti et al. 2008, 2012a). This method demonstrated high tolerability, fewer allergic adverse events, and improved patient compliance. Epicutaneous administration, using patches with grass pollen extract, was found to be safe, well-tolerated, and induced symptom improvement compared to a placebo, with benefits appearing to be dose-dependent (Senti et al. 2012b).

15.2.2.6 Cytokine Inhibition and Phosphodiesterase 4 Inhibitors

In the context of allergic inflammation pathogenesis, inhibiting cytokine functions holds therapeutic promise for allergic rhinitis (AR). Interleukin 33 (IL-33) plays a crucial role in inducing Th2 cytokine production in various immune cells, indicating its potential to contribute to Th2 cytokine-mediated allergic inflammation. Both animal models and human studies have suggested a potential role for anti-IL-33 treatment in allergic rhinitis. In an experimental animal model of allergic rhinitis, anti-IL-33 treatment demonstrated significant improved skin condition, reductions in nose-scratching events, and decreased eosinophil numbers in bronchoalveolar lavage (BAL) (Kim et al. 2012; Haenuki et al. 2012).

Phosphodiesterase 4 (PDE4) is a crucial intracellular enzyme responsible for the breakdown and deactivation of cyclic adenosine monophosphate (cAMP). Found in various inflammatory cells, inhibiting PDE4 results in increased cAMP levels, effectively suppressing inflammatory processes. Roflumilast, an approved PDE4 inhibitor for maintaining severe chronic obstructive pulmonary disease, has shown effectiveness in allergic rhinitis. In a randomized, double-blind, placebo-controlled study involving 25 patients exposed to allergen challenges, roflumilast demonstrated positive outcomes (Schmidt et al. 2001).

15.2.2.7 Probiotics

Probiotics show clinical promise for allergic rhinitis (AR) management, particularly focusing on gram-positive genera like Bifidobacterium (Miraglia Del Giudice et al. 2017). Studies demonstrate symptom improvement and enhanced quality of life in both children and adults with seasonal or perennial AR through probiotic treatments. Combination therapies involving probiotics and standard AR treatments, as well as with allergen immunotherapy, reveal added benefits. Some studies also explore probiotic efficacy in experimental AR models, indicating reductions in symptoms and nasal eosinophil counts. Probiotics, as an adjunct therapy, emerge as a valuable and effective option for AR management, emphasizing the need for further studies, especially using validated AR models (Jalali et al. 2019).

15.3 Asthma

15.3.1 Current Management and Future Therapies

The overarching objectives for extended-term treatment involve attaining control over symptoms, mitigating the likelihood of subsequent exacerbations, preventing persistent airflow limitation, and managing treatment-related side effects. A holistic strategy encompasses non-pharmacological interventions, notably trigger avoidance, and follows a stepwise regimen involving escalating doses of medications, predominantly inhaled corticosteroids (ICS), often combined with a secondary controller. The initial step involves a β2 agonist, progressing to the inclusion of theophylline (for adults) or leukotriene receptor antagonists before resorting to systemic corticosteroids, if necessary. Mastering the technique of inhalation and assessing comorbidities are crucial elements in the asthma management process. Ongoing targets for type II asthma include IL-5, IgE, IL-5Ra, and IL-4Ra, with corresponding biologic therapies such as Omalizumab, Mepolizumab, Reslizumab, Benralizumab, and Dupilumab (Gour and Wills-Karp 2015). Omalizumab, the pioneering biologic therapy, is a monoclonal antibody against IgE. Approved for mild to severe allergic asthma in patients aged ≥6 years with exceeding 30 IU/L IgE levels, Omalizumab hinders the attachment of IgE to its receptor (FcεRI) on mast cells and basophils, thereby limiting their allergic reactions (Eyerich et al. 2020). Clinical studies have demonstrated a 25% reduction in asthma exacerbations and admissions in hospital in both adults as well as children. Omalizumab also shows promise in reducing virus-associated exacerbations, potentially by enhancing antiviral responses. It boasts a favorable safety profile, with a low risk of anaphylaxis (0.1–0.2) (Corren et al. 2009).

15.3.1.1 β2-Agonists Pharmacogenetics

The primary objectives for long-term treatment involve achieving symptom control, minimizing the risk of subsequent exacerbations, preventing persistent airflow limitation, and managing treatment-related side effects. Hydrophilic short-acting β2-agonists (SABAs), including terbutaline and albuterol, have effects that endure for 4–6 h. In contrast, lipophilic long-acting β2-agonists (LABAs) like formoterol and salmeterol offer a duration of 12 h or more. SABAs bind weakly to the β2-receptor, resulting in their shorter action duration. Albuterol is widely used and available in various forms, including oral formulations, and metered-dose inhaler (MDI). Levalbuterol, an isomer of albuterol, produces comparable effects. LABAs like formoterol and salmeterol, commonly used, offer better symptom control and improved pulmonary function compared to placebo or albuterol. Salmeterol, acting as a partial agonist for β2-receptors, is not appropriate for managing acute asthma symptoms but is effective for maintenance therapy. The U.S. Food and Drug Administration (FDA) recommends combining long-acting β2-agonists (LABAs) with inhaled corticosteroids (ICSs) for moderate or severe persistent asthma. Monotherapy with LABAs is associated with an elevated risk of asthma-related death (Pearlman et al. 1992). Combination therapy with a LABA and ICS is preferred for patients aged 12 years and above, and it is also recommended for patients aged 5 years with severe persistent asthma or inadequate control on step 3 care (Education and Program 2007).

15.3.1.2 Leukotriene Receptor Antagonists or Theophylline

Zafirlukast and montelukast, effective cysteinyl leukotriene receptor antagonists, exhibit significant inhibition of exercise-induced bronchoconstriction and responses to inhaled allergens. Clinical trials demonstrate improved lung function, reduced reliance on rescue bronchodilators, and evidence of decreased eosinophilic airway inflammation when combined with as-needed β2-agonists. In the UK, these antagonists are licensed for use in patients with persistent symptoms despite inhaled corticosteroid treatment. They allow a reduction in corticosteroid dose without compromising asthma control. However, the comparative effectiveness of leukotriene antagonists versus increasing inhaled corticosteroid doses remains unclear. Some studies suggest that long-acting β2-agonists may result in greater improvements in asthma control, while others indicate comparable benefits with additional anti-inflammatory effects from leukotriene antagonists. Theophylline, traditionally used as a bronchodilator, is considered at lower doses to minimize side effects. Combined with low-dose inhaled corticosteroids, theophylline provides comparable asthma control to higher corticosteroid doses, potentially offering slightly greater lung function improvements. In contrast to long-acting β2-agonists, theophylline may possess anti-inflammatory activity, suggesting a role in specific patient subgroups. Further research is needed to elucidate the relative effectiveness of these therapies in different asthma populations (Patel et al. 2005; Licari et al. 2014).

15.3.1.3 Corticosteroids

Inhaled corticosteroids (ICSs) have proven highly effective in reducing airway hyperresponsiveness (AHR) in both adults as well as pediatric, correlating with improved disease outcomes. Studies, inculcating the Childhood Asthma Management Program (CAMP), reveal significant benefits in terms of improved lung function, reduced exacerbations, and fewer prednisone courses. The extended use of ICSs has shown significant efficacy in preventing morbidity and mortality among asthma patients, as evidenced by a case-control analysis indicating a tenfold reduction in the likelihood of fatal or near-fatal asthma in individuals using chronic beclomethasone dipropionate. Furthermore, consistent utilization of low-dose ICSs has been linked to a decreased risk of deaths related to asthma (Choby and Lee 2015).

15.3.1.4 Anti-IgE Treatment

Several biological medications targeting IgE synthesis or signaling pathways are in development, offering potential anti-IgE treatments. Lumiliximab (IDEC-152), a primatized monoclonal antibody against FcεRII, has demonstrated the ability to block IgE synthesis in human B-cells and reduce serum IgE levels. Ligelizumab (QGE031), a humanized monoclonal IgG1κ antibody with high avidity for IgE, has been investigated for its pharmacokinetics, pharmacodynamics, and safety. Monoclonal antibody-12 is another candidate for anti-human IgE therapy, potentially aiding in the extracorporeal depletion of IgE and IgE-bearing cells from peripheral blood. However, Omalizumab is currently the only licensed biological anti-IgE agent for human use. Omalizumab, a recombinant DNA-derived humanized IgG1 monoclonal antibody, was approved for severe persistent allergic asthma in patients aged >12 years with specific criteria, and later expanded to include patients aged 6–12 years with the same indications. It selectively binds to human IgE and received approval from regulatory authorities, such as the US Food and Drug Administration (FDA) in 2003 and the European Medicines Agency in 2005 (Laffer et al. 2008; Presta et al. 1993).

15.4 Conclusion

Allergic rhinitis (AR) is a globally prevalent condition, and severe forms significantly impact patients' lives, causing limitations in day-to-day routine, reduced quality of life, and sleep disturbances. The most effective treatment is customized based on the severity of allergic rhinitis (AR). Mild to moderate symptoms, particularly without substantial nasal obstruction, respond well to second-generation antihistamines, like rupatadine and bilastine, which offer improved side effect profiles. Persistent nasal obstruction is effectively addressed by corticosteroids, with topical intranasal formulations being preferred due to their efficacy and minimal systemic side effects. In cases of inadequate response, short courses of oral corticosteroids or nasal decongestants may be considered, but their use is limited to avoid potential adverse effects. For predominant runny nose, intranasal anticholinergics like ipratropium bromide are indicated due to their impact on cholinergic mechanisms. Patients with insufficient response to drug treatment may benefit from allergen desensitization through specific immunotherapy (SIT), which has demonstrated efficacy even in drug-resistant cases. Future perspectives for enhancing AR treatment involve the development of combination. The optimal approach involves combining medications like antihistamines and topical corticosteroids into unified formulations. Additionally, there is ongoing research into employing anti-IgE therapy, specifically Omalizumab, for severe cases of allergic rhinitis. Explorations are also being made into immunologic treatments targeting crucial elements in allergic inflammation, such as cytokines and toll-like receptors.

The escalating occurrence of severe asthma and cases resistant to existing medications necessitates the development of new therapeutics. Recent advances focus on understanding immunologic signaling, genetic aspects of airway inflammation, and immune system plasticity, paving the way for innovative treatment options. Ultra-long-acting β-agonists aim to enhance dosing frequency and adherence. Preclinical work on "dissociated corticosteroids" shows potential in maintaining efficacy while minimizing systemic side effects. The focus on developing targeted therapies for type 2 inflammation has grown, with particular attention to alarmins—epithelial-cell-derived cytokines that play a pivotal role in activating ILC2 and Th2 cells. Tezepelumab, which targets TSLP, has demonstrated efficacy in reducing asthma exacerbations in patients with mild asthma, irrespective of blood eosinophil levels and FeNO. Another approach in clinical development targets IL-33 directly or through its receptor (anti-ST2). An innovative approach involves optimizing the airway delivery of monoclonal antibodies, with ongoing development of a nebulized biologic therapy targeting IL-13 in animal models. Within the domain of small-molecule antagonists, Fevipiprant, acting as a PGD2 type 2 receptor antagonist, has exhibited potential in diminishing eosinophilic inflammation and reducing smooth muscle mass in individuals with moderate to severe asthma. The imperative development of biomarkers is emphasized to identify appropriate patients and foresee and track their response to biologics. These progressions collectively signify a hopeful trajectory for forthcoming asthma therapies.

References

Alevizos M et al (2013) Rupatadine inhibits inflammatory mediator release from human laboratory of allergic diseases 2 cultured mast cells stimulated by platelet-activating factor. Ann Allergy Asthma Immunol 111(6):542–547

Bangalore Thimmadasiah N, Joshi TK (2020) India: country report on children's environmental health. Rev Environ Health 35(1):27–39

Bousquet J et al (2006) Severity and impairment of allergic rhinitis in patients consulting in primary care. J Allergy Clin Immunol 117(1):158–162

Brożek JL et al (2010) Allergic Rhinitis and its Impact on Asthma (ARIA) guidelines: 2010 revision. J Allergy Clin Immunol 126(3):466–476

Casale TB et al (2006) Omalizumab pretreatment decreases acute reactions after rush immunotherapy for ragweed-induced seasonal allergic rhinitis. J Allergy Clin Immunol 117(1):134–140

Choby GW, Lee S (2015) Pharmacotherapy for the treatment of asthma: current treatment options and future directions. Int Forum of Allergy Rhinol 5:S35–S40

Corren J, Casale TB, Lanier B, Buhl R, Holgate S, Jimenez P (2009) Safety and tolerability of omalizumab. Clin Exp Allergy 39(6):788–797

Education NA, Program P (2007) Expert Panel Report 3 (EPR-3): guidelines for the diagnosis and management of asthma-summary report 2007. J Allergy Clin Immunol 120(5 Suppl):S94–S138

Eyerich S, Metz M, Bossios A, Eyerich K (2020) New biological treatments for asthma and skin allergies. Allergy 75(3):546–560

Gibson TB, Fendrick AM, Chernew ME (2012) Cost-sharing and productivity. National Bureau of Economic Research

Gour N, Wills-Karp M (2015) IL-4 and IL-13 signaling in allergic airway disease. Cytokine 75(1):68–78

Grubbe RE, Lumry WR, Anolik R (2009) Efficacy and safety of desloratadine/pseudoephedrine combination vs its components in seasonal allergic rhinitis. J Investig Allergol Clin Immunol 19(2):117

Haenuki Y et al (2012) A critical role of IL-33 in experimental allergic rhinitis. J Allergy Clin Immunol 130(1):184–194

Horak F (2010) Antiallergic and vasoactive drugs for allergic rhinitis. In: Allergy frontiers: therapy and prevention. Springer, pp 51–61

Horak F et al (2009) A placebo-controlled study of the nasal decongestant effect of phenylephrine and pseudoephedrine in the Vienna Challenge Chamber. Ann Allergy Asthma Immunol 102(2):116–120

Jalali MM, Soleimani R, Alavi Foumani A, Ganjeh Khosravi H (2019) Add-on probiotics in patients with persistent allergic rhinitis: a randomized crossover clinical trial. Laryngoscope 129(8):1744–1750

Kim YH et al (2012) Anti-IL-33 antibody has a therapeutic effect in a murine model of allergic rhinitis. Allergy 67(2):183–190

Konradsen JR et al (2020) Intralymphatic immunotherapy in pollen-allergic young adults with rhinoconjunctivitis and mild asthma: a randomized trial. J Allergy Clin Immunol 145(3):1005–1007

Laffer S et al (2008) A high-affinity monoclonal anti-IgE antibody for depletion of IgE and IgE-bearing cells. Allergy 63(6):695–702

Lee Y-G, Lee P-H, Choi S-M, An M-H, Jang A-S (2021) Effects of air pollutants on airway diseases. Int J Environ Res Public Health 18(18):9905
Li AM et al (2009) Leukotriene receptor antagonist in the treatment of childhood allergic rhinitis—a randomized placebo-controlled study. Pediatr Pulmonol 44(11):1085–1092
Licari A et al (2014) Current recommendations and emerging options for the treatment of allergic rhinitis. Expert Rev Clin Immunol 10(10):1337–1347
Linton S, Burrows AG, Hossenbaccus L, Ellis AK (2021) Future of allergic rhinitis management. Ann Allergy Asthma Immunol 127(2):183–190
Milgrom H et al (1999) Comparison of ipratropium bromide 0.03% with beclomethasone dipropionate in the treatment of perennial rhinitis in children. Ann Allergy Asthma Immunol 83(2):105–111
Miraglia Del Giudice M, Indolfi C, Capasso M, Maiello N, Decimo F, Ciprandi G (2017) Bifidobacterium mixture (B longum BB536, B infantis M-63, B breve M-16V) treatment in children with seasonal allergic rhinitis and intermittent asthma. Ital J Pediatr 43(1):1–6
Mösges R et al (2018) Short course of grass allergen peptides immunotherapy over 3 weeks reduces seasonal symptoms in allergic rhinoconjunctivitis with/without asthma: a randomized, multicenter, double-blind, placebo-controlled trial. Allergy 73(9):1842–1850
Mösges R et al (2019) Subcutaneous immunotherapy with depigmented-polymerized allergen extracts: a systematic review and meta-analysis. Clin Transl Allergy 9(1):1–15
Mygind N et al (2001) Mode of action of intranasal corticosteroids. J Allergy Clin Immunol 108(1):S16–S25
Nielsen LP, Mygind N, Dahl R (2001) Intranasal corticosteroids for allergic rhinitis: superior relief? Drugs 61:1563–1579
Paramesh H (2018) Air pollution and allergic airway diseases: social determinantsand sustainability in the control and prevention. IJP 85(4):284–294
Patel P et al (2005) Randomized, double-blind, placebo-controlled study of montelukast for treating perennial allergic rhinitis. Ann Allergy Asthma Immunol 95(6):551–557
Pearlman DS et al (1992) A comparison of salmeterol with albuterol in the treatment of mild-to-moderate asthma. N Engl J Med 327(20):1420–1425
Presta LG et al (1993) Humanization of an antibody directed against IgE. J Immunol 151(5):2623–2632
Sastre J (2010) Molecular diagnosis in allergy. Clin Exp Allergy 40(10):1442–1460
Schmidt BMW et al (2001) The phosphodiesterase 4 inhibitor roflumilast is effective in the treatment of allergic rhinitis. J Allergy Clin Immunol 108(4):530–536
Senti G et al (2008) Intralymphatic allergen administration renders specific immunotherapy faster and safer: a randomized controlled trial. Proc Natl Acad Sci 105(46):17908–17912
Senti G et al (2012a) Intralymphatic immunotherapy for cat allergy induces tolerance after only 3 injections. J Allergy Clin Immunol 129(5):1290–1296
Senti G et al (2012b) Epicutaneous allergen-specific immunotherapy ameliorates grass pollen–induced rhinoconjunctivitis: a double-blind, placebo-controlled dose escalation study. J Allergy Clin Immunol 129(1):128–135
Shreffler WG (2011) Microarrayed recombinant allergens for diagnostic testing. J Allergy Clin Immunol 127(4):843–849
Simons FER, Simons KJ (2011) Histamine and H1-antihistamines: celebrating a century of progress. J Allergy Clin Immunol 128(6):1139–1150
Varghese M, Glaum MC, Lockey RF (2010) Drug-induced rhinitis. Clin Exp Allergy 40(3):381–384

Correction to: Understanding Allergic Airway Diseases: Contemporary Treatment Paradigm

Parteek Prasher, Mousmee Sharma, Sachin Kumar Singh, Ronan MacLoughlin, Kavita Pabreja, and Kamal Dua

Correction to:
P. Prasher et al. (eds.), *Understanding Allergic Airway Diseases: Contemporary Treatment Paradigm*,
https://doi.org/10.1007/978-981-97-1953-2

The original version of this book was inadvertently published with errors.

1. Chapter 3 was published with incorrect affiliation of the authors "A. Dwivedi". The correct affiliation should read as "Department of Pharmaceutics, Maharishi Arvind College of Pharmacy, Jaipur, Rajasthan, India".

2. In Chapter 5, the last name of author "Lokesh Nager" was unfortunately published with an error. The correct last name should read as "Lokesh Nagar".

The above errors have been corrected with this erratum.

The updated version of these chapters can be found at
https://doi.org/10.1007/978-981-97-1953-2_3
https://doi.org/10.1007/978-981-97-1953-2_5

P. Prasher et al. (eds.), *Understanding Allergic Airway Diseases*,
https://doi.org/10.1007/978-981-97-1953-2_16

SPRINGER NATURE

GPSR Compliance

The European Union's (EU) General Product Safety Regulation (GPSR) is a set of rules that requires consumer products to be safe and our obligations to ensure this.

If you have any concerns about our products, you can contact us on ProductSafety@springernature.com

In case Publisher is established outside the EU, the EU authorized representative is:

Springer Nature Customer Service Center GmbH
Europaplatz 3
69115 Heidelberg, Germany

The manufacturer's authorised representative in the EU is Springer Nature Customer Service Centre GmbH, Europaplatz 3, 69115 Heidelberg, Germany. If you have any concerns regarding our products, please contact ProductSafety@springernature.com

Printed and bound by CPI Group (UK) Ltd, Croydon, CR0 4YY
01/10/2026
02225862-0006